MAD SCIENTIST, IMPOSSIBLE HUMAN

MAD SCIENTIST, IMPOSSIBLE HUMAN

An Essay in Generative Anthropology

Andrew Bartlett

The Davies Group, Publishers

Aurora, Colorado

Mad Scientist, Impossible Human: An Essay in Generative Anthropology.

Library of Congress Cataloging-in-Publication Data

Bartlett, Andrew, 1960-
Mad scientist, impossible human : an essay in generative anthropology / Andrew Bartlett.
pages cm
Includes bibliographical references.
ISBN 978-1-934542-35-4 (alk. paper)
1. Philosophical anthropology. 2. Language and languages--Origin. 3. Human beings--Philosophy. 4. Science--Philosophy. 5. Religion--Philosophy. I. Title.
BD450.B335 2014
128--dc23

2013047136

0123456789

for Joanne

"...take me down... to where you hide..."

TABLE OF CONTENTS

ACKNOWLEDGMENTS

I thank the Educational Leave Committee at Kwantlen Polytechnic University for granting me a two-semester leave for 2006-2007, during which much of the early research for this book was performed and the earliest drafts of chapters 2, 3, and 4 were composed. I thank the 0.6% Faculty Professional Development Fund at Kwantlen for the grant that covered a single-course time release for the spring semester 2011, enabling me both to work on the chapter about *Blade Runner* and to employ three student readers to critique the earliest version of the complete manuscript. I thank those student readers—Janet Eastwood, William Fast, and Jenine Sarchet, participants in my English 3380: Studies in Popular Writing (The Frankenstein Myth) (fall 2010 at Kwantlen)—for their work. Eric Gans read through the first versions of the chapters on Wells and Capek and gave encouraging feedback. Brian Bartlett (St. Mary's University), Ian Dennis (University of Ottawa), Andrew McKenna (Loyola University), and N. P. Kennedy (Kwantlen Polytechnic University) performed readings of a second revision of the complete manuscript and presented me with a diverse body of responses, going well beyond any call of loyalty. I take sole responsibility for the current edition of the work.

C. S. Lewis defines friendship in *The Four Loves* as that bond shared by those who answer *Do you see the same truth?* with a *Yes*. I have many friends to thank. Among those whom I have known for decades, I must thank Ernest Bauer (Montreal), Nancy Bauer (Fredericton), Geordie Haley (Halifax), and Paul Headrick (Vancouver) both for their affectionate loyalty and support of my creative impulses. As a member of the Colloquium on Violence and Religion, I thank William A. Johnsen (editor of *Contagion*), who delayed this book by commissioning a review article on the work of Eric Gans; Ann Astell, Sandhor Goodhart, Britt Johnston, Jeremiah Alberg, and Pablo Bandera. The 2005 meeting of COVR in Koblenz (organized by Ann Astell) with its theme *imitatio Dei* started my preoccupation with the Victor Frankenstein figure who plays but does not imitate God. Those stalwarts in the Generative Anthropology Society whom I have not already named – Peter Goldman, Marina Ludwigs, Adam Katz, Stacey Meeker, Richard van Oort, Matthew Schneider – need to know how much I value their intellectual labor and spiritual camaraderie. To Richard, I owe a special debt for the help he gave me when the Vancouver reading group *Sparagmos!* was active. What my intellectual life would be now if Ian Dennis had not appeared at the University of British Columbia in September 1995, professing knowledge of an interest in the work of Rene Girard, I will never know. I owe Ian for many kindnesses and for his loyal labor as Secretary-Treasurer of the Generative Anthropology Society. To Eric Gans, for his openness and generosity of spirit and tireless persistence, I offer my thanks. I am privileged to know both the work and the man.

I thank my parents Lester and Marjorie Bartlett for their love and care. They blessed me with five siblings, all of them always easy to love, all of them ceaselessly encouraging: Bruce in Beirut, Gail in Katmandu, Brian in Halifax, Carol in Toronto, Mark in New York – thank you. When I was in my late thirties, Joanne Horwood, Torben Christensen and Michaela Christensen embraced me as a member of their family. The happiest years of my life, I have lived in their embrace. Joanne, Torben, Michaela – thank you.

****

Quotations from Claudia Novak's 2007 translation of *R.U.R (Rossum's Universal Robots)* appear with the kid permission of Catbird Press.

Excerpts from *Future Noir: The Making of Blade Runner* by Paul Sammon appear with permission from Harper Collins Publishers.

Regarding the transcribed-from-subtitles excerpts from *Blade Runner*, grateful acknowledgment is made to the Blade Runner Group for their communication that they have no issues with my use of the dialog in such a manner, within the limited scope of publication of this book.

For where two or three are gathered together in my name,
there am I in the midst of them.

— Matthew 18:20

ONE

The Frankenstein Myth, Scientism, and Generative Anthropology

Four Stories, One Formula

For most experts, a myth is (among other things) a story. To introduce the Frankenstein myth as I have come to understand it, let me tell four stories. Each encapsulates the action of one of the four artworks this book will advertise, celebrate and illuminate. I say *advertise,* for one of my goals is to make you want to read, view, see, and study these artworks for yourself. Literary analysis should create an appetite for experiencing first-hand the literature under examination. I use the term *artwork* as it can cover a gothic novel of the romantic period, a modernist scientific romance, an experimental European drama of the 1920s, and a Hollywood movie of the 1980s. Be warned: to hear the four stories consecutively will remind the familiar (or inform the uninitiated) that to absorb the impact of the Frankenstein myth is to be spectator at an anguished, bloody tragedy. This myth belongs to the genre of science-fiction horror.[1] If we understand the Frankenstein myth, we understand it has opened a view onto a fantastic world that we are lucky not to inhabit and exhibited a scene of violent origin that we are fortunate not to claim as our ancestral home.[2]

Here is the precedent-setting model on which all other versions will be based.

In the late eighteenth century, a privileged young man, the first-born son of a distinguished family of Geneva, studies chemistry at the University of Ingolstadt. There, he discovers the "secret of life"—a process by which he can bestow life on inanimate matter. Enchanted with the prospect of fathering a new race to rival humankind, he fashions a gigantic human-like body out of sewn-together pieces of corpses, but the moment he successfully animates the thing is the moment he finds it utterly, intolerably ugly. Disgusted, he abandons it and collapses into self-pity and illness, to be rescued and taken home by his best friend. Meanwhile, the animated Creature develops independently in the wild, slowly gaining consciousness and language.

It learns, however, that humans cannot abide its presence. Disappointed when the family of a blind man it has befriended rejects it in violent horror, the Creature, now experiencing itself as a Monster, confronts the fact that its ugliness is an insuperable obstacle to contact or exchange with humans. It travels to Geneva to find its maker, but impulsively murders the man's younger brother and frames a domestic servant for the crime. The Creature confronts its maker and demands that he fabricate a female companion to allay the pain of its intolerable solitude, promising to abandon the regions of humankind with the companion. Full of reluctance and loathing, the young scientist travels to the remote Orkney Islands and works on the female thing, but destroys it before completing it. Taking revenge, the Monster murders the scientist's best friend and (on their wedding-night) the scientist's bride. Seeking his revenge in turn, the scientist pursues the Monster into the frozen northern regions, strangely assisted in the chase by the Monster itself. Physically wrecked, he is taken on board an icebound ship and befriended by an admirer who has himself been on a scientific expedition to reach the North Pole. To this explorer, the cursed experimenter tells his tale; then he dies, neither regretting his creation of the Monster nor repenting his neglect of it. The Monster appears at the death bed, laments its cruelty to its maker, and promises to immolate itself in solitude on the ices further north.

Many of you may recognize this as a retelling of the gothic novel *Frankenstein, or the Modern Prometheus* (1818) by Mary Wollstonecraft Shelley. The young Genevan is the ill-fated Victor Frankenstein; his young admirer on the ice-bound ship, Robert Walton, from whose transcriptions the text of the storytelling is taken; the hapless bride, Elizabeth; the best friend, Henry Clerval. The Monster in Shelley's account goes unnamed, just as he must remain nameless in this book.

Here is a second story.

Toward the end of the reign of Queen Victoria, in the 1880s, a biologist in England who practices vivisection becomes the object of public scandal when one of his victims, a grotesquely flayed dog, escapes from his laboratory. Taking as his assistant a heavy-drinking bohemian gentleman who must likewise leave England under a cloud of scandal, he sets up an experimental station on a remote island of the South Pacific. There, the biologist continues to practice experiments in vivisection: he aims to demonstrate the plasticity of living forms. Specifically, by severing and grafting pieces of animal flesh without anesthetic, he seeks to fashion a human-like creature, one beyond pain. His multiple experiments fail, but he persists. The animal-human hybrids, referred to as Beast People, receive a rough education in a rudimentary "law" that makes them obey the two Englishmen and gives them elementary powers of thought and speech through mind-control mechanisms that rely heavily on hypnosis. A third English gentleman, a victim of shipwreck, arrives on the island, rescued by the bohemian gentleman from being cast adrift a second time. However, troubled by the cruelty of the vivisectionist, believing the experiments are reducing human beings to animals, the third man flees in terror and, when cornered, tries to provoke a rebellion among the animal-human hybrids. The vivisectionist explains that he is not animalizing humans, but humanizing animals. The rebellion is quelled. The Beast People, however, have been inspired by the third gentleman's alien suggestions. One has seemed to go wild from tasting rabbit blood; the bohemian man regrets having brought rabbits onto the island. The Beast People now rebel of their own volition: the vivisectionist is murdered, his assistant killed. But the third gentleman survives. He lives alone, a single human among the Beast People, for several months, witnessing their slow reversion back into full animality, their loss of the "law" and of speech. By luck (a lifeboat without survivors drifts onto the shores of the island), he returns to England. He has been traumatized by the monstrosity he witnessed on the island, so he cannot shake a haunted conviction that the English people are too much like the macabre hybrids the vivisectionist created.

The above is a retelling of H. G. Wells' *The Island of Dr. Moreau* (1896), one of a series of much-loved scientific romances that earned Wells renown as a founder of science fiction. The vivisectionist is Dr. Moreau; his bohemian assistant, Montgomery; the traumatized survivor, Edward Prendick, the author of the posthumously discovered papers that become the source, once edited by a nephew, for the transcription of his first-person narrative.

The next account presents the action of the play in which the word *robot* as we most often use it—to refer to a mechanical humanoid—first appeared.

A young woman actively campaigning for human rights visits the factory on a remote island where the products marketed all over the world by a company called R.U.R. (Rossum's Universal Robots) *have long been manufactured.*

She hopes to incite their rebellion, but the charming company director and his colleagues, telling her about the technological practices of the founders of the company, demonstrate—to her embarrassment—that the automata are not human. They feel no pain, fear no death, experience no emotion, show no taste, and possess no identities. They are efficient, predictable, bizarre, unsouled organisms in human shape—nothing more. The Comic Prologue ends with the company director falling in love with the activist. Ten years later, the curtain opens on Act One. We learn from dialogue that in the intervening years, the young woman, having married the director, has suffered childlessness and has continued to believe the robots' odd near-humanness is wrong. *She persuaded the company physiologist to endow the robots, in secret, with elementary human feeling. A terrible unintended consequence of that secret alteration is that the robots worldwide have waged resentful war on humankind. Having annihilated all but a tiny remnant of the species, the militarized humanoids now come to the island. In Acts One and Two, we witness the company executives and the lady gathered in her sitting room, surrounded by a menacing, numberless army of robots outside the windows, standing still along the shoreline. The humans mournfully accept their impending deaths, exchanging opinions on the causes of the disaster, trying desperate last measures. None works. They are all executed, save one bricklayer. In the final act, the bricklayer is powerless when the leader of a committee of robots demands that he perform surgery, cutting him open in a quest for the irrecoverable secret of their manufacture. After failing in this gruesome task, the bricklayer meets two young-seeming robots, male and female, who seem to possess a sense of personal identity: they express a willingness to die self-sacrificially one for the other. He interprets their gestures as a sign that the future of planet earth might contain something like human beings.*

That is a re-telling of the action of *R.U.R. (Rossum's Universal Robots)* (1922), the brilliant originary robots drama, the best-known work in the English-speaking world of the great Czech writer Karel Capek. Hellena Glory is the young woman, Harry Domin the director, Dr. Gall the physiologist whose tinkering gives the robots resentment, Alquist the bricklayer, and Damon the leader of the committee robots. There is no idea about robot-human competition that this prescient play does not contain.

Now let us look at our fourth (and last) definitive version of the Frankenstein myth.

In a rain-soaked Los Angeles of the distant future, the Tyrell corporation, housed in a Mayan-like pyramid that towers above an urban wasteland, genetically engineers and sells biomechanical humanoids named replicants, whose owners exploit them (as soldiers, labourers, prostitutes) in the off-world colonies. Five have mutinied and travelled to earth. A blade runner is a policeman whose job is to execute replicants on sight. One such executioner, bullied by his ex-boss into hunting the rebels, visits the pyramid to meet with Tyrell and a beautiful woman. At Tyrell's request, he administers an eye-scanning test designed to differentiate replicants from the humans they resemble closely. The blade runner is shocked to learn that, even though she herself does not know it, the beautiful woman is a replicant. When Tyrell disabuses her, she visits the blade runner for a second opinion, but exits his apartment in fearful despair when he insists too bluntly she is nonhuman. Back to work, the blade runner tracks down and kills one of the rebels who has been hiding as an exotic dancer, shooting her in the back on a crowded street. A tall, powerful male replicant happens to witness the killing, beats up the blade runner, and prepares to kill him, when the beauty from the Tyrell corporation appears, putting a bullet through the attacker's head. She and the blade runner return to his apartment; after a quiet interlude and some pushiness on his part, they end up in bed. He tracks down and executes his third victim, another female, who has been hiding in the apartment of a genetic engineer who has worked for the Tyrell Corporation. Meanwhile, the replicant leader, tall and blond and menacing, has bullied the Tyrell employee into helping him infiltrate the pyramid towers. He confronts Tyrell, but when the genius claims he has no power to extend the four-year lifespan of the creature, the lead replicant murders his maker, crushing his skull between his bare hands. After he returns to the employee's apartment, the replicant leader pursues the policeman in a game of cat-and-mouse violence, following him onto a rain-soaked rooftop. But at the last minute, he saves his human rival from slipping and falling to a terrible death. He sits down in the rain; he speaks a few poetic lines, bows his head, and dies, his time expired.

Many may recognize in this account an admittedly skeletal version of the action in the Hollywood film *Blade Runner,* directed by Ridley Scott. The executioner policeman is Rick Deckard (played by Harrison Ford), the leader of the replicants Roy Batty (Rutger Hauer), the beautiful replicant Rachael (Sean Young). Joe Turkel played Eldon Tyrell, the genetic engineer murdered by Batty in the most despairing moment of the movie. I will approach *Blade Runner* as an exemplary post-modern instantiation of the myth.

Let us now ask the question outright: *what is the Frankenstein myth?*

At the core of these works and of any other that qualifies for inclusion in the canon of artworks promulgating and preserving it, one fundamental event shapes the action: *a mad scientist plays God by trying to re-enact the origin of the human, and disastrously fails.* Victor Frankenstein; Dr. Moreau; the Rossums (father and son), with their corporate heirs Domin and Gall; the genetic engineer Eldon Tyrell—for a while, the physiochemical knowledge they possess and the technology they put to work has them ruling their insular worlds. All these works feature a powerful character who acts out a singular desire. Insane in its extremity, it is the desire to create a new set of creatures a little like humankind but better than humankind, a new race or species enough like the human to be its rival but sufficiently distinct to outdo it—outperform it, outlast it, outlive it. Acting out that desire, the scientists defy, by ignoring, those religious traditions that place a single Creator (in the Judeo-Christian tradition, God) as an agent at the origin of the human. Not all, but certainly those mad scientists in and of the Frankenstein myth desire to prove a point: humankind is not good enough, and materialist science can do better at making humans than the agents sanctioned by religion (or anthropology, or education, or tradition) have ever done. In every case, *not good enough* entails something like *not sufficiently impervious to suffering.* If humankind is not good enough, then by implication God has botched the job. The mad scientist has determined to repair, to complete, or to replace God's substandard handiwork. The phrase *plays God* captures that rivalry. If the version of the myth is post-Darwinian, to the extent that a creator God is not held responsible for human inadequacy, then the scientist will try to do better than evolutionary processes have done.

The scientist *fails disastrously.* The failures are at once horribly on display and finely honed for heart-rending. In Shelley's novel, we watch the serial killing of Victor's young brother Ernest, the framed servant Justine, the loyal friend Clerval, and the oblivious fiancée Elizabeth; we learn of the self-destructive miseries of Victor, not to mention the ineradicable aloneness of the Wretch he creates; we witness the implacable hatred of maker and monster, with its undeviating escalation into a chase to the death in the icebound regions of the North pole. In Wells' scientific romance, the failures are signalled by the screams of the animals suffering under the knives of Moreau, the neglect of the Beast People chanting their rote laws, the chaotic murders of Moreau and Montgomery, and the traumatization of Prendick. In *R.U.R.,* Capek stages nothing less than the annihilation of humankind and an idiotic end to history, marginally softened by expressions of regretful sentiment and the demented hopes of the bricklayer Alquist in Act Three. *Blade Runner* shows us the frustrated misery of the enslaved replicants, the disorientation of Rachael, the murder of three humans, the execution of four replicants, and the untimely demise of Roy Batty. The scientist who plays God by trying to re-enact the origin of the human fails disastrously. To avoid such disaster, the myth teaches, we name his error and fear his contempt for the human as it is and always has been.

By *myth* I do not refer simply to a false belief, detached from a story. For our purposes, a myth is a story with deep and wide appeal to many like-minded people, more radiantly profound in content and more stubbornly necessary to a cultural group's coherence than a legend, folktale, or fairy tale. Like vampire stories or stories of zombie apocalypse, its largest contemporary competitors, the Frankenstein story raises the problem of *modern myth:* is not myth primitive by definition? Should we not think of *myth* as a category belonging to preliterate, non-scientific cultures? Because it allies itself with some combination of the irrationally primitive and the capriciously supernatural, the *mythic* appears to be in tension with modern rationality and modernity's faith in our ability to know Nature as a realm of law-like regularities and a territory plentiful with raw materials subject to technological control. Ironically, the Frankenstein stories present themselves as mythic most obviously because the human scientist, although not God, seems to possess god-like powers of creation: the materials are thus fantastic in their naïve supernaturalism. Relying on Robert A. Segal's concisely dizzying survey of the prodigious variety of theories of myth, I would say the definition of myth presupposed in this book harmonizes with those of Mircea Eliade and Bronislaw Malinowski insofar as for them, the category *modern myth* is not a self-contradiction. A myth is a story that a culture possesses so as to codify and make manifest its deepest beliefs, those beliefs it celebrates as self-evident. Segal does not refer to the work of Eric Gans, whose generative anthropology forms the analytical framework I put to work in this study. For Gans, a myth is at least ". . . narrative discourse, [and] an attempt, not to commemorate an event, but to tell a story that is to render plausible the existence of cultural categories" (*End* 142). Gans teases out a peculiar feature of modern myth: ". . . to the modern reader . . . the mythical narrative is inevitably devoid of literary value" (*End* 236). The direct referent of the *mythical narrative* lacking literary value is the transcription of a "primitive" narration of a myth. As a modern one, the Frankenstein myth is esthetic, not religious. It cannot be severed or detached from the specific literary, dramatic, and filmic texts that embody and represent it; when it is so abstracted, the leftover skeleton will be, as Gans hints, "devoid of literary value." In other words, the text of a primitive myth narrated by an aboriginal person and recorded by an ethnographer *is* the myth; the text of a *modern* myth, by contrast, must necessarily be an intellectual abstraction assembled from a combined set of literary or esthetic productions. The Frankenstein myth as we come to understand it will function to "render plausible the existence of cultural categories": it will render plausible the idea of *sacred, shared human origin* over against the reductionist idea that humans can be made of nothing but mindless matter.

It might seem needless that another book about the Frankenstein myth should appear in the marketplace of ideas. A number of valuable multi-text studies have been performed over the years—by psychologist John Cohen (1966), literary critic Chris Baldick (1987), historian of science Jon Turney (1998), pop culture historian David J. Skal (1998), theologian Elaine Graham (2002), cultural historian Susan Tyler Hitchcock (2007). What might justify an additional examination of Frankenstein phenomena? Few would quarrel with the assumption that the myth concerns more or less directly *what it means to be human*. Thanks to the way of thinking provided by the philosophical anthropology of Eric Gans, which I introduce below, a refreshing perspective on the meaning of the Frankenstein myth has now become possible. One of my assumptions is that we have become so familiar with it that we have lost our sense of its truth. It is a knife that needs to be sharpened, a statue that needs to be cleaned of grime and polished to a glow, a dust-damaged painting badly in need of restoration, a story whispering to be retold in bold new accents, in a newly confident tone of voice, viewed from the new perspective of a minimal anthropology. The bluntness, the grime, the sleepy underestimation induced by excessive familiarity must go. This book argues that they result mostly from interference caused by two trends visible in much of the academic critical literature on these and like texts in recent decades.

The first trend visible in the critical literature is this: the description of the myth's central figure as a mad scientist who plays God by making "humans" has fallen out of favour in recent decades. This book seeks to restore the mad scientist figure to its full brilliance by establishing that the Frankensteinian scientist is the dramatic embodiment of *scientism;* scientism, moreover, is alive and well and contaminating the intellectual air we breathe. To avoid excessive repetition, I will switch the label at times to *vain scientist:* the attempt to re-enact the origin of the human is a project carried out in vain (bound to fail); at the same time, the projector is distinguished by an incorrigible vanity (overconfident pride in scientific powers, conspicuously disproportionate self-regard). I will define scientism below, after I have spelled out my working definition of the Frankenstein myth. For now, let me suggest that one presupposition of this book is that the power of an idea is not always proportionate to its seeming sophistication. Here is an idea: *the Frankenstein myth is the esthetic revelation of the horror of scientism.* It may well do for us. The semi-explicit lesson of the myth, its tacit creed, is that humankind neither can nor should, whether now or at its origin, be re-created with models bound by the experimental procedures and mathematical formulae of materialist science alone.

The second trend pertains to discourse both inside and outside the academy. Both in the critical literature and in moral attitudes popularly taken in the postmodern era, a tendency called *victimary thinking* prevails. The victimary approach, with the best of compassionate intentions, makes the mistake of aiming to humanize the victims of the mad scientist, thereby obscuring the fact that the "humans" fabricated by the mad scientist are not quite human. As one cannot be a little bit pregnant or a little bit dead, one cannot be a little bit human; so the victims of the mad scientist are not human. To the extent that the victimary tendency operates, the horror of the Frankenstein myth diminishes. It becomes increasingly difficult to provide an accurate description of the moral error represented by the mad scientist. Attempts to appreciate the otherness of those beings I name *impossible humans* also begin to trip up. In its resistance to victimary thinking, this book will respectfully beg to distance itself from more than one commentary on these works.

Defending the *Mad Scientist Plays God* Formula

As a description of the central action in the Frankenstein myth, the formula *a mad scientist plays God by trying to re-enact the origin of the human, and disastrously fails* may inspire a few signs of hesitation and skepticism. However, the formula constitutes a thesis, a working definition, so an open rejection now of its purported obsolescence is best. It is not uncommon for the professional critic to betray embarrassment at the supposedly invidious oversimplification carried by the phrase *mad scientist.* The attitude might be paraphrased like this: *I am not going to disrespect the scientific establishment by crediting a cartoon vision of the bug-eyed cackling want-to-rule-the-world materialist, and I am certainly not going to kowtow to theistic pieties. Are we not yet ready to admit that there really is no alternative to scientific naturalism as a way to human self-understanding? There is real racism, real sexism, real social inequality oppression in the world—but real mad science?* Certainly, I concede that to re-apply the label is a little like taking out of the wardrobe a suit that one has not worn in a long time. But I contend that the fashion the suit represents has never really passed away, and if we put it on again, it will fit surprisingly well and feel good. The formula makes a lot of powerful sense. One might observe for starters that *mad scientist* carries an appropriately double valence: its connotations of "mad" as insane (irrational, delusional) mingle with those of "mad" as angry (deeply resentful, violent in attitude).[3] The notion that violent anger is the way toward irrational violence certainly fits with the Frankenstein myth: the mad scientist is so angry at the seeming failures of "God" that he tries to remove God from the universe by making a new, better set of creatures, freed of any entanglements in the humbug of God. On the face of it, the wish to re-start human history should strike most of us as crazy: is there nothing now in place on which one might build, are things so terrible that a totally new humankind is needed? The mad scientist alone in his private laboratory is not the public revolutionary inciting the desolate mob, but something of a like unilateral passion for destruction is there. Furthermore, one might observe that if the mad scientist represents *scientism,* not science, then we should feel less hesitation. There should be no embarrassment in a wariness of scientism.

Another reason to stick with the *mad scientist* concept is the fact that the figure has continued to appear in many artworks in the decades since *Blade Runner* and the formula fits them, too. Recent literary production includes many serious dramatizations of heirs of Victor—mistaken, hubristic, and destructive scientists trying to re-make the human. In the category of acclaimed literary fictions, one could include *Vienna Blood* (1999) by Adrian Mathews, *Oryx and Crake* (2003) by Margaret Atwood, *Wetware* (2003) by Craig Nova, Maurice Dantec's *Babylon Babies* (2005), and *Never Let Me Go* (2006) by Kazuo Ishiguro. In the realm of popular fiction, I would include titles such as Robin Cook's *Mutation* (1989), Tom Hyman's *Jupiter's Daughter* (1994), Ken Follett's *The Third Twin* (1998), John Darnton's *The Experiment* (1999), F. Paul Wilson's *Sims* (2003), the five-volume retelling of *Frankenstein* (2006) authored by the best-selling Dean Koontz and his co-writers, Michael Crichton's *Next* (2006), and Peter James' *Perfect People* (2011). In the movie industry, one could include Steven Spielberg's *A. I. (Artificial Intelligence)* (2001), Michael Bay's cloning thriller *The Island* (2005), and Vincenzo Natali's *Splice* (2009). Not least, Nick Dear's dramatic adaptation of *Frankenstein* produced by the National Theatre in London and directed by film-maker Danny Boyle (of *28 Days Later* and *Slum Dog Millionaire* fame) enjoyed huge audiences. When I venture generalizations about the *mad scientist* or the *impossible human* (about the *vain scientist, artificial human, or metahuman*), please know that they frequently apply not only to the four canonical works under intensive scrutiny but also to the works in this much larger set. These works too merit membership in the association of textual bodies sustaining and promulgating the Frankenstein myth. Their critical acclaim and market success would indicate that the mad scientist figure has not lost its esthetic appeal or ethical relevance. The Frankenstein myth is alive and well, not dead and buried; it circulates in the world; it is not locked in a laboratory deep freeze.

A mad scientist plays God by trying to re-enact the origin of the human, and disastrously fails. Skeptics may also hesitate over the phrase *plays God.* When Leroy Augenstein published *Come, Let Us Play God* (1969), a study rich with descriptions of dilemmas in medical ethics and social planning, dilemmas explored in a tone that may strike us as alien in its exuberance, the expression had a certain novelty. It is almost a point of honour for critics working with these texts to hint in passing at their knowing boredom with the *playing God* metaphor. And yet outside the domain of literary study, there is a substantial scholarly literature both about the metaphor itself and its real-world referents (Bayertz; Evans; Gordon Graham; Howard and Rifkin; Heyd; Peters; Nossal and Coppell;

Splicing Life; Verhey). As with the phrase *mad scientist,* this book aims to rejuvenate the expression by resituating it in a context where it might stand decisively for the wisdom it implies: something is *wrong* when the scientist carries out such experiments, attempts to remake the human from scratch, as if the communal past could be erased and history be re-started. Likewise, something is wrong with scientism—as I define it, the reduction of anthropology to biology. I have explored the metaphor at length in a study of its uses in the discourses of genetic engineering ("Accusations"). For now, a brief analysis of three of its levels may help.

First, the phrase *playing God* will not denote for us simply a rule against "interfering with the sacred laws of Nature." If we conceive of God as the designer of the laws of Nature, then the scientist accused of violating those laws may be accused of *playing God.* Even if you know *how* to animate a body from pieces of sewn-together corpses, even if you know *how* to create a sex-robot or a designer baby, so the argument goes, you should know *that* such procedures are obviously wrong. They should seem as (obviously) wrong as the laws of Nature you must understand to perform them have (obviously) become knowable. The objection to such an argument is this: one now has the problem of explaining why God would permit the specific laws and truths of Nature to be discovered and thus potentially violated, if God had intended, when creating Nature and us, the law never to be broken. The objection points to a certain confusion between cosmological and moral law, and the difficulty of deducing moral laws from laws of Nature. The physical law of gravity, the chemical laws that determine which substances are poisonous to human bodies, or the biological laws that govern which sex acts risk pregnancy, are indifferent to humans and knowable in a way that moral laws are not—those moral laws (for example) forbidding us from throwing a violent enemy down a stairwell, shaking arsenic into a tyrant's soup, or exuding contempt for the feelings of a partner in erotic play. Laws of Nature are indifferent to humankind; humankind differs from Nature in trying to have moral laws. Accusations of playing God that refer to violating sacred laws of Nature produce little of value. Several practices that many people now consider acceptable—inoculation, or organ transplantation; " . . . in vitro fertilization . . . embryo manipulation, embryo donation and surrogate pregnancy" (Kass 139)—were pioneered by scientists eager to share the possible benefits of their technological discoveries but accused at first of "playing God." We shall see that scientism is objectionable mostly not because of what laws of Nature it upholds as relevant to rocks and trees, insects and nonhuman mammals, but because of what laws it proposes for the conduct of human affairs. Scientism, like playing God, is easier to grasp as a violation of moral or anthropological truth than as a violation of the "laws of Nature."

Let us now move the metaphor up one level, from that of a purported identity between natural and moral law to that of an involuntary autonomy of the moral, cut off from the guidance of Nature. Given certain understandings about facts of Nature and the cosmos, particularly facts about our human bodies, we are faced at this level with extremely difficult moral questions in debates on such matters as euthanasia, reproductive technology, population control and ecological management. Competing interests clash; the primary fact at this level is that no matter what moral decision you make, somebody is going to suffer. There is no way out of the situation in which somebody, perhaps even the innocent, will be hurt. At this second level, *playing God* means something like *pretending to an impossible moral wisdom,* a wisdom we imagine God possessing (if God exists). We can universally agree that we wish we did not have to make such difficult choices, for we agree that the rightness or wrongness of each choice could be calculated with serene confidence only from an all-knowing, all-seeing, all-benevolent perspective such as God is believed to enjoy. For that matter, imagining even God finding such choices easy to make is next to impossible, at least as unhelpful as imagining God (cf. Rescher, *Limits* 36). This reluctant exercise of a moral authority that we wish we were not obliged to exercise is different from the violation of a supposedly transparent law of Nature. Our mental burden is the incalculability of the costs and benefits of a decision that must be made without the guidance of any absolutes. Now, take notice: the vain scientist in the Frankenstein myth is not forced by unfortunate circumstances into an ethical dilemma that he himself has not made. When the vain scientist decides to re-make the human, he is not choosing between alternatives. On the contrary, he is creating disasters from nothing other than the wellspring of his ambition. No necessity forces Victor to make the monster; no necessity compels Moreau to cut and paste his Beast People, or Rossum the older or younger to create the robots; Tyrell synthesizes his replicants of his own volition, not because he cannot do otherwise. These performances have no cause other than an imperious desire to exercise a technological know-how in a morally vacuous way. Scientism is wrong-headed not because of the scientific discoveries it makes about the cosmos but because of its casting aside of often ancient religious truths about the human. Such include the fact that intersubjective structures in human exchange are real despite their non-observable nature; the imperative of moral reciprocity is a human universal despite its irreducibility to the physical; and the goodness of compassion is indispensable to human society although the value of such goodness cannot be reduced to a mathematical calculus of costs and benefits.

A third level at which the *playing God* metaphor operates is that captured by our working formula: to play God to is to try *to re-enact the origin of the human.* This book thoroughly endorses the metaphor in this sense and intends to elaborate upon it. The claim that a mad scientist *plays God* includes an argument against scientism's disrespect for moral intuitions. But I take another long, serious step. Just as important is the level at which God is imagined, believed, or theorized in the abstract as a Being who accompanies humankind at its origin and whose tacit presence sustains the mysteries of human difference indefinitely in history. At this level, we become free to observe with special care the signs of the mad scientist's strange desire to re-create humankind, to make a new human race, in an implied rivalry with a creator whom the scientist hates and whose central position the scientist usurps. That hate-motivated usurpation (or a certain conspicuous indifference to God that amounts to the same attitude) operates ironically even when the scientist is an atheist: *ironically* because one should not bother acting out a hatred for that which one believes does not exist. I will elaborate on the anthropological idea of God below, but a preliminary observation belongs here: in the Frankenstein myth, we witness creation scenes in which "God" is violently displaced and replaced by the mad scientist creating impossible humans. When we see Victor animate the sewn-together body that will henceforth be the abandoned, solitary Monster; when we hear Moreau's victims of vivisection screaming in pain for hours; when we hear Domin tell Hellena the tales of old Rossum's trying to prove the nonexistence of God and young Rossum's creating the perfect worker by "chucking out" everything human from his re-design; when we witness Tyrell's attempt to make Roy Batty feel indifferent to the violence he has performed in his desperate mutinous quest, we are witnessing creation scenes. But such creation scenes make up amoral inversions of any scene in which we might imagine a creator participating in the emergence of an originary community of humans,

humans to whose goodness and flourishing the creator is tending.[4] In spectacular contrast, the relevant scenes in the Frankenstein myth display an oblivious scientist who himself stages a moment of creation to rival those of religious creation myths. At this level of re-enacting origins, the metaphor *playing God* should seem all but irresistible as a descriptor of the initiating action in the core story.

The Frankenstein myth is a creation myth not for the age of modern science, but for the era in which a materialist scientism contemptuous of humankind becomes a viable contender for public attention. If we think of the human community as united in a concrete moral equality, and think of "God" in the most minimal terms possible as the one spiritual Other of humankind—even if the God we thus acknowledge seems one historically passive, one beyond representation, one who offers no guarantee other than that of Being an absolute witness to the human drama—then we can sense how tormented and trapped are the monsters, Beast People, robots, replicants, and clones whom the mad scientist has, in his bad creative work, molded and sculpted and shaped for underthought purposes. True, we might resent God for a kind of seeming passivity in the face of historical human suffering (no one, not even the most faithful, is immune to such resentment). But it is significant that the freedom such purported passivity has permitted us as real humans (not impossible humans) strikes us as preferable to the torturous restrictions that mad science builds into the lives of its creatures.

Resisting Victimary Attitudes

A second trend in the scholarly work on these four texts—the trend of fuzzy confusions engendered by victimary approaches to the myth—is not ubiquitous in the criticism, but certainly significant, perhaps dominant. I will consistently be distinguishing real humans from impossible humans. You may soon ask, whence my impudent self-assurance in assuming that the long-suffering Monster in Shelley's novel does not qualify as human? The Beast People walk upright and talk and seem to have thoughts: why can we not classify them as human? Why believe the replicants in *Blade Runner* are not human? But my answer to the question—*Are the creatures of the mad scientist playing God the creator, humans?*—will remain, *No.* They are humanlike, yes; they are impossible humans, yes; humans of our kind, no. Now it is victimary thinking that makes my distinction seem invidious. It is victimary thinking that prompts questions and doubts about the nonhuman status of the mad scientist's creatures and that inspires the eagerness to protect possible victims such as those of the mad scientist by granting them "human" status, perhaps in the belief such a gift might enlarge our sympathy for them.

Briefly put, the victimary, a category applied intensively by Eric Gans, is the ethical response to the revelations of Auschwitz and Hiroshima.[5] It follows a moral model rooted in a climate of sensitivity to the human capacity for exclusionary violence. This climate of sensitivity is not irrational; it is, to a large extent, culturally productive in the best ways; in any case, it is all but inescapable, particularly in the academic circles where workers in the social sciences and humanities frequently must justify their research by referring to victims who might be helped by its findings. The risk is this: victimary thinking implements a model that tends toward reducing all social relationships to those of *persecutor and victim,* on the model of the Nazi and his victim in the death camps or the model of those who dropped the bombs that annihilated Hiroshima and Nagasaki as against their victims. Victimary thinking is sound and good in its desire to minimize human violence; where it errs, where it risks escalating rather than containing violence, is in the temptation it creates for the defender of victims to rush in self-righteousness toward taking up the cause of the victim while creating solidarity in resentment against the persecutor. The self-righteousness depends on a prior conviction that the victim is self-evidently such and the persecutor easily identifiable, the difference between them absolute, transparent, unquestionable. The unpleasant fact, however, is that to wrestle with moral questions in the postmodern era not infrequently means to be in situations where one cannot unambiguously sort out victim and victimizer, oppressor and oppressed, strong and weak, blameworthy and blameless. The causes and sources of human suffering are terribly many. As with most ethical touchstones, victimary thinking serves better to monitor one's own conduct than to motivate one's moral indignation against a rival. The thing that commends victimary thinking to us—the absolute desire *not to victimize others* and the absolute desire *not to be violent*—is the very thing that makes us vulnerable to its perversion. For the desire to spot and blame the violence of others is always easier to sustain than the discipline required to step back, to notice and to take ownership of our own violence (especially the temptation to accuse *others*).

Complicating matters further is the fact that to seek attention by claiming victim status for oneself is more common a form of behaviour in the postmodern era than it was prior to Auschwitz and Hiroshima, when people lived in a moral climate where it was considered more dignified to be private, perhaps even ashamed, of one's victim status. We do well to observe that the prevalence of victimary attitudes in postmodern culture is something of an historical anomaly, itself an ethical climate that may be dissipating. In any case, as much as I will persist in defining the nonhuman status of the victims of the mad scientist in the Frankenstein myth, just as certainly I will detail the sufferings of Victor's monster, the Beast People, the robot Damon and the replicants in Scott's movie. The problem is that in the Frankenstein myth, even though the victims of the mad scientist suffer terribly, they are far from being innocent themselves. The situation of Frankenstein stories is modern; it is complicated by horizontal envy and counter-accusation.

In the context of victimary attitudes, the presupposition of this book boiled down to a slogan is this: *if you want to understand the Frankenstein myth, let the impossible human be nonhuman and let the mad scientist playing God thus be seen to fail.* For ironically, if we argue that the mad scientist's creations are fully, truly *human,* then we have presupposed that the scientist has successfully imitated and outdone God. If there is no real difference between our realm of being and that of the Monster, the Beast People, the robots and the replicants, then we are depriving ourselves of the most serious objections to be made against the scientistic usurper as one who has violated our moral intuitions in arrogating to himself the Creator-like privilege of setting the design limits for the experiential reality of a human-like creature. If he has in fact succeeded *in the first place* as the creator of a fully human realm, then in an important sense the realm to which the vain scientist relegates his creatures cannot be considered any worse than ours in its ontological outline.

Simply at the factual level, this obliviousness to the failure of the vain scientist to create a *fully* human creature strikes me as a big mistake. The horror for Victor's monstrous creature is that his world is utterly *unlike* our human world; the horror is that the world to which Victor has confined him is dominated by the oppressive omnipresence of Victor as the one who made him ineradicably ugly. Quite the same oppressiveness has ever and always trapped the Beast People of Moreau, has confined the robots in *R.U.R.* and condemned the replicants in *Blade Runner.* They do not enjoy the freedoms, abilities, and terms of existence that ordinary humans enjoy. How could they, having been made as they were in a grotesque parody of divine creation or evolutionary emergence? How? The fundamental fact about the ontology of these creatures inside their fictional worlds is that of the omnipresent oppression in the "design" that the mad scientist has built into their restricted and tormented existences. Certainly we readers may suffer as humans, but we do not suffer just as these impossible humans do. The point of each story that contributes to the Frankenstein myth, properly appreciated at this more-or-less literal level, is one with the point of an invitation the myth extends to us as modern people. It is an invitation to stretch ourselves imaginatively to try to take literally the horror of what it would be like to be a monster such as that made by Victor Frankenstein or a replicant such as those manufactured by Tyrell. But to perform that stretch means not to comfort ourselves by deciding to call these characters *human* and insisting on their similarities to us; on the contrary, it requires the more difficult work of trying to imagine what it would be like *not* to be us.

Therefore, the slogan *let the monsters of the mad scientist be monsters* reminds us that it is probably a misappropriation to "humanize" the victims of the mad scientists inside their fictional worlds by re-classifying them from the outside (from our world as readers). I would suggest that nothing is gained by our waving the banner *human* over their painful careers. On the contrary, the harder we try to humanize the impossible humans in the Frankenstein myth, the more injustice we do them. Let Shelley's Wretch be the wretch he uniquely is, let the Beast People be beast people, let the robots be robots and replicants be replicants. They speak to us from a place we cannot go on the other side of the tragic curtain that separates their fate from ours. They speak to us, more specifically, from a nonhuman place that we could not wish to be.[6] To become conscious of the full impact of the Frankenstein myth requires more than disapproval of the experimentalist; it requires more than sympathy for his monsters. To become conscious of the impossible human as other requires us to sense the horror of having been made under such weird, unthinkable conditions. The coinage *impossible human* is designed partly to foreground that ineradicable difficulty in our experience as readers and viewers: it is impossible fully to identify with them. They are *impossible* in the sense that we cannot deal with them without trouble; we will never be relaxed, at ease, or unthreatened in their presence; they wear us down with their difference from us. As the vain scientist fails in re-enacting human origin, so we must fail in fully identifying with his creatures. ("Impossible" also serves as a reminder of the restrained science-fiction fantasy in each version of the myth: in reality, no human can succeed in re-enacting the origin of the human even to the partial extent the mad scientist does, despite the liquid spillage of such fantasies into real-world scientism.) Therefore, to explore wide-eyed the worlds of Victor, Moreau, R.U.R. and Eldon Tyrell is to be assaulted, overcome with feelings of disorientation and disgust. It is to be torn between the desire to understand the impossible human's painfully inspired resentment of us, and the fear of their alien bottomless violence toward us. When the experience is complete and we close the book or turn off the DVD player, we should be newly conscious of what it means to be human. We discover what it is like for us to find it *impossible* to embrace them as wholly human, no matter how we try, because we have understood their ineradicable difference from us.

More Criteria for Counting as a Story that Builds Up the Frankenstein Myth

I have sketched the two polemical edges of this study. One is the claim (contesting the views of those fed up with what they believe to be clichés) that to formulate the core of the Frankenstein myth as a *mad scientist plays God by trying to re-enact the origin of the human, and disastrously fails* is to permit a new releasing of its esthetic and ethical power. The other claim contests the domination of victimary thinking in the post-modern academy. It suggests it is best to describe the non-human status of the victims of the mad scientist rather than deny and obscure it. Four more lines I draw now will further map the inquiry this book is undertaking. First, an *anthropocentric* criterion—the mad scientist must create a human-like creature in his madness—differentiates other stories with overreaching scientists from those contributing to the Frankenstein myth. Second, I classify the myth archetypally as a tragedy of shock and horror. Third, I impose an *historical* criterion: the myth originates with Mary Shelley's text, and the six basic narrative components of her story must appear in all its variations. Fourth (accenting one of the six narrative components), I privilege a *dialogic* criterion: it is not enough that the creature be human-like; the story must also include an explicit dialogue, however brief, between impossible human and scientist-creator, in response to the Judeo-Christian tradition of human-God dialogue against which the Frankenstein myth sounds its deepest reverberations. Frankenstein is to the Monster as Satan is to Job.

The anthropocentric criterion requires that the scientist attempt to make human-like creatures by some scientific method or another—animation by electricity (Victor), grafting of tissues (Moreau), synthesis of fantastic raw biological materials (Rossum), genetic engineering and biomechanics (Tyrell), genetic engineering (Atwood, Mathews, Nova) or cloning (Ishiguro, Darnton, Bay). This criterion constitutes an official notice that there are numberless stories where scientists brazenly show the proud moral indifference of Victor and his heirs, but engage in projects that have nothing to do with the creation of humanoids or an attempt to re-start history. The maker of the dinosaurs in *Jurassic Park* (Spielberg) belongs to the category of the mad scientist as a genetic engineer who takes terrible risks, but he does not make humanoids. The scientist in *The Fly* (Cronenberg) is an overreaching experimentalist whose errors have the terrible effect of turning him into a nauseating agglutination of *homo sapiens* and housefly; interestingly unlike the Frankenstein myth, he makes of *himself* a victim—in the Frankenstein myth, it is always the *impossible human* who is the founding victim. The scientist in Wells' *The Invisible Man* likewise does the most harm to himself; he is not an heir of Victor Frankenstein. Many mad scientists seek "divine" powers only of destruction: those who seek to wage war with thrilling new weapons (atomic, biological, chemical) are Frankensteinian in their morally oblivious arrogance, but have nothing to do with making metahumans. David J. Skal's *Mad Science and Modern Culture* (1998) surveys the copious variety of scientists whose projects involve Frankensteinian hubris and incaution, but not the goal of re-enacting creation. For this study, a true Frankenstein story must feature a vain scientist who plays God by trying to re-enact the origin of the human. Not just any scientific hubris with disastrous unintended consequences will do.

In defining the Frankenstein myth *archetypally,* I impose the criteria that a story contributing to the myth must exemplify what Northrop Frye in *Anatomy of Criticism* (1957) defined as the tragedy of shock and horror. Frye delineated a subtype of tragedy characterized by "a world of shock and horror in which the central images are images of *sparagmos,* that is cannibalism, mutilation, and torture. The specific reaction known as shock is appropriate to situations of cruelty or outrage" (222–23).

That such images play a central role in the Frankenstein myth is beyond question: Victor cutting cut up corpses to make an eight-foot tall body that he will electrocute into wordless consciousness; Moreau carving up animals and making them into Beast People in what he calls the bath of pain; the production and sales team at R.U.R. manufacturing millions of robots from vats of vein, muscle, bone and tissue; the Tyrell corporation growing replicants from recombined DNA to sell them as slaves in outer space. Frye places an oxymoronic *villainous hero* at the center of tragedies of shock and horror: "In such tragedies the hero is in too great agony or humiliation to gain the privilege of a heroic pose; hence it is usually easier to make him a villainous hero" (222). Clearly, Victor and Moreau and Harry Domin and Tyrell are only villainous heroes, if heroes at all.

Meanwhile, the fateful performances of Shelley's Wretch, the Beast People, the Committee Robots, and replicant Roy Batty qualify as responses to "agony or humiliation." In the Frankenstein myth, the Victor-figure will condemn his impossible human to performing as a "villainous hero." This archetypal criterion has the valuable consequence of excluding many tales of artificial humans in which torture, bloody surgery, and violent revenge play no role. The many robot stories of Isaac Asimov, for example *The BiCentennial Man,* made into a memorable Hollywood film (Columbus), dramatize the fates of artificial humans but certainly do not qualify as versions of the Frankenstein myth. Likewise, the feminist science fiction *He, She, and It* (1993) by Marge Piercy, examined at length in Elaine Graham's *Representations of the Post-Human* (2002), tells the tale of a team of scientists who create cyborg humans. The relations between creator and creation in Piercy's text are not dominated by violent conflict; on the contrary, the male cyborg proves exquisitely skilled in love-making, a pleasure to his female human master. It does bear noticing that in both examples even the good-natured cyborg comes to an untimely end: he chooses to commit suicide. To that extent, even in such cases of modestly benevolent humanoid creation, the vain scientist fails to create a being who can enter seamlessly into the world of human history. He fails to re-enact the origin of the human.

A third criterion is historical or chronological. Mary Shelley's novel is the founding work of the Frankenstein myth. It is true that there are precursors in legends of the man-made or synthetic human. In *Human Robots in Myth and Science* (1966), John Cohen details many such talking statues and strange automata, such as those created by Hephaestos in

Homer's *Iliad* (15–25). The Jewish legend of the Golem resembles the story of the Frankenstein myth to the extent that a humanoid creature is made by humans and its makers lose control of it; but there the similarities end, given the atheistic cast of the creators in Frankenstein stories. There are affinities between the Frankenstein story and the Faust legend, given the shared theme of *forbidden knowledge;* the central project of Faust, however, has little to do with re-enacting the origin of the human and, like the Golem legend, it takes place in a context not of atheistic materialism but of theistic supernaturalism, in a universe populated by angels and demons, ghosts and spirits, and framed by the guarantee of divine judgment in an afterlife realm (see Reichardt 136–37). The story of Frankenstein is different from the legend of the Golem, the Faust legend, or tinkering that produces mechanical chess-players, ducks, or dolls. Nor would Somerset Maugham's *The Magician* (1908) belong to the myth: although it tells of the creation of a homunculus, it has nothing to do with re-enacting the origin of the human. It is about magic, not scientism. A few significant differences between the Faust and Golem traditions and the Frankenstein myth are worth noting. Mary Shelley was working in a context where Enlightenment atheism was worthy of some public respect, despite the minority status of its adherents. She was writing in a context where biology, although as yet unnamed, was well on its way to becoming a science in its own right, with (for example) the evolutionary speculations of Erasmus Darwin, the grandfather of Charles, echoed in the text of *Frankenstein; or, the Modern Prometheus.*[7] And not least, although the expression *playing God* had not yet been invented at the time of her writing, Mary Shelley herself casts the work of her character Victor Frankenstein as a project that expresses an intentional rivalry between Victor and the creative centrality of the God of Judeo-Christian tradition.[8]

My fourth, final criterion for inclusion of an artwork among those building up the Frankenstein myth is the presence of a narrative sequence consisting of six narrative kernels or plot-points: [1] *experiment; [2] partial success; [3] defect; [4] dialogue; [5] violent revenge; [6] death.* Most decisive among these is the dialogue between mad scientist and impossible human. But let me give a quick survey. First, the mad scientist discovers facts about the natural world that enable him to perform powerful and impressive experiments. The demonstration that his humanoid-creating experiment has succeeded is one of the must-do moments. Even if we try to fit our four texts to Shelley's model of the moment of animation, there is considerable continuity between them. In *Frankenstein,* there is only a sigh and a shudder, as the creature opens one yellow eye (1.4.85–85). On Doctor Moreau's island, the cries of the tortured puma at a certain moment sound to Prendick like human screams; that phonic shift marks a successful quasi-humanization (9.50). In *R.U.R.,* old Rossum wants to usurp the creative throne of God; later, Rossum the son decides to chuck everything not related to useful machines—in fact to stop playing God, to make robots rather than people—and the production of mechanical meta-humans follows *en masse* (Pro. 7–9). In *Blade Runner,* the interrogations of Leon and Rachael under the Voight-Kampff machine are a postmodern substitute for the indignities of the operating table: if you can identify the replicant as replicant by such psychological testing, then you are permitted to kill it.

Second, the partial *success* of the mad scientist's project complicates matters in the story. This element guarantees the Frankenstein myth is opposed not to science, but to scientism, for the partial success of the scientist is to some extent admirable. The readers or audience members must be able to understand the temptation the vain scientist has failed to resist: the scientist alone cannot be our scapegoat. If the myth would expel any party, it would expel mad scientist and impossible human together as one misbegotten, doubled unit, cursed twins. Mad scientists demonstrate genuine theoretical expertise (in sciences such as chemistry, biology, zoology, genetics, neurology, physiology) and practical technological prowess (using the scalpel, calibrating the equipment, splicing the DNA). Victor's made-out-of-corpses monster comes to life and survives; tested as an experimentalist, Victor passes. Moreau's Beast People look like humans and practice something like a primitive religion; to the extent that Prendick is fooled into thinking them human at first, they "prove" Moreau's hypotheses about the plasticity of living forms. Rossum's robots are amazingly successful as consumer products; they are powerfully efficient machines. The replicants in *Blade Runner* likewise "work" as well-made machines are expected to work; Tyrell is a corporate demigod, for his neo-Mayan pyramid home symbolizes political power and financial independence beyond the most maniacal dreams of avarice and tyranny.

These partial successes, however, are followed by the revelations of a *defect* in the design of the artificial human. The weirdest here is the ugliness of the Monster as it appears to Victor once it is animated; why Victor did not see that defect before the animation is a profound enigma in the text.

For Moreau, the devolution of each of the Beast People into animality is the sign of their defectiveness; he has not succeeded in creating a humanoid beyond pain and fear. Capek's robots become defective when Gall secretly endows them with a capacity for resentment of robot-human difference. Tyrell fails to anticipate the emotional growth that his replicants undergo, which opens the possibility of rebellion; therefore, he programs them with false memories and builds into them a four-year lifespan.

We need not linger over the violent revenge (fifth) and death (sixth) plot-points. More important for our purposes is the fourth plot-point, the dialogic exchange. The story must, to qualify for membership in the canon of texts belonging to the Frankenstein myth, include dialogue between the impossible human and the mock-creator: they must meet, and the Victor figure must be required to account for himself, however briefly. When the artist who constructs a vision of the Frankenstein myth represents this moment of the impossible human asking his maker about his suffering and the reasons for it, the artist is working in the context of Biblical revelation and the Jewish and Christian traditions of dialogue between God and humankind. The lament of the impossible human at the moment of its deepest despair will resemble the lament of Job.

> Why didst thou bring me forth from the womb?
> Would that I had died before any eye had seen me,
> and were as though I had not been,
> carried from the womb to the grave.
>
> Job 10:18–19 (RSV)

The question is, *why have you made me in the first place?* And to that question, the mad scientist—a tawdry substitute for the God whom Job queries—never has a satisfactory answer. Often, he has no answer at all. Believers or atheists, we must concede a grotesque disproportion between the explanations, rationalizations, excuses and alibis of Victor, Moreau, Harry Domin and Eldon Tyrell, and the richness of a theological tradition that includes at least the *possibility* of a good Creator in dialogue with humankind. I am thinking here of a Creator who could, at one level, step back from his handiwork and see that it is good.[9] I am thinking here of a Creator who is believed to have taken on the form, in another of his Persons, of a suffering servant, God Incarnate certainly not as victimizer of humans but as victim of humans and revelation to humans of their violence. No mad scientist comes even remotely close to genuine rivalry with such models of divine creativity, whether the One who can call his works good or the One who becomes divine self-giving incarnate in victimized human form. Victor Frankenstein cannot substitute for the figure of the self-giving Crucified.[10] To repeat, Frankenstein is to the Monster as Satan is to Job. The phrase *plays God* feels appropriate when we recognize the disproportion and distance between the mad scientist-impossible human as a set of two, and the set consisting of the Creator and humankind engaged in genuine dialogue.

Scientism as the Reduction of Anthropology to Biology

This book proposes for your meditation the thesis that *the Frankenstein myth is an esthetic revelation of the horror of scientism.* I propose a new definition of scientism as *the reduction of anthropology to biology.*[11] But I begin by taking the precaution of spelling out a point already made. I wish to avoid a certain *myth* in the sense of false belief. Paisley Livingston outlines it in the relevant pages of *Literary Knowledge: Humanistic Inquiry and the Philosophy of Science* (1988): the complex of confusions arising from a conviction that "science equals scientism" (107).[12] I do not in this book invest in that form of postmodern relativist skepticism which questions the objective knowledge the natural sciences have produced and which expresses suspicious contempt for the scientific "establishment." I do not reject external realism, the idea there is a reality external to human mind. On the contrary, I hope to avoid any such anti-scientific attitudinizing of the type interrogated by Paul Gross and Norman Leavitt in *Higher Superstition* (1994). I do not consider modern science a social problem as does Sal Restivo. Perhaps scholars in the humanities should be embarrassed by the Sokal hoax, rather than mounting defenses that deny the special productiveness of that set of philosophical presuppositions, epistemological methods, and institutional practices that fall under the rubric of modern science.[13] I agree with Livingston's claim that the instrumental reliability of much scientific achievement is something the framework relativism of the postmodern skeptic makes it difficult to explain (73–79; see also Rescher, *Limits* 38–39, 42). In short, this book is not itself, nor does it interpret the Frankenstein myth as, an attack on science. Furthermore, with respect to the perennial debate between the "two cultures" of literary creativity and scientific research (Snow),

I do not claim an identity of the type of authority that legitimates the kinds of knowledge that we get from everyday observation, sacred texts, works of art, and economic practice (on the one hand), and the type of authority that legitimates the peculiarly powerful knowledge produced by the empirical and mathematical methods of the natural sciences (on the other). The types of authority are not identical; it is vain to wish one could make them so.

Nevertheless, there is a point at which skepticism about the findings of scientific naturalism is permissible; and not only permissible, but commendable; and not only commendable, but in our time ethically pressing. That is the point at which the fact that human beings do not belong to Nature in the same way that other things belong to Nature gets to be ignored, overlooked, downplayed or obstinately denied.[14] I hope that most readers will find themselves reminded of this fact after a few reflections. When the levels of reality named by *the anthropological* are reduced to the level of the biological and the physiochemical, there the illegitimate, unreasonable ideology of scientism begins to knock off track the humane, reasonable institutional practices of science. Scientism is an ideology: that point has been made in different ways by the existentialist humanist William Barrett (xv), by the philosopher of biology John Dupre (*Disorder* 167), by the Heideggerian philosopher Frederick Olafson (xv) and others. When the natural scientist begins to study those parts or signs of the human being that are not obviously reducible to the physical body, everything changes: everything must change. The facts that have been provided by physiology and medicine seem valuable, for we know ourselves as creatures of flesh, blood, bone, muscle, skin, hair, nervous system and braincase buzzing with feelings from joy to misery. Defenders of scientific naturalism, therefore, rather than pointing to the less obvious benefits of modern hydrology or mechanics, often point to those accruing from modern medicine when they seek to lessen the hostility of science's opponents. Even respecting the great advances medicine and technology have made in the last 100 years, however, we might keep in mind a truth that the philosopher Nicholas Rescher points out: to alleviate suffering is not to guarantee happiness; a human can be at once perfectly healthy and utterly joyless (*Human* 186).

When we get to the level of biology, to present as a factual claim that human animals do not belong to Nature in the same way other things belong to Nature seems more questionable. One thinks of the oft-repeated dictum of the prominent contributor to the modern synthesis in biology, Theodosius Dobzhansky: *Nothing in biology makes sense except in the light of evolution.*[15] The problem with the truth that humans belong to Nature like everything else does is the part of that truth corrupted by the popular influence of sociobiology, with its numerous suggestions to the effect that *humans are just like other primates, only more intelligent.* On the other side, there is a certain strand in cognitive science, in particular the computational model of mind that dominates much neuroscience. This other camp betrays an ambition to explain everything human with suggestions to this effect: *human minds are nothing but sophisticated computers evolved to maximize reproductive fitness.* Now we owe it to ourselves to value and respect knowledge; the ethical question is whether we value and respect each form of knowledge in the appropriate way and to the appropriate extent. I am not alone when I suggest reluctantly that sociobiology and cognitive neuroscience make it their goal not just to explain, but to explain *away* our knowledge of the ordinary experiences that we have in interpersonal exchanges as somehow less valuable than the knowledge of ourselves that we are expected to have once we have been thoroughly tutored in the moral, religious, and anthropological implications of such reductionist biology and neuroscientific physicalism.[16]

Given the prestige far too often granted to hypotheses proposed by those who do sociobiology and neuroscience, the Neuromania and Darwinitis that Raymond Tallis courageously challenges, we might learn to ask more often and more insistently the question: *what non-trivial anthropological fact can the theory of evolution tell us about ourselves—about me, myself, I, and you, yourself, reader?* Consider: you and I evolved from animal ancestors. I would not be writing this and you not reading it if our primate parents had not had sex, if our primate mothers had not conceived and carried and given birth to us, as did their biologically determined parents before them; we are therefore (grand conclusion!) bipedal evidence of the *reproductive fitness* of our animal ancestors, human and non-human. So let us applaud this conclusion to the drama that has brought us onto the scene, accepting its lesson that we ought to live in gratitude to the cosmological reality of the maximizing of reproductive fitness that our ancestors both ultimate and proximate incarnated. What follows? It means that if we are—to take one religious tradition—Christian believers, we now feel obliged to reinterpret the Holy Bible, especially the first few chapters of Genesis, and to formulate a more nuanced theology to accommodate a revised chronology for the much earlier appearance of the human on the planet. (That has already been done, one might note, more than thoroughly and more than once.) But what else? What properly anthropological, ethical, social truth follows from the revelation that *we humans evolved as other animals have evolved?*[17]

I put the question a little rudely because in this book, it is the *reduction of anthropology to biology* that I am taking as a polemical target, not from a position of theological obstinacy or rationalist prejudice, but rather from the position of one who has been emboldened to own his intuitions about four instantiations of the Frankenstein myth tested against the originary hypothesis of Eric Gans. Let me clarify that I use the word *anthropology* to name a far broader field than that belonging only to departments of anthropology in contemporary Western universities. When I say *the anthropological,* I mean *the human* as liberally conceived as possible. I mean not just archaeology, ethnography, paleontology, but the study of everything humans do that animals do not: *anthropology* is the pursuit of an understanding of everything to do with humans as acting, believing, desiring, historical creatures endowed with consciousness and language, able to engage in linguistic exchange, economic exchange, and technological home-making. Studies that occur in many disciplines would count as anthropology thus conceived: studies in the interpretation of sacred texts, religion, mythology, and theology; in philosophy (ethics, esthetics, epistemology); in linguistics, literature, literary criticism, theatre history, art history, critical theory, music, sculpture, architecture, dance; in humanistic psychology, in sociology, economics, and political science; in the history of religions, peoples, nations, minorities, ethnic groups, ideological movements, and political parties. Simply to walk through such a list is to be reminded of the size of the veritable oceans of knowledge in which we swim that have no immediate reference to the categories of physics, chemistry, and evolutionary biology. How did it happen that those three should have been awarded the pedestal on which they now stand as smiling-mime purveyors of all ultimately authoritative models for inquiry into the human? I would even suggest that the *reason* humans do physics, chemistry and biology is ultimately that they hold the promise of helping humans with practical problems that belong more immediately to other realms: physics for the social needs answered by architecture, chemistry for the personal needs answered by medicine and pharmacology, biology for the ecological care expended to preserve habitats and homelands for humans and nonhumans alike. (Meanwhile, I do not quite intend by anthropology in this wide sense the politicized research programme currently given the label "cultural studies."[18])

Scientism begins when materialist reductionism is applied to human beings. In *The Wonder of Being Human: Our Brain and Our Mind* (1984), the neuroscientist Sir John Eccles and the philosopher of psychology Daniel N. Robinson present a sketch of five troublesome constellations of belief, with this comment following the sketches: "It is our view that these five truisms of today's folk philosophy have much to do with the dissatisfaction and aimlessness infecting modern life and that they deny the modern citizen the pleasure and happiness of the human adventure" (6). I agree. In a series that will include relativism, materialism, evolutionism, and environmentalism,[19] Eccles and Robinson place scientism first in line.

> 1. *Scientism.* Where Scripture was once that court of last recourse in which every claimed truth had to plead its case, science is now taken to be the ultimate arbiter. It is difficult today to discover any significant proposal for government, social life, interpersonal relations, education, morality, or personal happiness that is not defended in "scientific" terms or criticized for being "unscientific." There is now a near-synonymy between "true" and "scientific" and a general conviction that what is not "scientific" cannot be true. Thus does the layman turn to the "behavioural sciences" and the "social sciences" for directions in matters of daily life the way he might turn to physics and chemistry for direction in his understanding of brute nature. As he understands things, the realm of thought is filled by only two finds of entities, scientific *truth* and utterly subjective *opinion.* There can, therefore, be no valid basis for opposing anything that science has not legislated against or for defending what science has not established. (4)

There is much to commend in this nontechnical definition: its implicit acknowledgment of an historical- narrative tension between religion (which has always been with us) and modern science (which has not);[20] its claim that appeals to science too often stop dialogue and thought by inducing the layman to shrug at his own half-informed opinion and leave passionate inquiry to others. Scientism, by attributing to a tiny minority of people the possession of authoritative knowledge, encourages a kind of idle cowardice: *Maybe "they" know; I don't; but having my opinions is better than having a fight with the experts.* Meanwhile, the thrice-repeated scare quotes around *scientific* signal that Eccles and Robinson object not to science, but scientism: some claims pretending to scientific status claim the privilege legitimately. Others borrow the prestige mistakenly: the scare quotes around "behavioral sciences" and "social sciences" query their claims to the property "scientific"; the methods that produce reliable knowledge when studying aerodynamics or mammalian digestion or the genetics of tadpoles have at best limited applicability to the linguistic, interpersonal realm that human beings inhabit.

Meanwhile, in *Scientism: Science, Ethics, and Religion* (2001), Mikael Stenmark makes the point that the way of science itself simply does not encompass all ways of knowing: "There are domains of knowledge outside of and independent of science. We have at least the domains of observation, introspection, self-reflection, memory, language and intention" (33). Human beings have bodies but they are not *just* their bodies; humans are *more* than their bodies (as, for that matter, many animals and plants may be considered more than material objects). Humans are creatures made up of intentions, beliefs, desires, hopes, regrets, histories, identities, and projected futures. More primarily, humans are made out of the exchanges linguistic and economic they have with other humans, without which exchanges they would never be counted as human at all. The problem begins, therefore, when the philosophical presuppositions of *materialist reductionism* that have been so successful in the natural sciences, in particular physics and chemistry, are taken up imperialistically and imposed upon human beings, those odd animals who spend so much time exchanging words and things in scenic configurations.

By materialist reductionism I refer partly to the "substantive reductionism" that Richard H. Jones in his magisterial study *Reductionism: Analysis and the Fullness of Reality* (2000) describes as follows: "The reductive process for materialists is always the same: the reality of a whole is identified with its parts or bases, the reality of those parts with their parts, and so on, until all that is left are what are obviously the physical bases. Whatever composes rocks is ultimately all there is of reality" (25). But I refer more specifically to what Jones calls *structural or causal* reductionism. It may well be true that the only stuff in the universe is particles of matter in fields of force, but that does not mean that the stuff is always structured the same way. Indeed, Jones acknowledges that substantive reductionism is ". . . probably subscribed to by virtually all scientists and philosophers (or at least by those who are not religious believers)" (26). However, the "structural variety of ontological reductionism" (26) goes to more drastic lengths than substantive reductionism. Because it is important for us to be clear about this doctrine of structural reductionism, I quote Jones at some length.

> Just as reality is composed of only one stuff, so too reality's structures are flattened into one type of structure. All properties at apparently higher levels of complexity have to be explained as effects of the properties and interactions of the presumably simpler structures of matter at the most basic level. Mental phenomena are reduced to physical processes or their products. Biological properties and processes are merely the products of molecules and atoms . . .
>
> . . . .
>
> . . . substantive reductionists can also be structural antireductionists who accept biological and mental causes—they merely claim all of reality is made of one substance, regardless of the number of types of structures with which it is structured. However, for structural reductionists to be true to their reductionist vision, they must go further and argue that only the lowest subatomic level of physical structure provides all the real causes at work in any phenomenon. (26)

Substantive materialism, therefore, does not necessarily entail scientism. The philosopher John Searle, for example, subscribes to a particular form of biological naturalism without thereby trying to explain away consciousness or language with computational and mathematical models. Raymond Tallis declares himself an atheist humanist; his commitment to celebrating the uniqueness of human consciousness and to describing the ontological difference it makes does not require that he include in the description those structures a theologian would. Eric Gans, the founder of generative anthropology, insists only than any self-respecting, coherent anthropology must own the ontological reality created by the exchange of signs. Each of these three may be comfortable with *substantive* reductionism; none would be content with a thoroughgoing *causal-structural reductionism.* By contrast, adherents of scientism as the reduction of anthropology to biology profess structural reductionism on top of substantive reductionism: they believe interpersonal human linguistic consciousness can be *reduced* to something other than itself, usually a mindless evolutionary algorithm (see, for example, Pinker 30). Humans are ultimately nothing but animals, and animal minds are nothing but cognitive processors: they are machines designed by the impersonal activity of evolutionary process or they are "objects" not caught up in the paradoxes of *all* human representation. Human representation, I contend, *includes* the representations of scientists. It is scientism to believe that science—when it takes the human being (person, subject) as its object—can escape the paradoxes of representation to which *all* knowledge is subject. The reduction of anthropology to biology is the denial of the *irreducibility* of the structure of human sign exchange; it is the obliteration of what I will be describing below, following Eric Gans, as the scene of representation.

Scientism is a reductionist ideology based on a metaphysical position neither required nor necessitated by the practices and institutions of scientific research itself. Jones helpfully provides the clue that a reliable indication of a thinker making a reductionist move is the presence in a statement of modifiers such as *nothing but, no more than, only, just, merely, simply.* Some examples of reductions that he provides are these:

> Optics has been reduced to electromagnetic theory. Albert Einstein's discovery of '*E=mc2*' shows that energy and matter (mass) are simply two forms of the same 'stuff.' Chemistry is nothing but physics. Macroscopic objects are really just structured collections of microscopic objects. Life is nothing but an effect generated by inanimate material. Humans are nothing but robots controlled by our genes. (14)

To speak about reductions, therefore, is to presuppose *levels of organization* and to move from a higher to a lower level: psychological phenomena we experience at the level of our minds are reducible to brain activity; brain activity is reducible to the level of physiological mechanisms dictated by our evolved bodies; the causes that make the observable physiological body move are reducible to microscopic phenomena knowable only through organic chemistry; the chemical is reducible to the physical; the physical level of subatomic particles is reducible to that of the even smaller phenomena measured by quantum mechanics.

When is a reduction dangerous or mistaken, then? Jones opens his book with one of the clearest examples of a scientific reduction I have found. It is worth using as a source of templates, to help us sharpen the idea of the reduction of anthropology to biology.

> In science, a familiar example of a reduction is the explanation of heat. When we explain why one object is hot simply by pointing out that it is being heated by another hot object, we only explain why that one particular object is hot—we have not explained the phenomenon of *heat* itself. At best, this story of explanation would lead only to a chain of hot objects, one heating another. To avoid an infinite regress and thus to explain heat itself, we have to refer to something which itself is *not hot*—that is, something neither hot nor cold but something to which the concept of "heat" does not apply. For gases, scientists have advanced such an explanation in terms of the movement of molecules: the molecules themselves are not hot entities, but their movement generates heat. The temperature of a gas is reduced to the average kinetic energy of its molecules. The temperature rises when energy causes greater molecular motion. Heat is not *caused* by molecular movement but rather *is* simply nothing but the movement of molecules. There is nothing left to explain about heat after its reduction to phenomena (molecules in motion) that are not themselves hot. In short, heat is completely explained by reduction. (13)

If one restudies this passage and substitutes *the human* for *heat,* then one opens oneself to a view of the way reductionism might be a problem at the level of anthropological reality. For example, try this out.

> *To avoid an infinite regress and thus to explain humankind itself, we have to refer to something which itself is* not human*—that is, something neither anthropological nor biological but something to which the concept of "human" does not apply.*

I beg the reader to recall this appropriation and reformulation of Jones' form when, below, I explain the way that generative anthropology refuses to reduce its model of human action to anything less than a mindful, minimally conscious, autonomously structured linguistic event. The point of Eric Gans' originary hypothesis, we shall see, is that we must *include ourselves* in the model of human origin that we hypothesize and thereby refuse to reduce the ontological level that we inhabit to anything less than that one which genuinely language-using animals might occupy. For another example, try these out.

> *There is nothing left to explain about the human after its reduction to phenomena (unconscious brain activity and neurological computations) that are not themselves human.*

> *There is nothing left to explain about humans after their reduction to phenomena (evolved organic energy transfers designed to maximize reproductive fitness) that are not themselves human.*

The "laws" of reductionist evolutionary biology and its sister sciences (sociobiology, evolutionary psychology, cognitive neuroscience infected with eliminative materialism) are not themselves human; that is why such biology as the science of the human-as-object reduces the human to something *other than itself.* Scientism is the reduction of anthropology to biology.

When the neuroscientist wishes to reduce conscious subjective experience to brain activity observable through mechanical laboratory means, he (thinks he) is reducing it to something that is not itself human (something not subjective)—thus reducing *mindful linguistic activity* (that which is human) to *brain activity* (something that is not itself human). When the expert in language evolution wishes to reduce language to an animal communication system, he is reducing *symbolic exchange in the presence of other humans on a scene of paradoxical imitation and rivalry* (that which is human) to *instrumental information exchange that might help to maximize reproductive fitness* (something that is not itself human).

A qualification for directness: science is often reductionist, but not all reductionism is scientistic. Much scientific work is based on observation and description of *what is,* accepting it as *what is,* without seeking to reduce it to something made of parts smaller than those that appear. Think of birdwatchers, stargazers, child-care workers; geologists, oceanographers, ecologists. Reductions themselves are often indispensable to knowledge seeking: many of the benefits of science are results of the real-world success of reductionism as an epistemological attitude and reductions as experimental or descriptive strategy. Generative anthropology (as we shall see) from one point of view is openly reductionist: it shares its reflections on the basis of a limited number of explicit conceptual presuppositions about what the human is and does. From another point of view, generative anthropology is anti-reductionist because it defers to the authority of other disciplines and certainly does not assert ownership of the magical solution that will dissolve the social and political problems of humankind, whether through a naturalistic or a historical determinism.

To bring the Frankenstein myth back into view, let us pause. The figure of the vain scientist is not irrelevant to our contemporary world because scientism, deliberate or inadvertent, does now influence public discourse inside and outside the universities (Evans; Midgley; Tallis; Rose and Rose; Stenmark). Its influence is most visible in the frequency and ease with which thinking people of good will, including (sadly, strangely) many workers in literature, are either semi-consciously accepting or explicitly announcing beliefs to the effect that *humans are nothing but animals* and *humans are nothing but computers.* In other words, many people are more or less comfortable with the idea that we are ultimately nothing but animated bodies or robots or beast people or genetic machines: we are entities that can be fully known by the exercise of the methods of natural science exclusively. There is little scandal today in the blithe insinuation that creatures much like us could be in principle, if only science could complete its work of getting at the material determinants of our experiences and "behavior," manufactured. (Few seem aware the ubiquitous noun "behavior" has connotations sinking toward the scientistic: if behavior is naturally determined, human action is naturally free. Humans who exercise freedom are not *behaving* like army ants or red deer; they are *acting* on scenes of history.) If only natural science could complete its work, then the fantasies in the Frankenstein fantasy could become realities: we could have love and sex with robots (Levy); we could make designer babies; we could become elegant cyborgs with multiple super-prostheses, or transfer an entire personal consciousness in a microchip, or make a humanoid in something like the way Eldon Tyrell makes his beautiful replicants. Even the potentially undesirable side effects of such realizations now often slip off our anthropological radar screens. That the undesirable possibilities are not imagined in any detail is more evidence that the cloudy climate of tacit scientism has a great anesthetic power to put our humanizing intuitions to untroubled sleep.

Now by contrast to scientistic anesthesia and semi-idle dreaming, the misery-inducing consequences of fully paid subscription to such beliefs are just what the Frankenstein myth tries to show, in shocking color. One reason to wrestle with and re-experience Wells' vision in *The Island of Dr. Moreau* is that, approached a certain way, it raises powerful doubts about the tenability of the claim *humans are nothing but animals:* it does so because Wells has enabled us to share at a distance the subjective experience of a man who has been *traumatized* into accepting such a harmful delusion. One reason to engage again with *R U.R. (Rossum's Universal Robots)* is that, approached a certain way, it exposes (how prophetically it exposed!) the bankruptcy of the claim *humans are nothing but computing machines* by dramatizing the difference between a mechanical humanoid and a fully human creature attached to others through language, scenic conflict, and shared historical memory—even if the end of history is biological extinction. My claim that scientism is widely accepted today is a large-scale, extra-textual claim; it would need the support of its own book. This is an essay in textual analysis, neither a journalistic investigation of public opinion nor a philosophical treatise on the limits of scientific naturalism. Thus I will confine my assertions about the relevance of the Frankenstein myth to this climate of opinion to those occasions where the works themselves provoke such assertions. But I proceed boldly with the thesis that Shelley's gothic novel, Wells' science fiction, Capek's play and Scott's movie examined together through the lens of a particular philosophical anthropology may restore to us a felt awareness of the wrong of scientism.

The particular philosophical anthropology I will put to work is named generative anthropology. It was founded by Eric Gans, professor of French and Francophone Literatures at the University of California, Los Angeles, and first set out in his book *The Origin of Language* (1981). Gans has refined, extended, and developed his ideas in the decades since then.[21] A numerically modest but intellectually immodest group of scholars interested in exploring and testing Gans' various hypotheses has grown over the decades.[22] In the remainder of this study, to avoid redundancy I will use the terms *generative* anthropology, *minimal* anthropology, and *scenic* anthropology interchangeably to refer to this new way of thinking. Perhaps the most important point now to be made about Gans' thought is this: his is an anthropology *founded on language.* In that, it has something of the flavour of post-structural critical theory and something of the fondness for the paradoxes of representation thematized by Jacques

Derrida. Ultimately, however, Gans' thinking is far less skeptical and far more compatible with common sense and scientific epistemology than is Derrida's thought. Gans' work is, some argue, a respectful completion of the theory of the more famous Rene Girard, once his teacher. It completes Girard by putting the problem of representation at the center of its thinking in a way that Girard does not. With these remarks behind us, let me launch into the difficult territory where we try to take seriously the following suggestion: we cannot reduce anthropology to biology because *we cannot reduce human language to anything other than itself.* In other words, an anthropology founded on language must be explicit in its presuppositions about what language is and does. One might say that it is the paradox of human language that it (simply) (simply?) cannot be reduced to any experiential phenomenon other than itself. In that, experience of language is a little like experience of God, and a lot like our experience of being together with each other as *nothing but humans.* Can we think our togetherness as humans without thinking of language? I doubt it. So let us begin.

Studying to Say Almost Nothing of the Origin of Language

As Stephen Anderson reports in *Doctor Dolittle's Delusion: Animals and the Uniqueness of Human Language* (2004), the original 1866 constitution of the Linguistic Society of Paris "is famous for explicitly prohibiting the discussion of matters concerning the origin of language at the society's meetings" (18; see also 308–309). Members were persuaded that allowing such discussions would open the floodgates to too many hypotheses consisting of too much speculation backed by too little evidence. During the heyday of structuralism, it was not cool to knit one's brow and ask a question of origin: diachronic transitions pushing the rock up the hill from one state of affairs to the next were out; synchronic structures floating above or buried beneath states of affairs were in.[23] In recent decades, the intellectual pendulum has swung back at least a little. Since 1985, the Language Origins Society has met over twenty times There are now over a dozen titles in the Oxford University Press series *Studies in the Evolution of Language.* The field of inquiry into the evolutionary origins of language has its intellectual superstars, among them Derek Bickerton, Terrence Deacon, James Hurford, and Michael Tomasello. In the academy, the swinging back of the pendulum follows partly from the prestige of sociobiology, aggressively launched by E. O. Wilson in the mid-1970s, and its more genteel offspring, evolutionary psychology. It owes something to advances in cognitive linguistics, comparative psychology, and genetic paleontology that have provided bases for the mounting and testing of hypotheses. Outside the humanities, it is no longer the consensus that inquiry into the origin of language must be irresponsibly speculative. Some hypotheses are more plausible than others; the field flourishes, thick with ideas and energetic with debates (Aichison). As for members of the reading public busy outside the walls and gates of universities, interest in the mystery of what forces worked in evolution to produce the oddity of humankind has never declined.

Inside dwindling departments of literary study, things are different: thinking about the human universal is scorned; thinking about differences between humans, adored; the more elusively subtle the difference, the better. The object of this inquiry is a set of four artworks representative of the Frankenstein myth. Literature is made of language; one might expect that thinking about language might happily include thinking about its origin. However, the question of language origin particularly and questions of philosophical anthropology generally create awkwardness today (Wells and McFadden, "Introduction" 2; Good, *Humanism* 22). Explicitly to ask *what it means to be human* in contemporary literary-critical circles is felt to be at once naïve and presumptuous. We are not discouraged in the humanities and sociological-political study from asking what it means to be Ukrainian, Canadian, heterosexual, queer, conservative, liberal, European, African, adolescent, elderly, empowered, homeless, migrant, corporate, or any number of sub-categories of the human. But *the human* is out. Literary criticism is sometimes intensely archival and thus safely evasive of big questions, at other times courageously politicized but safe in an express commitment to an identifiable clientele. *The human* cannot quite be a clientele. There is no embarrassment in theorizing the *post-human* or the *transhuman,* for these signifiers permit one to ally oneself with dreams of technological body-extension or the ecosensitive shaming of consumerist excess, but *the human* is out-of-date, untenable, atavistic. Most professional readers of literature and cultural critics are suspicious of the deployment of any figure suggestive of the human universal, as something bound to exclude and thus to victimize (Gans, *Signs* 190–96). The effect of the flight from such questions among literary critics has been negative. The public has quietly fired (metaphorically speaking) the literary scholars unwilling to ask the big questions and replaced them by non-literary scholars quite willing to provide bold answers: the public now listens to sociobiological discourse, evolutionary pscyhology, cognitive neuroscience and the like when it comes to pronouncements about *the human*. The idea that texts—sacred or secular, religious or esthetic texts—might convey genuine knowledge of the human animal has become almost an antique proposition.

Apart from the resistance to generalizing about humans in literary critical circles, there is another obstacle to face, this one inclusive of the reading public. Many educated people of good will today resist the suggestion that the possession of language makes humans significantly different from other animals. The valuable work of animal liberationists seems to have had the unintended consequence of obscuring a certain fact. The fact is that animal communication systems do not compare well to language (Bickerton; Cooper 1–3; Deacon; Dupre, *Darwin's* 5, 68–69, 72; Eccles and Robinson; Lowe 177–83; Napoli; Tomasello). Certainly, we might observe that animal studies not trapped by behaviorism have established the fact of the richness of animal minds. Nonhuman animals have memories; they feel pain and express emotions; they make short-term plans and decisions (Bavidge and Ground; Grandin; Midgley, *Beast*). Nonhuman animals display evidence of mental states to which one can attribute something like semantic content (Hurford 58; 155–56). All the same, language differs fundamentally from the communication systems of nonhuman animals. I have loved my West Highland terriers; I can express myself to my terriers and they express themselves to me. But I cannot have with them anything like the symbolic exchanges and verbal conversation I can have with a grown human of ordinary ability, or even an ordinary four-year-old human. We can use words to think about dogs (let us this instant: *Lassie; bloodhound; pit bull; wolf*), but dogs have never been discovered using words to talk about us. And dogs will never use words to talk about us, unless in some far-off future where science fiction becomes reality—unless genetic, organismic modification of the type performed by Margaret Atwood's vain scientist Crake or Craig Nova's mad scientist Briggs has been successfully realized. That humans cannot converse with nonhuman animals as they can with one another should be evidence enough to get educated people of good will to admit a fundamental difference between language and animal communication systems. But it is not.[24] So how can we proceed with a minimal anthropology founded on language if many readers deny that language makes humans significantly different from animals?

For generative anthropology, the originary language act differentiates between the object in a sacred center and the human-performed abortive gestures of appropriation that designate it from the distance of a profane periphery. The humans are aware of the presence of that object and of the presence of each other on such a scene. The evolution of language, even at its earliest stages, is thus the evolution of religion or ritual. It helps to reflect that nonhuman animals do not exhibit patterns of behaviour we could in any way describe as religious. Even if in some sense higher mammals and birds have something we might be willing to call "language" by analogy, they do not have ritual scenes with centers and peripheries, nor is their play equivalent to ritual performance. The originary function of language was the sacralisation of material objects on scenes of spiritual ritual activity: the objects in the center were made sacred, as represented and consumed. No ethologist has observed animals engaged in patterns of ritual behavior where the object of joint attention is a thing they mutually name. Only humans gather in formations where they focus their attention on central objects and share representations of those objects with religious awe or esthetic interest verbally expressed.

It might also help, from the pole of "science" as opposed to "religion," to meditate on the link between our language and our technological domination of the planet, between language as that evolutionary difference which gave us the brains we have and thus our technological powers everywhere so manifest. The not-irrational fear of ecological disaster has spread by emotional contagion among us so deep and wide, and has come to be so universally experienced (I am tempted to say *religiously* experienced), that a denial of the scenic configuration at the source of our mental powers has become a nervous habit. Sometimes it seems to me we speak as if we believe that murmuring semi-consciously the self-humbling mantra *we are animals, we are no better than animals, we are nothing but animals* might by magic reduce the harms created by our difference. But the mantra and its tacit self-loathing do animals no good. Dolphins and chimpanzees, parrots and prairie dogs, honeybees and elephants (among others) all have fascinating and surprisingly brilliant communicative abilities. It does not follow that they are ever going to use language like ours and the brain powers that come with it to discuss the problem of greenhouse gases and plastic garbage in the Pacific, let alone find ways to reduce them. To confront the brute reality that humans alone have language and the special intellect it brings is to take a long step not away from, but toward, moral responsibility for "the environment."

However, there is a relevant way, meanwhile, in which the tenacious desire to claim that we are "just animals" makes sense. A full expression of the intuition at the root of this tenacious desire would find words to absorb the Darwinian truth that in the primate past our hominid ancestors were animals without language; and *then later* they—we—were animals with language. But how can we now theorize the space and time named by *then later?* Insofar as reminders of our embodied animality impose upon us a consciousness of the evidence for natural selection and thus a consciousness of our obligation to confront rather than shrug away the mystery of language origin, they serve a valuable purpose. The desire to take the reality of pre-linguistic animal minds into account in our ontology (rather than a clinging to the fallacy that that animals somehow have language) quite correctly acknowledges the reality of those animals who, at the origin of the human, must have been the first to experience something like that which we now experience when we use language.

In the space created by the fact that our earliest ancestors were once nonhuman animals, there, the intuition that we remain *nothing but* animals, belonging to and one with the rest of Nature or creation, should take its place. Such a space is the site of the originary hypothesis of Eric Gans.

I beg the reader to pause to consider what has just been said, out of respect for animals, including those who were our first ancestors, and respect for the ties that bind them to us and us to them. Once in the history of this planet, certain animals like us—sufficiently like us to warrant our shared faith that they were human like us—must have had an experience like that we have when we use language. But the first nonhuman animals to have it must have been *nothing but* animals before they had it. It is a model of that transition as event that a theory of the origin of language must provide. If the hypothesis about the evolution of language does not describe what was happening as an event in the minds of the very first *animal* users of human language, then what is it about? If such a hypothesis does not describe what it felt like, what mental event was occurring for them, then it is avoiding the question of our shared historical origin. This idea—that we must think of the origin of language as such a minimal, memorable shared human event—is the originary hypothesis of Eric Gans in a nutshell.

Now Gans' work, while respectful and aware of empirical anthropology, is a philosophical anthropology, an abstract model, a new way of thinking. Although external evidence from outside philosophical anthropology must be relevant to the question, the mystery of language origins cannot be dispelled by empirical evidence or positivist findings alone. The problem is conceptual rather than empirical (van Oort, *End* 29–30). The real and the hypothetical are not mutually exclusive: a hypothesis about the character and nature of a problem makes a real, concrete difference in how we approach the problem. Any model of originary human language events will necessarily be speculative, but to own the *speculative* in this context is not to embrace inconsequentiality or submit to a regime in which all ideas are equally, randomly credible because nobody knows anything. On the contrary, to apply the qualifier *speculative* here is to acknowledge that such a hypothesis must conceptually shape the empirical evidence rather than be inductively compelled by it. Most powerful new openings to knowledge of the real begin in the *speculative,* otherwise known as the *creative.*[25]

To own the speculative quality in theses about language origin does not mean that anything goes. It means only that empirical research alone cannot provide a sufficient model of the earliest human language event; more specifically, it cannot insofar as we must think of it from the inside as *an event in which we might have participated.* If we do not think of it as an event in which we might have participated, then it is something happening to them, our ancestors. A moment's reflection reveals we mean *them but not us.* And if not us, then it is not the origin of language as *we* experience language. We locate the originary human animal on the other side of the boundary of linguistic consciousness, while remaining comfortably seated in our directorial beach chairs with *humankind* stamped on our backs. The difficulty—it is a serious difficulty—of thinking the event that moved us from then to now (and moved them to us) thereby gets side-stepped (Gans, *Science* 6–10). We draw a blank; we avoid the question; we decide politely to agree that *further research is needed* when we have no good idea, really, of what we are looking for.[26] We will not have formulated a hypothesis about where we came from because even before beginning the inquiry, we will have determined not to include ourselves in the hypothesis formulated. We will have agreed not to speculate on an event in which we might be obliged to include ourselves.

Not to lose sight of the Frankenstein myth, recall that its key event tells of a human origin that cannot possibly be true: a vain scientist playing

God the creator tries to re-enact the origin of the human, and fails. The Frankenstein myth takes up the question of human origin and gets in our face with it, with an answer that we should find shocking. Its answer is quite unlike the model proposed and tested in generative anthropology.

On That Which Necessarily Must Have Happened Accidentally

If a first problem in thinking about the origin and evolution of language is just owning up to the embarrassing weightiness of the question, a second is this: if we are to respect the theory of natural selection, then we must explain the emergence of language without reference to supernatural agents. We must think of the originary language event as a necessary accident: it must have happened, and yet, in a way, been accidental, *uncaused* by anything outside itself. (Note in that formulation both the refusal to submit to the imperatives of scientistic reductionism and the will to propose the autonomy that operates at the ontological level of *the human*.) The first language event must appear to us as something accidental: we can imagine the material cosmos as not having needed to include the event at all; on the other hand, at the same time, it must appear to us as necessary, for we simply cannot imagine the cosmos without this event having happened in it, since we are here and here we are as language users. We are in and of the cosmos; but cosmologically, there was then and there is now no *need* for humans. The cosmological irrelevance of humankind is part of that to which Gans alludes in his maxim *theology is good anthropology, but almost always bad cosmology* (*New Way* 55).

Language origin must have been accidental, not "Divine" or supernatural. Theology is bad cosmology when it refuses to concede to scientific naturalism that, given the establishment of Darwinian evolutionary theory, there is no need at all in Nature itself for the appearance of humankind. An irrational "faith" that puts God as an agent outside Nature only so that God can put us back inside Nature is *bad cosmology*. Thinking of God as such a (first) cause apart from humans does not help us along very much, at least not if we want to respect modern science. If theology argues that God made the cosmos a certain way, such that we were bound necessarily to emerge from the cosmos naturally as creatures with (unnatural) language, then it makes the Creator stand outside Nature so as to give us, inside Nature, our weird reality as language users. But a scientific model of language origin is expected never to invoke

God. A scientific hypothesis is expected to concede that the origin of language was accidental—indeed, there was no origin, there was only a gradual evolution built on imperceptibly small differences—insofar as we can imagine the cosmos as not having needed to include it at all. We can picture a world without us.[27] The existence of our species may be a random contingency.

And yet our possession of language seems necessary to our sense of ourselves in the universe. An originary language event appears to have been necessary insofar as we cannot imagine the cosmos without one having happened, to the precise extent that we cannot disown by removing ourselves from, disown by somehow stepping out of, our bizarre situation as language users in the cosmos. *Theology is good anthropology* because (unlike scientism) theology does not deny the bizarre uniqueness of the human. Theology, focused on making philosophical sense of religious discourse and sacred texts, is wholly comfortable with our tragicomic possession of language. Indeed, theology usually accounts for our possession of language by positing an extra-cosmological Creator whose design made us different. In this context, to say language is *necessary* does not mean that the originary language event was mechanically predictable in a Newtonian-LaPlacean universe of causes and effects. It does not mean that the design of the universe itself had to include us as language users, as part of its clockwork or its organic development or its inscrutable obedience to dramas of the gradually emergent. Nor does it mean that language origin had to happen because the first language users had no choice whatsoever and were entirely forced by matters in motion outside of their scenic situation to use language. *Necessary* in this context means rather that the reality of our using language makes it necessary for us to hypothesize its origin. An originary language event must have happened because we have language events now, and such events had to happen once for the first time somewhere. Do we wish to deny that we participate in historically counting language events? Here we are, reader, you and I, on the scene of representation, exchanging signs. I guess at your intentions, you guess at mine. Mutually educating guesses. How did we get here?

Here we are. To approach that *here* with a pause, with a certain reverence, is warranted in the context of an investigation of the anti-human quality of the scenic *here* in the Frankenstein myth. In the vain materialist practice that founds its scene, there is no *we* but *one:* the mad scientist alone hovering over the body of his not-yet-animated, not-yet-humanoid victim. Not unrelated is the fact that once we have language, it is possible for us to step outside of the transcendent reality we share as language users and to place the originary event in a distant picture-frame out there and back then, in a scene that does not belong to us now. It is possible to deny the spiritual equality that we share with other humans as language users and turn them into objects of materialist science alone, without the mediations of the scenic center as the originary hypothesis conceives of it. It is possible to do that in thinking about the "primitive" hominids who were somehow our biological ancestors, but not really us, members of our historical kindred. Maybe some language event happened to them as objects, but it does not have anything to do with us here and now as subjects. One might effectively translate the gradualist slogan *language just evolved; it did not happen at once* into a formulation like this: *An originary language event might have happened to them, but we do not need to think about it happening to us.* Materialist scientism, therefore, is free to cut the drugged-and-bound question of language origin adrift just as religious fundamentalism is free to smother it with theological pillows. Scientism does so by a different route, though. It does not tell a story of scenic interaction, an event between agents. It ducks out of the discomfort by agreeing that the first language event was "caused" by genes, or by "selection pressures," or by forces outside any hypothetical event in which we might imagine ourselves participating. The important thing for materialist science is that we not imagine *ourselves* as scenic actors in any punctual model of the evolutionary origin of language.

When Gans claims that the question of language origin is avoided because it is dangerous in the way that religious questions are dangerous, he speaks a profound and itself dangerous truth. The originary question risks violence because answering it commits us to a description of *the human,* and such commitments have ethical consequences. After Darwin, most intellectuals shy away from offering punctual models for human origin out of deference to the gradualism central to the concept of natural selection. Natural selection explains the emergence of species through the slow, gradual accumulation of small differences over long periods of time: to use Darwin's phrase, through "the full effects of many slight variations, accumulated during an almost infinite number of generations" (*Origin* 170; see also Mayr 18; 46; 99). Therefore, it seems to the self-respecting intellectual that a thesis of language origin ought not to take the form of a punctual hypothesis and propose a model of an event, even if merely as a hypothetical model to be tested against the available experiential data and our anthropological intuitions, our common sense. (Yes, common sense can deeply err, but so can experimental and statistical method; let not scientism terrorize us into a prohibitionist shame at disdaining to note any murmurings of our common sense.) To think of it as a single event, to reduce human language origin to one hypothetical model of one event, would mean violating a sacred principle of evolutionary theory.

However, inside biology itself, Stephen Jay Gould and Niles Eldredge's explanatory model of punctuated equilibrium has proposed some modification of the Ultra-Darwinist gradualist orthodoxy and found many adherents (Eldredge, *Reinventing*). Michael Tomasello's *The Cultural Origins of Human Cognition* (2001) starts with the problem of the empirical and statistical fact that human culture and language have appeared too quickly and in too short a time to be accounted for in terms of "normal processes of biological evolution involving genetic variation and natural selection" (*Cultural* 2). Richard Van Oort has drawn attention to the striking parallels between Michael Tomasello's scientific model of the scene of joint attention and Eric Gans' model of the scene of representation (*End* 52–68). Derek Bickerton's thesis in *Adam's Tongue: How Humans Made Language, How Language Made Humans* (2009) relies on niche construction theory, itself something of a challenge to the orthodoxies of eventless genetic gradualism. Bickerton's ideas about language origin are compatible in significant respects with the originary hypothesis of Eric Gans. It is in the context of the work performed by experts such as these, who question the prohibition that unilateral eventless gradualism has imposed on the formation of hypotheses about the origin of language, that I proceed without feeling like a demented heretic (see also Eric White 76). Meanwhile, a quick reflection on the nature of events as such might help reduce the fearsomeness of the shape that lurches toward us with *punctualism* tattooed in fluorescent blue on its rectangular forehead.

There is an ambiguity in the concept of the event: an event can be singular or general (Hempel 97). That the general form of a human event must have happened once somewhere for the first time gives no offense to common sense. We can speak of events as abstractions. *The hunter shoots an arrow* names an event; *the cook serves a dish* names an event; *the pilot drops a bomb* names an event; *the animals exchange signs on a scene to defer intraspecific violence* names an event. Numberless particular, historical instances could exemplify concretely these abstractions; each has happened numberless times in the course of human history. But no ontological heresy is committed in the claim that for each general event, a singular event must have happened, one we can think of as the general event happening the first time. If we can assume that any kind of event must have happened once in particular for the first time itself, then the question is why we feel uneasy when we are asked to think of a first *language* event, which is to say, a first human historical event, a mindful exchange of signs. And certainly, the hypothesis of an originary language event is as much about an abstract form that would have been repeated many times, as it is about its having happened once for the first time: generative anthropology does not really have any quarrel with Darwinian gradualism anyway (Gans, "Little Bang"). What we might call the *onceness* of any originary language event will have to be ontologically inseparable from its repeatability; the reality of its having been performed once for the first time will have to be inseparable from the reality of its having been repeated with uncountable variations ever since.

The Exchange of Abortive Gestures of Appropriation

One answer to the big question of what language events do, the answer represented by the originary hypothesis of Eric Gans, is that the first language event must have been equivalent to the first ritual event. Whatever happened must have been memorable to a community of sign users. Two or three must have been gathered in a collective act of naming, the invention/discovering of naming. Given the assumption that the origin of language is the origin of human history (of collectively remembered events), it seems sensible to connect it to the beginnings of ritual and myth. Ritual and myth are anthropological categories not reducible to the biological. To make a connection between the origin of ritual and that of language might seem to be asking for trouble, for ritual implies cultural paraphernalia (costume, musical instruments, dance equipment), and myth implies a narrative discourse about an already-established deity. If ritual and myth already are there, there is language; and we have presupposed the existence of that which we set out to explain the coming-into-being. The challenge is to construct a model for a linguistic memory that could refer to next to nothing, and yet function as both a precursor to a minimal form of linguistic interaction that might plausibly involve ritual action, and as a precursor of the symbolic exchange that might evolve into the sentences exchanged in mythic narratives.

Before the discovery/invention of language, a number of ecological and biological elements would have had to be in place. Our prehuman ancestors would have been members of a social species, able to recognize kin relations among conspecifics as chimpanzees do. They would have been endowed with sophisticated cognitive abilities: the ability to navigate, to exchange information with indexical calls, and to perform in the surprisingly "intelligent" intra-group competitions for dominant positions that chimpanzees display.[28] The one new and radical idea that Gans brings to the inquiry is an insight he borrows from Rene Girard. The insight is a model that posits the pragmatic paradox in human social interaction as that of mimetic desire: the tension between imitation and rivalry, or the tension between what we might call the mimesis of learning and the mimesis of conflict. In positing the pragmatic paradox of rivalry and imitation as the situation in which the symbolic sign was invented/ discovered, Gans hypothesizes that our prehuman ancestors possessed not just any degree of a capacity for imitation, but an unusually high, extraordinarily intense capacity for mutual imitation. Now on the one hand, insofar as this capacity would have increased learning and co-operation, it would have been a benefit to the group: it would have been "selected for." On the other hand, insofar as the peculiarly intense imitative behavior would have increased the frequency of occasions for conflict between members of the group, it would have developed into a disadvantage. In moments of intense appetitive conflict for food objects and other objects, such over-intense imitation must have been destabilizing.

Imagine two or more of our prehuman ancestors having happened upon the natural affordance of an object of appetitive interest, equally hungry and prepared to appropriate it, equally intense in their mutual imitation of the gestures or cries (or both) that express—all mammals can express themselves—that appetitive interest. That is to imagine a purely biological mechanism at work: the mechanisms of appetite, hunger, indexical communication and behavioural imitation. As long as such animals faced with such moments of conflict were protected by a biological mechanism (such as a dominance hierarchy) from letting the conflict get "out of hand," they would not have needed language. The food object would have remained nothing but a food object (nutritional material); it would have remained nothing but a (biological) thing. There would never have evolved any need for a sign, a name, or a word to represent that thing. The alpha animal breaks the pattern and eats, and then the subordinate animals wait their turns and eat. As long as our hominid ancestors only used indexical signs and regulated conflicts exclusively by means of dominance hierarchies, they would have remained prehuman (in a significant sense, not *really* our ancestors: biological ancestors, but not our first historical ancestors).

Gans invites us to consider the plausibility of the hypothesis that language was invented and discovered in the exchange of an *abortive gesture of appropriation*. He hypothesizes that the first signs of language were ostensive gestures designating a central food object in the presence of the object, but understood as substitutes for the object itself. We still have experiences reminiscent of the primordial scenic ostensive, when a few shout *fire!* or *man overboard!* The ostensive is a symbol used in the presence of its referent to designate that very referent: for generative anthropology, designation and presence are the fundamental categories of language. The originary abortive gesture of appropriation would have designated and named the object as a scenic center of desiring hesitation, rather than pointing to it indexically as an object of appetite. Hypothesise again a group of animals–us as animals, endowed as described with the set of biological capacities sketched above–surrounding a food object and intensely interested in it, pointing toward it and moved by appetite. For now, the gestures are indexical and unidirectional. For now, we speak not of desire, but of appetite; for now, not of symbolic signs, but of indexical signs. (Desire presupposes the symbolic; indexical signs of appetite are not arbitrary symbols.) In the pragmatic paradox of rivalry and imitation, the situation is unstable: the very indexical gestures or cries that increase our animal appetite (contagious imitation) are those that that increase our animal conflict (escalating rivalry). By contrast now, imagine that one of the prehumans in the group notices—by accident?—his indexical gesture as an object of perception different from the food object. The gesture gets noticed as a thing in itself. It cannot be stressed enough that the initial cognitive processing of the new noticing of the *abortive gesture of appropriation* must have been very brief, short in duration. It would have been enough that (if it had been viewed from the outside) to attach to it the adjective *accidental* would not be inappropriate. Nevertheless, it has become for an instant an *abortive* gesture of appropriation. There is a hesitation; a hesitation is there. Indexical pointing has given way to a minimal awareness of a gesture that aborts appropriation. The animal's consciousness oscillates between the food object-as-referred-to-by-the-abortive-gesture on the one hand and the abortive-gesture-as-referring-to-the-object on the other hand. Stop to imagine that oscillation of attention.

Exchangeability is the human difference. As long as one prehuman alone has the experience of hesitation or awareness of the abortive gesture of appropriation as a thing in itself, there is as yet no language. There is no language because there is no exchange of symbolic signs. Not until symbolic signs are exchanged, not until two or more animals pay a minimal attention to each other's gestures as things exchanged as *substitutes* for the food object, do those indexical calls mutate into the signs capable of symbolic reference. Following Terrence Deacon's revision of Charles Peirce's semiotic categories, generative anthropology defines a sign that performs symbolic reference as opposed to indexical reference, as one that *refers in the first place to other signs* (Deacon 83; 99). This is the perhaps the hardest thing to grasp: properly linguistic signs refer *in the first place* not to objects in the world (they do not provide information about referents alone); rather, they refer to other signs (they defer violence through representation, creating the scene of representation). The sign is performed not simply in the presence of its referent, but in the presence of other signs like itself. In the originary language event, one newly mindful human animal imitates the abortive gesture of appropriation and attends to the similarity and difference between his gesture and that of his conspecific. The imitated animal imitates the sign made by his imitator. The imitator imitated, imitates the sign of his imitator. There we have a minimal model of symbolic reference and linguistic consciousness.

It is a model of signs referring exclusively to an extra-linguistic referent (in this particular case, a food object). When another participant notices the first abortive gesture of appropriation—imitates, as Michael Tomasello would say, the *intention of* the other prehuman's abortive gesture of appropriation—then we have symbolic reference, a language event, the cognitive precursor of a ritual event. The specifically human-exchanged sign is *about* other signs first; it is instrumentally referring to the food object as well, but second. Symbolic signs' referring to other signs first before referring to things is the reality that generates transcendence from immanence.

Now when in such situations we exchange abortive gestures of appropriation, we have not permanently lost sight of the food object: certainly not, for appetites and animal rivalries and behavioural energies wind up to the highest possible intensity. But we have discovered/invented a temporary substitute for it. We cannot eat the abortive gesture of appropriation, but we can experience the gesture's power to *defer our intraspecific tension* on the scene, where we hunger for the same material thing, where the impossible tension between rivalry and imitation dominates. Our attention can be focused on the exchanged abortive gestures of appropriation themselves as part of a scene structured by those abortive gestures; we can become aware of the scene itself, signs exchanged on the periphery but intensifying our desire for and resentment of the central object. This event is *the naming of a central object, structured by the paradox of rivalry and imitation.* All of us, we use a recognizably similar aborted gesture of appropriation, a name, but each of us is aware that the name (the gesture) is different from the central object named; and each is aware that my use of the gesture is not identical to yours: my hand is not your hand, my vocalization is not your vocalization. In the beginning, human consciousness of the symbol itself being exchanged does *nothing but* briefly substitute for the consciousness of the material object of appetite (the reduction is deliberate). That consciousness of scenic hesitation is consciousness of the originary sign. The shared consciousness of the exchange of signs preceding an exchange of things is the essence of a human event. Such consciousness transfigures human events into *actions* as distinct from *behaviour*. Notice that in this creative moment, the scene is *shared*. By contrast, in the originary moment of the Frankenstein story, there is no sharing in the relation between the solitary mad scientist and the mock-human, meta-human bodily objects in his doomed re-enactment of the originary.

At the same time, a certain pressure at work related to involuntary animal fear—here I do reduce the event to a quasi-determined outcome of ecological necessity—would have been very intense. These prehumans would have been feeling the well-known animal emotion of an intense *fear: fear* of bodily harm and death.[29] The organism's immanent submission to the imperative of bodily self-preservation would have contributed. In the moment, we fear intensely the instability of the situation: the appetitive reach for the central object is just as much an aggressive reach. In the originary event, we fear violence; we fear death; it is a crisis that must either be deferred by symbolic exchange or resolved by the differential structures of animal dominance behaviours. Prehuman appetite mediated by the exchange of signs is converted into desire; prehuman aggression mediated by the exchange of signs is converted into violence. The originary religious illusion (the future of which, despite Freud, will endure and endure) is located here: the origin of language that is created by *nothing but* the pragmatic paradox of rivalry and imitation leads to a certain understanding that is not "objectively" the fact. The central food object is not in reality itself alone the *cause* of the exchange of signs. The "real" cause is the whole paradoxical situation. It is a situation in which the only way to change it is to exchange peripheral signs as substitutes for the one central thing; the central object itself is just one part of the scene. But that one part becomes infinitely significant. In re-presenting the central object and creating a difference between it and the exchanged gestures that substitute for it on the periphery, originary humans create an immaterial difference between *sacred center and profane periphery*. In a sense, this awe-inspiring and fear-saturated difference between sacred and profane, center and periphery, abortive gesture of appropriation and terribly desirable, violence-generating object becomes the differential basis of the gradually-emerging experience of the idea of the One central being other than the community (us on the periphery). And insofar as the originary humans experience this illusion or delusion, they do not experience the hesitation as something they choose but rather as something predetermined by a scenic center they resent and fear. In this sense, the model of the event is a model of determinism rather than free choice: the originary humans had no choice but to represent the center with signs in order to become human. The choice was (paradoxically) a necessity, and the difference between profane periphery and sacred center would have been experienced as submission to an all-powerful central will (submission to the will of the gods, or God).

The scientistic thinker might leap on my recourse to the noun *delusion* as evidence for the *nothing but* of a materialist ontology. The problem with such a leap is the fact that humans who exchange signs experience the structural situation in which they perform such exchanges as *real,* a real structure, a structure irreducible to parts at a lower ontological level. Human language is not animal communication; symbolic reference cannot be reduced to indexical reference simply because the signifier for each subsists at the level of material reality (recall the distinction between *substantive* and *causal* reductionism). We experience the structure of the scene of representation as something real, something as real as hunger or thirst, gravity or magnetism. The reductionist, of course, is free to go ahead and attempt to explain away such experience. The claim I am staking in the name of a minimal anthropology is that such an attempt is the essence of scientism: scientism is the reduction of scenic, linguistic anthropology to materialist biology. For generative anthropology, there is no good reason to explain away the fundamental status of reciprocal exchange of *representations* in an ontology of the human that deserves to be named an ontology *of the human*. A sign is not itself just a thing; a sign is a relation between things. Likewise, human beings are not describable as *nothing but things*. The human being is irreducibly "in" a relation between scenic actors*:* we always exchange signs with each other, and those signs refer to something other—or for those not hostile to a minimal religious sensibility, an Other—different from us in a scenic center we share.

To review quickly: conflict with others on the periphery increases animal fear and generates human violence. The pragmatic paradox of rivalry and imitation means that to imitate a mutually reinforcing, intensifying appetite is at the same time to imitate a mutually intensifying, reinforcing aggression. The increase in intraspecific tension creates an intense fear. Therefore, the originary event both converts animal appetite into desire, mediated by the gestures substituting for the scenic center, and converts animal aggression into violence, mediated by gestures substituting for the scenic center. The central object becomes a sacred thing; and when the food object is consumed, it will be consumed with an aggression more fierce and powerful than the aggression that the nonhuman primate would have exercised. The exchange of signs re-presenting the central food object sacralises the food object. The central object is now conceived in mind as wholly Other to the peripheral sign-users; if the sign intends desiring hesitation (representation) prior to appropriation, it equally intends desiring violence (more than animal aggression) once appropriation begins. Before ritual consumption, we exchange the name of the *whole* central object; during ritual consumption, we exchange the name of the central object broken (torn, fragmented, divided) into parts we consume. This consumption will be *more* aggressive than it would have been before such events. Violence is born with the sacred.

It is prudent now to emphasize a point already made above: the exchange of signs in this model *precedes* the exchange of things. For humans, representation precedes and structures consumption; indeed, the human event itself is about consuming in a different, new way that involves language and the sharing of named things. The naming of the whole object before consumption operates in a kind of practical continuum with the naming of the parts of the object divided, distributed, and consumed. This continuity allows the performances even from the earliest, most minimal exchanges of abortive gestures of appropriation, to resemble a ritual act. First, the action includes the minimal difference between sacred center and profane periphery. Second, the participants in the nascent community remember the central object through its sacred name, in a minimal religious experience. We can think of such minimal memory as the first historical memory, the first to generate a space of linguistic transcendence in which the name "has" a reality different from its referent. There are philosophical critiques brought to bear against the materialist transmission models of communication so loyally defended by reductionist cognitive psychologists and neuroscientists, critiques pointing out their conceptual incoherence (Olafson; Tallis). In its own way another such critique, Gans's originary thinking is not a "transmission" model of linguistic exchange. It hypothesizes that linguistic exchange at the origin created the transcendent space in which humans operate, agents irreducible to mindless, material "transmissions" of cosmic "information." The origin of language is the origin of ritual; ritual is not about material truths alone, but about truths sacred to human community—in particular, the sacrality of language as that which makes us *different* from all other animals, because we know—we know—we are doing something different from languageless animals and inanimate machines when we exchange signs. It is that knowledge (our knowledge) that scientism, in many different ways, tries to take from us.

Gans calls the originary experience of sign exchange and sign-altered consumption *the deferral of violence through representation.* The fundamental function of language and culture is the deferral of violence through representation. When generative anthropologists speak of the scene of representation, they mean the fact that humans organize their intentional actions in and on such scenes, exchanging and manipulating objects situated in scenic centers. But each such scene has a dimension irreducible to materialist categories. There are for us, millennia later, almost numberless examples of scenic interaction: rites of birth, marriage, initiation, and mourning of the dead. We enjoy mealtimes, scenes with food at their centers; and the food is surrounded by symbols. We participate in religious or political ceremonies, scenes at the center of which objects of worship or admiration represent sacred or authoritative beings to the participants. We "get into" sporting events, scenes with center ice, pitcher's mound and home plate, the diving board, the bull's eye. There is no human exchange that is not scenic. In our technologized world, the virtually scenic dominates. A conversation between cellphone users, one in Iceland and one in Mozambique, takes as its center the virtual space into which each interlocutor speaks in turn. Human events have the structure now they had at their historical origin. Nonhuman animals have not evolved such structures; they do not need such scenes as we breathe in and perceive without even noticing; they do not by themselves occupy a transcendent reality teeming with thousands of such collectively remembered events. In the words of Kenan Malik: "Humans have learnt to learn from previous generations, to improve upon their work, and to establish a momentum that has taken us from cave art to quantum physics and the conquest of space. All animals have an evolutionary past. Only humans make history" (167).[30]

In the context of this book's exposure of scientism, it merits emphasis that our earliest ancestors chose to use the sign. Without losing sight of the fact that this choosing would have been the smallest possible mental event (we have no quarrel with Darwin's belief in small differences as the engine of evolution in this respect), we must imagine they chose it. Even for our very first ancestors, the decision to defer intraspecific violence and violence against the sacred central object was a real decision. As a model of will, the originary use of language is inseparable from a process of learning. Agency means learning to represent scenic objects as a sign user equal to other sign users, rather than being doomed to eat only that which your purportedly biological (evolutionary, genetic, neurological) destiny *causes* you to eat. The object drained of sacrality, beauty, erotic and economic value is *nothing but* a non-negotiable cause of pre-determined conflict subject to *mechanisms*. The object considered that way can have no human history. The object transfigured by its symbolic status on the scene of representation becomes a thing about which we can reason: human linguistic communication is nothing but the representation of our resentments and desires and the reasons by which we seek to explain them. The scientistic biologist fixes the human in the material world as a user of indexical signs pointing only to a food object "out there," exchanging "information" about objects "out there." But the scenic object is in here, inside our circle, with us on the scene of representation, between us. This minimal anthropology with its faith in free agency determined by conflict over sacred centers on the scene of representation opens the human to everything that we can learn about objects through the exchange of symbols that represent it before we consume it "in here."

Scientism, both revealed in the fictions contributing to the Frankenstein myth and strangely persistent in the real world, with its hurried-up desire to reduce humans to the predictable, manipulable, or at least purely material objects of biology (or chemistry or physics) obliterates the scenic event so as to deny agency and shut down human history.[31] For scientism, our experience of language is defective with folk psychology. For scientism, we have never had anything meaningful to say about objects because we ourselves (in materialism, "selves" are illusions) are nothing but objects. We have never gotten beyond indexical horizontality, the use of indexical symbols to refer to features of the environment as other higher animals refer to things. For scientism, there is no transcending the material. We are material. The human is reducible to the material because the fact of language makes no real difference.

Our earliest ancestors would have found themselves in situations similar to the (literally) originary scene over and over again. Genetic, neurological, and physiological advantages would have been "selected for" as language and the brain slowly co-evolved in the niche we had created as language-users. Slowly, we would have been freed from limitations on learning imposed by the confinement of competitive behaviours to the inequalities of pre-determined dominance hierarchies. Nonhuman animals do not need symbolic reference to mediate their knowledge of the world, because they do not need it to resolve their conflicts. Humans needed symbolic reference, and still need it, to mediate intraspecific conflict. Most of human history shows, the dream of a society of pure co-operative pacifism aside, that innovative learning flourishes where social competition between equals flourishes, and it dies where brute anti-social inequality squashes the equal freedom to compete. Learning requires competition between reciprocal equals. Ritual would have evolved "naturally" from the repetition of the minimal event of naming a central food object. It is beyond the scope of this book to describe in detail Gans' brilliant dialectical account of how ostensive designation in the presence of the scenic object would evolve into the imperative (from invocations of the object in its absence). Nor can I detail his account of the dialectical origin of predication, a discovery equivalent in momentousness to mindful ostensive naming itself (Gans, *Origin* 98–196; *Originary* 62–85). However, since we are investing in the metaphor *playing* God, we do need to consider Gans' anthropological idea of God.

For Gans, from collective memories of the central locus of the scene, memories supported by sign exchange, the intuition of "God" or a sacred Being transcendent to and other from the human community evolves. When Gans speaks of the anthropological idea of God, God as the referent of the first human sign, he means the imaginary divine Being that would have been equivalent in its transcendent Otherness to that being which monotheism now names "God." The originary humans' evolving cultural awareness of that Being would have gradually emerged as a memory. After the physical food object had been destroyed, distributed, consumed in a "meal" at the conclusion of each ritual repetition of the event, the Being itself would have been understood as beyond representation, the center-as-such conceived as a Being. Experiences of God would have become possible when the empty central space occupied by the consumed food object gradually became associated with the collectively remembered and therefore spiritual presence of a Being.

> For the idea of God is the idea of what subsists in the physical being's absence, and this supratemporal subsistence of the scenic center with respect to the temporal being that fills it is a direct consequence of the originary experience of representation. The sign can only designate what occupies the center of the scene, and the being of this center, the center-as-being, is what we call God. (*Originary* 38)

The originary hypothesis questions "revolutionary atheism" by relocating God to the human realm. At the same time, it makes the idea of God as necessary to our historical self-understanding as a model of language events is to it. Humans may not always "believe in" God (obviously, to us modern skeptics), but they will never cease trying to forget God.

> We retain the idea of God without necessarily believing in it because of the indispensable persistence of the communal ground of the scene independently of the individual members of the community. The nonbeliever may be spared the undignified examination of

> what one "really" believes. Once the idea of God exists, it cannot be forgotten; and once it has been forgotten even for an instant, human culture is already engaged in the process of secularization of which the contemporary atheist is the final product. Our hypothesis attempts to convince the [atheist] only that because the idea of God, to which anyone is free to deny belief, is coeval with the origin of humanity, the process of this forgetting can never be concluded. Even if someday not one believer remains, the atheist will remain someone who rejects belief in God, not someone for whom the very concept is empty. (*Originary* 42–43)

Let me repeat for emphasis: *the process of this forgetting can never be concluded.*[32] One way to understand the reason that Victor Frankenstein, the vivisectionist Moreau, the robot-making Rossums and the god of biomechanics Tyrell all fail in playing God the creator is to grasp the fact that their projects are founded on the presupposition that "the very concept [of God] is empty" and the process of the forgetting has long since been concluded. They believe that a willful ignorance of our scenic origin as moral equals will do no harm. They believe that something other than exchange on the scene of representation made us human in the first place: their models of human creation or origin are exclusively materialistic.

Over against this materialism, I suggest that the "being of God" in a minimal anthropological sense will continue to be "present" whenever language events are occurring. Imagine any few humans exchanging signs mindfully, rather than, to make the point clear, attacking each other in wordless hate—in biologically determined or robotically mechanical hate. They who exchange signs are to some extent (a real extent, however incalculably minimal) realizing the presence of the sacred Being whom their distant ancestors named—whom *our* distant ancestors began to name—at the origin of human community. The ordinary person's prohibition against *playing God* expresses the common sense fear of a forgetting of the intuition of moral equality that grounds human exchange in our linguistic transcendence of materiality. The same moral equality grounds our transcendence of that level of being to which nonhuman animals remain bound, although we share the world with them. When the mad scientist plays God by trying to re-enact the origin of the human, he is supposing the transcendent relations created by the sign can be understood in the flattened terms of immanent matter.

He assumes sociality does not *really* count; the power of laboratory and technology and expertise demonstrates the irrelevance of sociality. This book argues that the Frankenstein myth steadily teaches that such scientism is deeply wrong—"wrong" in the sense of inaccurate as a description of the real, and "wrong" in the sense of failing to recognize the ontological reality of irreducible moral equality between humans on the scene of representation.

Experience of the Object-as-Sacred: Revelation without Cognition

In my analysis of the Frankenstein texts, I will comment at times on the human being as object—as sacred, esthetic, erotic, or economic object. For example, the minimal economic is relevant to my thesis about the Monster's condemnation to exclusion from social exchange in Mary Shelley's *Frankenstein;* the object as esthetic is particularly relevant to my theorization of the mechanical in Capek's *R.U.R. (Rossum's Universal Robots*). It might seem relentlessly abstract to continue with an exposition of the originary hypothesis and the center-periphery structure in human events, along with fundamental categories such as resentment, desire, sacrality, significance, violence, and consumption. If you feel your patience being tested, skip freely over about thirty paragraphs to the section "Exchangeability and Desacralization" and thus speed up your movement toward our exploration of Mary Shelley's novel *Frankenstein* in chapter Two (or go directly there, if you like). You may return to the remaining sections of this long introduction when you please, or consult the index as necessary.

****

In the rest of this chapter, I will break down the scenic language event into its "moments" or "components" as a way of unpacking the minimal anthropology that inheres in this model of *the abortive gesture of appropriation* and the center-periphery structure it generates.

The central scenic object may be thought of as a concrete thing such as food, territory, weapons, treasure, or a person of interest (lover, parent, child, friend). It may be as abstract as a political office or a disputable mathematical equation. The scenic model of center and periphery is an heuristic, just as the minimal model of the originary event is an heuristic.

The minimal event as an experience of the sacred is best described as a pure revelation which may not even include "cognition." What is first learned about the object of appetite is only the sacred prohibition against appropriating it; the experience of that prohibition is the experience of the sacred. *Why can I not possess the center? It repels my desire infinitely; I resent its infinite power.* The originary religious delusion is the attributing to the central object a cosmological power that is in reality an effect of collective sign-exchange between humans. The first cognition would have had only the inaccessibility to appropriation of the object as its content. Resentment and awe (awe-ful resentment) of the misunderstood, repellent center would have dominated the minimal naming-feeding actions of our earliest ancestors, actions not yet fully ritual.

The experience of the object as sacred is paradoxical: the most communal knowledge is the least propositional and thematizable knowledge, the most inarticulately ostensive. The central object's inaccessibility to appropriation is one with the impossibility of its being known as a consumable good. To the extent that all members of the community can share a truth, inside the restricted space of this first moment in linguistic consciousness, the truth designates primarily the power of the center as such rather than desire on the periphery. It therefore risks meaninglessness. And yet, meanwhile, to the extent that knowledge of the prohibition is shared equally by all sign-users, knowledge of the object-as-sacred is the form of cognition most powerfully and purely communal. The sense that humans are equal before one creator or the one God finds its content here. We may affirm the spiritual Other of all humankind *is;* the Other may be (first and best) minimally described. The experience of the object as sacred—even the self-cancelling emptiness of this Other, as in the quest for enlightenment in Buddhist tradition—equalizes the moral status of those who have access to it. Where the presence of "God" exacerbates rather than defers crisis, or justifies collective violence in mindlessness rather than prompting mindful hesitation over the prospect of collective violence, we misapprehend "God." Where the presence of God defers crisis, God is. If we do not love each other on this scene—if we do not defer our mutual resentments—then God-as-love is not present.[33]

The experience of the object as sacred compels a uniformity of ritual performance and likeness of material signifier. Under the spell of the violent communal sacred, we know the center as a "locus of dispossession" (Gans) to such an extreme extent that our awareness of our exchanged signs does not include the center-periphery oscillation of esthetic experience. The success of ritual repetitions depends on the minimization of any private experience of the desirability of the central object, such as esthetic exchange affords. Ritual prohibits thoughts about the desirability of the sacred center as a locus open to *my* occupation; my private fantasies as a sign-user on the periphery are sacrificed to a unity of experience. Gans describes it this way: in the "stable imaginary structure of resentment" we feel ourselves each as only a "self on the periphery [that] is definitively alienated from the desired object" (*Originary* 119). The "force" of the sacred object is conceived as "independent of representation" (*Originary* 118): "Although the sacrality of the center is coeval with the sign that designates it (as the 'name of God'), it is experienced ('revealed') as ontologically prior to the sign, and is therefore *independent of the esthetic effect* that operates between the sign and the referent" (*Originary* 124) [emphasis added]. The sacred enforces a similarity of form in the symbols we exchange: we must sing the hymn at church the same way, cross the stage at convocation the same way, wait in line with other citizens to shake the President's hand the same way. If we deviate, we desacralize the object-as-sacred. A bride or groom who says at the crucial moment *I would prefer not to* rather than *I do* violates the communal sacred with a scandalous revelation of an alternative central object that up until that instant had been imagined only privately. If a rule that protects a central object of desire from violation is easily broken, the object has lost its sacred power. In the era when scientism becomes possible, during which the Frankenstein myth emerges and operates as a relevant cultural force, it is human being itself that has lost its sacred inviolability. In the era of the possibility of scientism, humans are free to try in vain to make themselves into *nothing but* material objects of technological manipulation.

Any knowledge that we might have of the object-as-sacred is therefore strangely minimal. Gans writes of the difficulty we must have in putting ourselves imaginatively in the place of our earliest human ancestors.

> We need to be able in principle to discuss the origin of language with its originators, for if no such dialogue were possible, we would face this origin as a natural rather than a human phenomenon. But when we thematize language and equate the origin of the human with that of the object of this thematization, we find it difficult to speak with those whose only theme is the sacred referent. (*Originary* 16)

The genetic determinist and the evolutionary psychologist, unlike the generative anthropologist, might for their parts be satisfied to *face this origin as a natural rather than a human phenomenon.* Those of us who would remain unsatisfied and challenge such a reduction by giving the originary hypothesis a try, stretching our anthropological imaginations, will still *find it difficult to speak with those whose only theme is the sacred referent.* True. Our earliest ancestors are difficult to imitate, minimal interlocutors. The referent of the experience of sharing the name of the object-as-sacred cannot be the referent of an esthetic imaginary, an economic evaluation, or a scientific investigation. It is untouchable, unknowable, not even there to be thematized beyond its being fearfully named as the unknowable center of a structural reality. Its referent accompanies the paradoxical incompatibility of our universal moral equality and our universal desire to be first in appropriating the center, conjoined with an ever lurking resentment of the impossibility of satisfying such desire on the communal scene.[34]

Sacred objects seem irrationally revered from a perspective outside any particular community, but comprehensible from the perspective of those inside. There is something at once alarmingly irrational and indelibly average about experiences of the object-as-sacred. When they are contrasted with experiences of the significance-drained cosmological object analysed and broken down into exchangeable parts, the irrationality becomes clearer. Bring into your memory the image of a place you love: the Manhattan skyline, St. Paul's Cathedral, the Parthenon, the Mayan ruins at Chichen Itza, Jerusalem, whatever you please. Viewed from the outside—by a human not attached to the place by communal experience—the place is nothing but a physical structure, neither to be worshipped as desirable destination nor protected by violence from the desacralizing intrusions of outsiders. However, viewed from the communal inside, the place is infinitely valuable; it may well seem worth defending to the death. On a different scale, it is the private sacred as mediated by the erotic that transforms a brick-and-mortar house into a home and changes the body of an intimate partner to whom we have promised love into the incarnation of a person transcendent.[35]

We today are unable to explain experiences of the sacred "scientifically" in the same way our first ancestors, not yet even speaking in sentences, would have been unable to explain their originary experiences of the sacred center. No such experience is explicable under a merely physical, chemical, biological, or neurological model. Brain scans reveal what is happening in our brains; but we are not brains; we are humans on scenes of symbolic exchange, who sometimes experience the sacred. What human does not hold some place, someone, somewhere, at least a little sacred? Who among us has *nobody*—a child, a parent, a sibling, a friend—whom we would protect if necessary by violent action? Does not everyone have a place to imagine fighting for, even if it be the place (ironically) where the Crucified once and for all shows us why in his memory we might wish to give up fighting and violence against each other? Experiences of the sacred are explicable only by an anthropological model that includes a description of the pragmatic paradoxes that produce such experiences. The originary hypothesis places such paradoxes at the start of its way of thinking about the human.

Experience of the Object-as-Esthetic: Imaginary Possession, Recognized Inviolability (To and Fro)

The object-as-sacred must remain non-individual, indefinite in its forbidden distance. Human learning about specific, particular worldly objects—food, clothing, landscape sites, precious metals, other animal bodies—is learning mediated by the object-as-esthetic. Esthetic experience permits our self-consciousness of the desirability of the central object as one specific thing to come into extended play: "Only through the esthetic experience of the center can the object be known in its specificity" (Gans, *Originary* 126). One can know the name of the prohibited apple without knowing the taste of the prohibited apple. To begin imagining its taste is to begin experiencing the apple as an esthetic object, oscillating between the name-of-the-apple (the apple imagined) and the apple-only-as-named (but really all the *more* there by virtue of being named).

The experience of the object-as-esthetic consists in a mental oscillation between "imaginary possession" and "recognized inviolability." At the pole of imaginary possession, the individual sign-user on the periphery imagines possessing the concrete central object to which the artwork ultimately refers (the unspeakably perfect landscape, a marriage with the likes of the wonderful hero or heroine, or that exalted state of which the Beethoven violin concerto speaks). One fantasizes consuming it as a real thing all by oneself, which moves in the direction of a knowledge of its specificity. At the pole of recognized inviolability, by contrast, one is reminded of the inaccessibility of the real-world referent of the artwork: one remains aware of peripheral others who are sharing in the exchange of signs of the referent (symbols substituting for the real thing). Partly the force of *others'* seeming satisfaction with the sign as a substitute for the referent motivates *my* satisfaction with the purely esthetic experience. It is comforting that everybody who listens to the recording hears the same "Here Comes the Sun" by the Beatles, and it would be absurd to lose pleasure in it because one was not there in the real studio when it was recorded. When imaginary possession is shared by many, the improbability of real possession is proportionately less distressing: the centrality of the referent of the artwork is mediated not just by the seemingly objective content of its concrete appetitive qualities, but also by the force of the manifest exchangeability of the collective representations of it.[36] In stopping to study the Mona Lisa in the Louvre, we contemplate the inaccessible, mysterious and inviolable referent who was Rembrandt's real-world model, but equally we contemplate those enigmatic formal properties in the painting which have made it worthy of the gazes of "imaginary possession" of countless viewers over the centuries: the painting is a thing in itself, the inviolability of the concrete circumstance of its real-world creation irrelevant.

The work of art must ultimately refer to something real, concrete, appetitive, edible, tangible, touchable, consumable (even the most non-representational painting, something by Jackson Pollock or Mark Rothko somehow does so)–with that, it moves toward conveying the *recognized inviolability* of its referent. And yet the work of art must ultimately refer to *itself* rather than to that referent, for its nature is to create in esthetic oscillation an experience of the unreal, the abstract, the non-appetitive and imaginative, rather than the corporeal prepared for consumption or the practical about to be used. Art's whole purpose is to be beautiful for the mind, not just the body, to enable "possession" of the inviolable referent not in the real world but in the imaginary realm of private contemplation. A food artist may tell you that the success of such art is all about the *preparation,* that which enhances imaginary possession; and yet that preparation depends on *recognized inviolability,* for the vulgar boor who paws her hors d'oeuvres before anybody has had a good look at the spread shows despicably bad taste, no manners. The as-yet untouched Thanksgiving dinner on the table is beautiful because it *refers* to the concrete meal about to happen ("recognized inviolability"); but it is also beautiful because it does *not* so refer, but rather becomes an object worthy of admiration apart from the fact we will eat it ("imaginary possession"). The imaginary possession is possible only before we consume (and thus violate) the inviolable referent; the duration of the time before is a result of the meal's "recognized inviolability." Indeed, there can be something a little ugly and disappointing, anticlimactic and vacuous about the carcass and grease, the dirty silverware and spotted tablecloth that follow the actual feast.

Art is certainly about the real object, but it is about imagining a realm where the real is more desirable than it could be in real life. Here is the paradox: that the world of art is about the real world makes it an invaluable substitute as imaginary possession; but the real world, once we recognize the inviolability of those specific referents in it that art makes desirable, becomes more valuable than the world of art alone. Photographs of my beloved that I take on travels enable imaginary possession of their referent even while reminding me of the inaccessibility of their referent. I fall in love with my real beloved more, not less, after she has become incarnate in a work of art—after she has become the referent of a sign that refers to her as herself (so that she is no longer *nothing but* her unrepresented, material self). Art transfigures and transforms that which would otherwise be *nothing but* the nutritive, the physiochemical, the materially advantageous. The esthetic, like the sacred, moves the object in a direction opposite to that taken by scientism: it makes things exchangeable and transcendent rather than reducing them to the mutely material and corporeally immanent.

Esthetic experience is inseparable from the waiting that makes the thing desirable. *When are we going to eat? How many more miles? Why don't you just ask her out?* The signs we use when designating the object as recognizably inaccessible are the signs by which we designate its desirability. How can we satisfy a desire if its object is prohibited? Meanwhile, how could an object be desirable if it were so utterly indifferent to all that none might block my access to it? An object of such indifferent accessibility might satisfy an appetite. But an object has esthetic value only if my experience of it oscillates, between imaginary possession and recognized inviolability; the recognition of that inviolability depends on my being attentive to others' attention to the *representations* of the object. Gans' hypothesis is that language began as an abortive gesture of appropriation. In the originary language event, there is the briefest of mindful hesitations between the gesture as *abortive* (recognized inviolability of the food object), and the gesture as one nevertheless *of appropriation* (imaginary possession).

Art is not religion; the experience of esthetic pleasure is not that of participation in the sacralising rituals of a religious community; the private experience of imaginary crisis and deferral in our engagement with artworks is not the worship of a god or God. The work of art produces its effects on those who experience it one-at-a-time; it is thus unlike the object-as-sacred, which works on everybody in the ritual scene all at once. However, as the social cement that binds people with shared tastes and passions for specific things together, and as the sphere in which modern people liberated from religious uniformity represent themselves as individuals—represent their beautiful, desirable, inviolable persons to other similarly desirable persons—the sphere of esthetic experience has displaced the experience of the object-as-sacred in secular modernity.

At the same time, esthetic experience is indispensable to the human; it is originary. We discovered/invented our humanity in situations of crisis, situations of deferral of those crises through representation; we continue to rediscover and reinvent our humanity in our exchanges of experience that invest in narrative, dramatic, visual, musical, kinetic, culinary, sartorial and other modes of esthetic deferral. Esthetic experience connects us to our shared origin in the scenic structure of signification. It connects us to the ever renewable means of the generation of transcendence from immanence. My personal, individual imagination (unlike anybody else's) has its origin in a collective imagination (the social structures of exchange by means of which the inviolability of the referent of the beautiful artwork is reinforced). Only a few in the world make contact with the real actor face to face, but many enjoy proximity with the movie star, the imaginary character in the film. Next, when linked to the manipulation of pieces of the divided-up sacred object (hands-on examining, whittling, polishing, tinkering), the experience of the portable privacy of the artwork will be mixed with a knowledge-seeking activity detachable from esthetic knowledge: the measuring, quantifying experience of the object-as-economic. Quantification opens the way to evaluation and scientific cognition. Scientific knowing is impersonal to the extent that we can appeal to the concrete referent more directly during its in-our-material-presence consumption than we can appeal to the inaccessible referent of sacred prohibition or the inviolable referent of esthetic oscillation.

Experience of the Object- as-Economic: Sacrificial Consumption, Economic Value

Originary language events, had they not led to re-organized consumption of the food object represented in the first place by the exchange of signs, would have been without practical value. Gans imports the term *sparagmos to* designate the moment at which the exchange of things takes over from the exchange of signs, the moment at which humans eat, consume, work upon and process the material objects they have represented. *Sparagmos* names the moment of violent destruction and consumption of the central object, its being torn up, broken apart, divided into the bits and pieces of necessary reduction. The mapping of territory that measures the land, the naming of particular kindred who are exchanged in marriage with members of other groups, the comparison of this tool with that one, or this ornament with that, resonate with the consciousness of the object as a concrete-thing-in-itself that is born with the sparagmos. Language events involve not signs alone (signifiers, signifieds, the *words, words, words* that may have bored Prince Hamlet); language events include referents (objects, things, the tangible, the concrete). In generative anthropology, reflecting on the sparagmos means reflecting on the way complete acts of representation eventually include sacrificial violence. The sacred and beautiful object-in-mind is "sacrificed" to the appetites of bodies; it is submitted to the "laws" of biological and economic necessity. We must eat; we must (more often than not) exchange one thing for another in preparing to eat, being able to put food in front of one another. In the sparagmos, the transcendent considerations of religion and art do not disappear, but they come into tension with the pragmatic considerations of economics and science. Neither as sacred nor as esthetic alone can the object satisfy hunger or serve biological need. It must have concrete (edible, nutritive, "fitness-maximizing") value as a cosmological object (although that will never be all or the only value it has).

The participants in ritual consume that which they represent. Consumption requires "violence": chewing, swallowing, digesting are experienced differently by humans as opposed to nonhuman animals because, having named the things we consume, we can think about them, and think about them almost endlessly. Animal consumption may be aggressive, but it is never, properly speaking, violent. As Gans learned from his teacher Rene Girard and as generative anthropology maintains, there is no violence without the sacred. Only humans know violence, for only humans share sacred centers that can be violated. Animals are not violent; aggressive, certainly; violent, no. Only humans do violence. And some violence, alas, is necessary to our flourishing. Meditation and fasting can purify, but *nothing but* meditation and fasting starves.

When we try to imagine our earliest ancestors who knew only the sacred name of a ritual food object and were living millennia away from later stages in language evolution (when imperative culture and then declarative culture, with its invention of predication, would add immeasurably to their cognitive capacities), pictures of shaggy *homo erectus* types passing pieces of rough dinner around the fire slide intrusively into view. What then? If we wish to take our kinship with ancient primates seriously and invest in the theoretical framework of Darwin and his heirs, then the picture should not embarrass us. On the contrary, we should be prepared to take something from such a context seriously as a minimal model for the earliest human events. It is worth pondering the fact that Charles Darwin felt touched to have descended from baboons but a little disgusted by thoughts of his descent from human "Barbarians" ("Descent" 253–54).[37] The point of the originary hypothesis is that our shaggy ancestors exchanging nothing but ostensive gestures in scenic configurations were significantly like us, because for them as for us—unlike other primates—scenic representation precedes consumption. And we can think of numberless dramatizations of the minimal truth that in human affairs, representation precedes and mediates consumption. For humans, invitations precede and conversations accompany the sharing of food, while the silent, solitary dinner is a potent symbol of asocial misery. Ants make their sandcastles, honeybees their hives, beavers their dams and prairie dogs their tunnels without prior symbolic exchange; but for humans, the blueprint precedes the building. When chimpanzees go to "war," even the most cleverly Machiavellian do not discuss strategy as we do.

To the extent that we do leave scenic objects alone and stop ourselves from consuming them, such things are objects we hold sacred. The citizens who love and enjoy the city park hold it sacred in a way that the developer who wants to build condos on it does not. The shopper who buys shoes with the intention of making them last a lifetime holds footwear sacred in a way that he who (content with planned obsolescence) buys cheap in the expectation of throwing them out does not. There is a ratio between an experience of the object as sacred or esthetic and an experience of the object as economic-and-scientific, as "nothing but" a resource to be exploited, raw material. It is the ratio by which holding the object sacred means protecting it from, and prolonging its significant protection from, the demystifying violence of consumption. To consume is to demystify. The fish in the frying pan is not the fish in the river. The new couch at home is not the one in the showroom. The morning after is not the night before, however pleased one might be by the reciprocation of a passion. To the extent that we refrain from the violent consumption of a desirable object and live wisely with the difference between the sign (the artwork) and the thing (its referent), to that extent we are sacralising it or appreciating its specifically esthetic value.

In an era of anxiety about *the environment* such as ours, claims implying the necessity of violent consumption may cause uneasiness. However, there is no signification without significant memory, and the sign is born thanks to the sacrifice of the referent; a language or ritual event remains incomplete until the "empirical" referent has disappeared. The economic moment in and of human events desacralizes, de-estheticizes, privatizes and individualizes: it permits (horror of horrors for the utopian ascetic) *appropriation*. But if there is no appropriation, there is no economy; no economy, no society. Thus to insist on economic reality is not to deny the real evil of one's having too much, the evil of excess (gluttony, sloth, envy, lust). Nor is it to say that there is not something truly admirable in the self-restraint of hermits, monks, nuns, prophets in the wilderness and moral pioneers who may never come back from the edge of their innovative separation from the social body. It is (however) to remind ourselves that even to such ascetics the following insight of Girard applies: "the romantic does not want to be alone, but *to be seen alone*" (*Double* 24).

The economic moment of the sparagmos is the place where we find *evil* in the minimal model of scenic events. Gans describes the willful ignorance of personal responsibility for one's participation in the violence of consumption as the model for all acts of evil: "The sparagmos, in which the exercise of violence toward the sacred center is accompanied by *the denial of individual responsibility for this violence,* is the model for all acts of evil, both collective and individual" (*Signs* 145) [emphasis added]. Consumption can seem the root of evil not because we have bodies (how could just being made of flesh be sinful?), but because we deny individual responsibility for that which we have done as embodied contributors to collective violence. The minimal model for human evil is *losing oneself in the mindless violence of the sacrificial mob,* losing oneself so as to be "free" not to take responsibility for one's dependency on the communal scene of representation-and-consumption. In its adherence to this idea, generative anthropology agrees in spirit with the anti-sacrificial ethic of the anthropology of the cross developed by Rene Girard and his colleagues.[38]

The communal experience of the object-as-sacred maximizes the visibility of the signifying gestures on the periphery. By contrast, violent consumption in the sparagmos minimizes the mutual visibility of exchangeable gestures. The object-as-economic converts imaginary into real possession; with real possession comes moral danger and the possibility of anesthetic anti-sociality. In the sparagmos, the satisfactions of individual consumption threaten to erase both the memory of the whole object-as-sacred and the memory of the imaginary object-as-esthetic. In consumption, we are each alone to some extent with the bodily sensations that we have. At the extreme, consumption is the highway to solipsism: nothing could be more real than this my perceptual and sensory reality, that which I know all to myself and by myself, now that I have my hands on the real thing. Maybe I am the only one in the world to feel this way; maybe I am the world. The glutton, the miser, the hoarder—and not least, the scientistic isolate contemptuous of scenic exchange and obsessed with the *materiality* of the object—all make their appearance now. Horror movies, obsessed with sadistic bloodiness and the dismemberment of flesh, get their grotesque imbalance from an overinvestment in the sparagmos detached from the sacred and the esthetic: that is why they are often condemned as obscene, in bad taste. Each of our four examples of the Frankenstein myth as tragedies of shock and horror has been thus condemned. They show the horror of a world where the human-object-as-demystified by consumption has all but annihilated thoughts of the human-object-as-sacred-and-beautiful; they show a world initiated not by a universally creative God but by a fragmented and fragmenting vain scientist.

Let us recapitulate the three moments in scenic language events so far described—the sacred, esthetic, and economic "moments." In the experience of the object-as-sacred, we know the scenic center as the One, untouchable, owned by all together and none individually. In the oscillatory experience of the object-as-esthetic, we each imagine (thanks to the sign) possessing the central object to our individual selves (we indulge in imaginary possession of it) while remaining aware that the sign alone cannot substitute for the thing (we are frustrated by the referent's recognized inviolability and we experience the sign's inadequacy to its object). By contrast, the experience of the object-as-economic converts sacred inaccessibility and imaginary possession (mediated by the esthetic) into real possession. We come to know the object-as-consumed at the price of knowing the central object's victimary status (because we have torn it apart).[39]

The purportedly beastly economic world of everyday human interaction—buying, selling, getting, spending, bartering, trading, auctioning, dealing—that world undercuts the aspirations of those devoted to a ritually-compact world of religious awe and mystery, a world in which the experience of the object-as-sacred, punctually re-produced in a community of like-minded believers, would reign supreme. Conversely, the reproaches directed by economic hardheads at artists and esthetes for their impracticality, unworldliness, escapism and the like are reproaches grounded in a consciousness of the object-as-economic, a consciousness seeking to restrain the privilege granted to esthetic cognition and seeking to discipline others to include the economic in a complete picture of human reality. Life without economic cognition of the object would not be human life. The experience of the object-as-economic gives us knowledge of the value of things in the social world of practical exchange—the reasonable price, the better business, the decent bargain; market reality. Human events are unthinkable without economic exchange. That is why (as we will see) the Monster in Shelley's story suffers so deeply: he is barred from participation in the economic.

I said above that the object-as-economic transforms imaginary into real possession. The conversion entails a cost: when the ideal object is divided and distributed, we must accept the economic fact that we each get only a piece of it. Nobody gets it all in human culture; nobody gets to be (to have the Being of) the divine central One. The imitation of the Inimitable is impossible; imitative rivalry with the Inimitable is self-destructive; that is why the expression is *playing* God. Playing God is not really possible, so the "humans" that the vain scientist creates are likewise impossible. Just as scientism believes it can successfully account for the anthropological by excising from the human-as-object its sacred, esthetic, and economic qualities, so does the vain scientist dream that he has acquired all the truth when he gets his hands on the uncut material stuff and gets his head around the "laws of Nature" needful to make a human. But alas,

Victor and his heirs learn (especially when the impossible humans take revenge) they had been given only a part of the truth about Nature. The mad scientist learns that despite his momentous discovery, he was only one more miserable economic actor on the collective scene all along, dependent on those from whom he believed he could separate himself, as if he could permanently occupy a view from nowhere while staying human on the scene of representation. Nobody can.

The Object-as-Cosmological: From Good (Minimal) Science to Scientism

Scientific knowledge is bound to economic exchange: the invention that works is the one that sells; the knowledge that proves instrumentally reliable in exchange after exchange, becomes priceless and universally desirable at the horizon. The scenic is practical because it presupposes the presence of other human beings; but when the scientific attitude degrades into scientistic dogma, that presupposition is abolished insofar as humans become purely *material* objects. The human is not really present if the human other is conceived as just a bag of bones or a sack of matter, conceived as a thing knowable exclusively by the atemporal laws of physics or chemistry, or the "universal acid" of ultra-Darwinian theory,[40] a thing without spiritual history. Participation in human community requires the exchange of signs; it requires the collective remembering of historical events. Apart from solipsism, being human alone on a scene is a metaphysical impossibility. When the scientific knowledge one seeks is perversely desired as knowledge of a thing that can have nothing to do with mutuality or exchangeability, then one is seeking to know the object as *nothing but* cosmological, divorced from the pragmatic-economic (let alone the sacred or esthetic).

The exchange of signs is irreducible to the exchange of things carried out as if representation and language had not already always transfigured their objects. One formula to help us unfold the ethical implications of the priority of the exchange of signs is this: "The exchange of signs defers crisis; the exchange of goods operates within the space of this deferral" (Gans, "Free Market"). The exchange of signs creates the transcendent space inside which our horizontal exchange of material things occurs; the priority of language is fundamental to the definition of the human operative in the discipline of generative anthropology.[41] To believe that we exchange nothing but material things when we exchange signs is to fail to rid ourselves of what I would call "the object delusion"—in a challenge to those enamoured of the book with a similar title by Richard Dawkins. Signs are not objects: our experience of language events is not reducible either to an experience of biological events, even less so to our experience of brain events. And to wish to reduce or to know our scenic experience that way is anyway to pursue the futility of mad science, in the first place. Why do that? Only humans use symbolic reference in exchanges mediated by a communal source, that source being one with the historical origin of human difference from the rest of the world we live in. The historical origin of human difference is one: both our one originary material difference from our animal kindred, and the originary difference from our one spiritual Other, by some of us named God.

Where is good science in language events? We can posit a model of the "originary scientist" in the individual quantifier who compares his or her piece or bit of the consumable object to the bits and pieces of others consuming. Only in consumption can we quantify scientifically. The idea of a *quantifiable* value in a whole, untouched sacred object seems bizarre: the nature of the object-as-sacred is to be all communally-owned quality, no divided-up quantity. And yet the originary object before consumption does have a quantitative value: the ironic quantity of being indivisibly whole, One. The signs exchanged prior to the sparagmos are "without value" in the sense that they do not *really* quantify; however, the necessity that a thing have exchangeable value (the nutritive, the concrete) explains why the object-as-sacred (religion) must find its completion in the object-as-economic and the object-as-quantified (science). A religion that does not help humans eat will not last. Sign exchange prior to the sparagmos names the inaccessible central locus conceived as a being; it thinks of the central object as the *revelation of* this inaccessible God, a Being whom (in our resentful awareness of the object's inaccessibility) we can never hope to cut up or subject to mathematization, let alone consume once and for all.[42] There will never be enough material "evidence" to "prove" that God exists, for God does not exist in the mode of a material object; only revelations of God exist so. The materialists who protest the inaccessibility of the Divine to sensory detection and physical measurement have always already mistaken its ontology.[43]

Originary science would have begun as an act of courageously desacralizing free will, in the earliest human refusal of the sacrality of the whole object. It would have taken the active shape of a refusal to accept as equal one's portion of the food object. This refusal has two aspects. First, it means the repudiation of any part as an equal to the whole food object. *My use of the symbol as name of the sacred whole does not equal my use of the symbol as name of this concrete part. Their referents differ: I must find a new name for my part of the object alone.* This refusal to accept the part as equal to the whole entails a rejection of the one name-of-God as an adequate, mentally satisfying sign; in that sense, scientific representation of objects has always suspected sacralising representation. For the originary scientist, the name-of-God would have been experienced individually as failing to name satisfactorily because of the manifest (perceptible, sensory) difference between part and whole: the parts in my hands are tangibly, "empirically" measured against the remembered sacred whole, and they are found wanting—different empirically from the (previously imaginary) whole. The scientist therefore thinks beyond the part toward the scenically centered whole, but in focusing on the part, must begin to discount the value of the object as untouchable, inaccessible, whole, and sacred to all participants equally on the scene.

Second, the refusal means a rejection of any acceptance of the portions of the food object *as equal to each other.* The acceptance of empirically unequal parts as equally satisfying socially (there lies the origin of the difference between arithmetical and geometrical equality) might well be the basis of universal morality. However, such an acceptance cannot be the basis of science, for science must reject the equivalence of sacred whole and profane part. Indeed, as we have seen, reductions are about breaking the whole down into parts ever smaller. Scientific thinkers measure differences in personal property: it helps to think of economic valuation and comparison as the originary science of Nature. Such a social science must have come first: division, partition, breaking down, inspecting, measuring, comparing, bits against bits, my portion of the no-longer-sacred-object as opposed to yours. Let us compare desacralized parts. Science names the real empirical differences between parts of the consumable object, as things accessible to perception unmediated by sacred and esthetic norms. Science names parts accessible to the sensations consumption allows.

Furthermore, scientific knowledge of the cosmological object differentiates humans as objects, one from the other: in scientific knowledge of human bodies as objects, our perceptible differences considered impersonally are taken into account at the expense of our universal sameness before the one Other whose presence might cancel such differences out.[44] Our bodies matter; our bodies differ; if we are all the same, it is only because we all have each one equally valuable soul or spirit or mind—equal access to the transcendence on the scene of *shared* representation of the One center. But in a radically empirical, necessarily desacralizing attitude toward the illusory whole, scientific cognition gives significance to parts in the service of representation; for science, representation is the production of meaningful, valuable, exchangeable differentiations. Notice, too, that such quantification calls for new language: we use the sign to name the "same" thing, while knowing that the things in reality are not the same, but parts different one from the other. The naming of parts is an act of science.

Science is knowledge of the concrete object as part of Nature, which means that we must locate science in a moment of ecologically disruptive action. We must eat, we must make shelters, and we must protect ourselves from predators. We get to know the object—not the imaginary but the real thing, not the idea but the thing "out there" in Nature—only thanks to the necessary violence of breaking it down. Edible roots pulled from the earth, berries taken from the bough, birds falling from the sky, fish flopping at our feet, trees falling in the forest, felled by us. Violence against the natural object is not pleasant, it is not easy; it takes work. Transformative contact with material objects calls for tools themselves crafted from natural materials, from the hand-axe to the particle accelerator, from the microscope to the movie camera. Those of us in the humanities who work with texts ought to acknowledge that knowledge is incomplete without the violence of the exchange of things, of consumption. If we resent modern science and technology for their aspirations to "objectivity" and believe that we can produce knowledge without doing such violence, if we believe that we can occupy a pedestal of pacifist self-withdrawal and operate in a world of representations, as if representations created objects ex-nihilo rather than doing *nothing but* representing them, we likewise wander away from scenic anthropology into the vacuum of free-fall insignificance. Generative anthropology would encourage a discipline of resistance to such anti-scientific moralizing grounded in contempt for the externally real. Paradoxically, interfering with Nature is the experiential foundation of most of our love for it, maybe all of it. Where we make our homes, there we interfere with Nature. But we love Nature and care for it only because it is our home.[45] The "whole" of "Nature" (unknowable anyway) is as vulnerable to the paradoxes of the sacred as any other particular object we experience on the scene.

Quantification gives power: counting, measurement, predictions of movement. At the same time—as the Frankenstein myth shows when scientism becomes thinkable at the dawn of the Industrial Age—the power that quantification gives can be terrible. The instrumental reliability of science gives access to destructive forces in a way the attitudes of religion do not. We cannot destroy the planet with prayers alone, but we might destroy it with prayers and weaponry, atomic or biological or chemical. The most dangerous religion is not the one with God on its side, but the one with the most advanced technological weaponry on its side. And the worst horrors of scientism follow from the object delusion taken to extremes. Materialist quantifying applied to humans, to the exclusion of any application of other ways of knowing, obliterates the scenic now as it would have done at the beginning.

The scientistic delusional wishes to reduce the human body-being to nothing but a cosmological object. Victor Frankenstein boasts that as he digs up corpses in cemeteries, he believes the bodies of human dead are only rotting matter. Dr. Moreau argues that sensations of pain are only meaningless data when he explains the serenity with which, in his quest to make a new species, he subjects the animals shackled to his operating tables to vivisection without anaesthetic. In *R.U.R.*, the younger Rossum succeeds in creating robots only after he manages to delete everything human from the recipe. The blade runner Rick Deckard, shocked to learn the replicant Rachael is not aware she has been manufactured, hears the biotechnologist Eldon Tyrell: "Rachael is an experiment, nothing more." The reductionism in generative anthropology, by contrast, is not the reductionism of scientism. It is our contention that to reduce human beings to cosmological objects alone is to do them the violence of scientism, to submit them to mad science. And generative anthropology can help to show how the Frankenstein myth dramatizes with impressive power the horror of that violence. Materialist quantification unaccompanied by knowledge of humans as sacred objects, humans as esthetic objects, humans as erotic and economic beings, is the obliteration of all the moments in historical events except that of the desire for quantification and consumption. Applying such materialism to human events is interpreting the world as if human beings are not actors on the scene of representation. It is scientism, mad science; it is the extreme opposite of originary curiosity satisfied and a minimally universal anthropology offered up for the taking or leaving.

Exchangeability and Desacralization

The Frankenstein myth leads us to apprehend an anthropological truth: to avoid madness (the open road to insanity, the slide into dehumanizing violence) we must conceive of the revelations of science as situated on the scene of human exchange. The scientist (because he is human) is never alone in the universe; he might think he can carry on as if he is alone with cosmological objects, but if he does carry on like that, he courts disaster. Scientific representations of objects are produced by humans; they ought always to be produced *for* humans, too—meaning with all humans virtually in mind.[46] The scientific discovery/invention that is *minimally desacralizing and maximally exchangeable* tends toward the practically good; it is compatible with minimal originary science. The scientific discovery/invention that is *maximally desacralizing and minimally exchangeable* tends toward something with the anti-social effects people name *evil:* to celebrate such discoveries in a moral vacuum is scientism. We might learn to think of evil as that which tends to the obliteration of exchangeability. Exchangeability presupposes sacralisation; sacralisation creates the space for exchangeability. There is no trade without a minimally sacred space between the traders. Scientism believes that we get to know objects exclusively through the exchange of scientific ideas. Scientism despises knowledge of the object as sacred; it despises knowledge of the object as esthetic; it despises knowledge of the object as esthetic or erotic; it despises knowledge of the object as economic, insofar as it does not care about the practical application of its discoveries. For scientism, all such imprecise, unquantifiable modes of knowing objects rank second-best or lower and must belong to the ultimately inconsequential realms of fancy and fiction, imagination and intuition.

There is a proportional relationship between the exchangeability and the sacralisation of an object. With the term *exchangeability,* I mean to refer to the way that some discoveries and inventions seem both reproducible from a practical point of view and desirable to people engaged in exchange. The exchangeability of a thing is visible. Ask about a device, product, or tool that results from a discovery: how easy is it to make and to distribute? Gold and silver are hard to mine, so their exchangeability tends toward the minimal and their sacralisation toward the maximal. Only an exchange-hating utopian would want a world where one could turn on a tap at home and see liquid silver pouring out to fill a flask. In fact, that is more or less the world in Capek's play before the robots learn to resent their lazy human owners and kill them all. The offensiveness of the plastic bottling of drinkable water hints at the possibly future spectacle of a maximal desacralization of water as a seemingly Natural, sacred thing available to all. The more scarce or precious the resource, the more likely it will be sacralised and the more limited will be its circulation in the community in the form of the exchangeable. Thinking of one's attitude toward oneself makes the desacralization-exchangeability ratio clearer: one's health, one's time, one's life—each is a finite resource. As the Faust in each of us learns, we have only one soul to sell to the devil. The older I grow, the more I know one life is all I have been given. There are hundreds of things I do the first time only once: jumping from a height into the water below, leaving my parents' home, getting into a fist-fight, receiving cash wages, saying *I love you* with my heart in my throat. Human events are not exchangeable, for they are not *things;* only the signs of them (verbalized memories, topics of conversation, and mementos for art) are exchangeable.

Science for the good creates things that humans all over the place need and want. There are technologies so exchangeable that we have forgotten their history and come to experience them as "natural": fire, cooking, hairdressing, the sounds of music, language itself as the ever-helpful naming of things. The tools developed by modern science alter irrevocably both us and Nature: technologies of medicine, visualization, mobility and communication. The invention that we hear about first as a rumour, see in the possession of neighbours, and then try out, fascinated by it in our own hands—who would wish away all such experiences of new goods on the market? *I tried it for the first time.* And yet there seems to be no technological expansion without a shrinking of cultural sacrality, without violence done to sacred centers. Insofar as historical change is driven by our increasing control over the natural world, moments in the technological past are episodes in the process of desacralization. When we handle any new technology, the residue of desacralization comes off in our hand. Railways and automobiles take out coaches and horses and carriages. The family dinner dies and the drive-through window triumphs. Bob Dylan plays his electric guitar and the crowd boos. Electronic mail destroys the slow art of handwriting letters.

To prove his point and re-enact the origin of the human, the mad scientist must make something much like a human body. When the human body is in question, the tension increases between a mindlessly laissez-faire privileging of exchangeability (if we the majority desire it, then it must be good for everyone) and a mindlessly paranoid prioritizing of desacralization (if we the majority are scandalized by its coming into circulation, then it must be a kind of thing bad for everyone). There is no shortage of moral indifference to technologies relevant to space pollution, the pine beetle, or shark fin soup. But when the marketing of a technological device that might affect the human body enters the picture, hot opinions proliferate. It is one thing to breed farm animals; it is another to breed human beings in cages or chains. It is one thing to train a dog or horse to obedience; it is another to brainwash the owner of a human body and thus make a "robot." The mad scientist operates in secrecy; factory farms do not welcome sightseers, and weapons development takes place under strict control. The overtones in the vocabulary of science conquering "Nature" increase in their potential for insidiousness when the scientist is conquering the "natural" human body. Reductions become personal when they dictate one's concept of the body. Are my emotions nothing but chemical reactions? Are my thoughts nothing but neurological events? Is my being seemingly endowed with free will nothing but a body every microscopic motion of which is determined by physical causes? Are my conscious thoughts nothing but *unreal* epiphenomena of *real* brain activity? Scientism sharpens its scalpel as it approaches the embodied human person.

This book will not pull out from the top hat full of our confusions a white rabbit solution to the ethical debates raised by technologies threatening dehumanization. I do propose that if we set the *exchangeability* of a technology against its power of *desacralization,* then the ethical dilemmas that the technology presents become more open to inspection, more concrete. The exchangeability question is this: what is the pragmatic possibility of all humans owning the object or knowledge in question? What kind of world would one inhabited by humans all having it then be? Would it be a desirable world? That is a first step. But the task is not just to ask whether most *other* humans find a given technology desirable, so that you can give up whatever you think and accept it as morally commendable if the majority answer is *yes.* Being human takes courage: majority rule can be majority insanity. Extreme libertarian consumerism (each peripheral individual decides that any object at all can be exchanged in the center) is nothing but such mechanical acceptance, the mirror image of extreme totalitarian uniformity (the center alone dictates what objects the peripheral individuals might exchange in it). Instead, the task is to imagine all humans who desire a particular object or knowledge actually possessing it. Imagine the wishes of everyone everywhere coming true, as those of the mad scientist in isolation come true. Is the world you then imagine one you would wish to inhabit? The ease of exchangeability of a scientific invention can be weighed deliberately against the intensity of desacralization that would follow its exchange.

The application of this ratio between exchangeability and desacralization helps us to grasp the relations between religion and science, which are certainly not relations one can think of in terms of simple conflict (on this, see Brooke's valuable history). Some science is minimally desacralizing and maximally exchangeable; it does not threaten religious belief. The minimality of fire makes it the symbol of something universally good. How to imagine humankind without fire? Medicine is good, the shaman or healer in the center when we get sick. And that which modern society expelled with the medical conquest of disease was a realm of occult powers: demons, witches, spirits, the gods often thought to be legitimating suffering by passing judgment in bodily injury and ailment. Once the microscope detects germs and viruses, why continue to believe in malevolent spirits and fateful curses: is modern science not preferable to that? Technologies of communication seem maximally exchangeable and minimally desacralizing: it is hard to see how devices that enable writing and translation, the preservation of text and the transmission of reports, from writing to paper to printing to the telegraph, telephone, television, internet and wireless iPad, could be so desacralizing that a world without them would be more hospitable to humankind.

However, at the scientistic extreme, modern technology creates devices that we find maximally desacralizing and minimally exchangeable. Unregulated experimentation on human bodies and minds is now considered suspicious, akin to mad science: research ethics boards screen proposals even for the calculation of public opinion from survey interviews, let alone projects testing drugs or psychiatric techniques on people. The nuclear bomb is a real-world incarnation of mad science; the bomb is minimally exchangeable and maximally desacralizing. It desacralizes human relations, putting all of us under permanent threat.

Technologies of nuclear weaponry remain under strict control in the international community; they must. Or consider—most relevant in the context of the vain scientist playing God the creator of the human—genetic engineering, in particular the technologies that promise to make it possible to perform human cloning.[47] In the wake of the cloning of Dolly the sheep, cultural fears were aroused to the extent that both the President of the United States and the European Community banned human cloning. The ban is an event we might describe as a quasi-sacred historical revelation. To imagine the technology of cloning as universally available–a world where to start the process of cloning oneself would be as easy as to make an appointment at a fertility clinic—is to imagine a world that has desacralized the human in a radically new way.[48] That human cloning has its cool, reasonable defenders is evidence that something like scientism is alive and flourishing, tenured and published. What we fear in human cloning on demand is ultimately the legitimation of the grotesque inequality between the clone and the donor, the original and the copy.[49] It is not unrelated to the inequality between the mad scientist and his impossible human victim, where the victim cannot enter human history as an equal to others already participating in it because his entrance has been excessively mediated by the real-world equivalent of a mad scientist.[50]

That humans find it desirable to minimize the exchangeability of the nuclear bomb, to minimize the freedom to sell one's body to researchers for cash, or to minimize access to cloning technology, follows the maximally desacralizing effect of each upsetting technology. If these examples appear to be extreme, that is because they are in reality. As an insidiously accepted scientism seeks to put human agency on the chopping block, we might wish to step back and murmur *no, thank you.* A strong grasp of the Frankenstein myth as the textual revelation of mad science will mean a rejuvenation of our knowledge of the limits of scientific knowing, a recovery of the irreducibly sacred in the human. Mysteries should not be multiplied beyond necessity. But language and consciousness should be at least a little mysterious and paradoxical. We do not wish to know a thing that can be completely known. Our philosophical wish to know ourselves is therefore paradoxical at its core, for it is a wish that evolved from religious experience. In the technical terms of generative anthropology, the truth of ostensive faith in the undeniable ritual presence of the central object evolved into the truth of reason in the undeniable absence of the referent of the predicated subject.[51]

The technical terms, however, may not be needed for all. One might say it this way: to declare that the human object (the human-as-object) can be known by the laws of physics or chemistry or biology alone, or even philosophical logic alone, is to draft an opening line in a nascent credo of scientism. It is to begin to become Frankenstein-like. A minimal common sense of the scenic sacred does us *nothing but* good.

Tortured Matter, Multiple Errors

Playing God presupposes the existence of a creative or willful agency external to and other than, separate and detachable from, the object on which that agency works. In the Frankenstein myth, that agency attaches itself to a recurrent image of interrogative torture. The persistence of that image implies something about the ethics of the materialism that infuses scientism. Science has always had an investment in the predictable and the controllable (Rescher, *Limits* 28, 76; 153); scientism extends that investment to a totalizing faith in the predictability of human "behaviour" that forgets the scenic reality of human exchange.[52] For much of what the scientist predicts of me, I may, having been informed of the prediction, choose to defy. From the separated externality of the God-like agency in the *playing God* metaphor, a moral inequality relevant to such considerations of predictability follows. It is thanks only to the advantages of knowledge that the mad scientist has over the not-quite-human objects of his manipulations that the potential for terrorizing in the mock-creation scenes of the Frankenstein myth holds. It may be that we have become so accustomed to the inequality between the mad scientist and his mock-human objects on the laboratory table that we fail to feel the full horror of that inequality. I invite you to step back with me and view the perversity of the scene, re-taking the picture with a sharper focus.

Those who come to Mary Shelley's novel only after having had their imaginings of the Frankenstein story formed by an acquaintance with one of the Universal or Hammer or other movie versions may well feel disappointment at the sparseness of visual detail Shelley supplies to help us picture the scene of the monster's animation.[53] The image of the mad scientist as torturer is presented more clearly in the films than it is in the original novel. The central picture is that of a humanlike body supine on a table, being exploited as an instrument—in effect, interrogated as an object—to *prove* the materialist doctrine that the human is reducible to the physicochemical, to prove that the human is reducible to the predictable and controllable. An artificial humanlike body on the laboratory table is being tortured as if it could confess that from it alone as matter, spirit might emerge. But matter has nothing to confess; inanimate matter does not exchange signs with itself or us. We speak; as humans, we exchange symbols with each other. Our common-sense fear of mad science is not unlike our common sense fear of torture, for both are demonstrations of the ultimate excess in dehumanization. If Victor animates the body of the to-be-Monster, his project of having created a man-thing will be complete: it will have proved that Victor could do it.

However, the "good" this proof does belongs totally to the vain scientist. It does the creature no good whatsoever, beyond bringing the thing into a pre-determined existence where its integration into human history is impossible. The success of the vain scientist is not success for the metahuman of the un-alive body about to be born. Once the scientist has proved his point that "it" can be done, he is more or less indifferent to the "human" he has created. Unlike the Gods of creation myths, mad scientists do not care an iota for the fated subjective experiences of the beings they have created.[54] Victor and his imitators must torture matter to try to get "the human" out of matter, but then they are done. Victor does not love the Monster as a parent might love a newborn child, nor does he bring his object into subjective being just by speaking as do the Gods of creation-from-nothing who speak the object into animate being with words alone. In *The Island of Dr. Moreau,* Wells intensifies the image of the supine body being tortured, with Moreau vivisecting animals deliberately without anesthetic in the belief that the pain they experience will shape their minds into the superhuman mental condition he wants them to exhibit eventually. In *R.U.R.,* Capek stays cheerfully distant from any such focus on the bloodily sadistic through three of the four acts of the play; but in the post-apocalyptic epilogue, the robots come demanding of the sole human survivor, the carpenter Alquist, to cut them open—*Cut!—Cut!*—in a search for the secret of life.

The scene of the mad scientist as torturer of not-yet-human non-bodies shows that the human cannot be made out of matter alone and that it was not so "created" at its historical origin. One cannot re-enact the origin of the human by just manipulating matter. And yet the mad scientist hates and defies that religious (anthropological) truth. It is horrible for us as readers or viewers to intuit that such torturing could go on forever, because it will never produce a *real* human; but the mad scientist refuses to accept the evidence of his senses and common sense, and so he continues, nonstop, trying, failing, trying. Whatever comes of such experiments will belong to the mad scientist first and the human community second. That is why I use the term *metahuman* from time to time as a variant on impossible human. *Metahuman* is meant to connote the difference between the false doctrine of the reducibility of human being to a materialist ontology and the true originary unity of humans in shared linguistic presence. The human originates in a memorable scenic exchange of signs; a sign is not reducible to a material thing.[55] The mad scientist believes in every case: *I can make a human being thanks to my manipulations of matter alone. I know the necessary scientific laws of matter.* He means the chemical laws (Victor), the physiological laws (Moreau), the quasi-biological (Rossum) or the genetic laws (Tyrell). *Furthermore, the laws I know are all I require to re-enact human origins; I do not need to know about religion or esthetic experience or economic value.* The errors contained in such a creed deserve enumerated exposure. I limit my closing exposition to three.

First, one creature alone is not a community. Aiming to re-enact the origin of the human, the vain scientist appears oblivious to the irreducibly social character of human origin. The one God of the Abrahamic religions and all creator deities in world mythology make humans so that they *exist in groups, n*ot so that they can spin out into atomic existences without mutual contact. One can cut and splice, stitch and pry, twist and torture matter as much as one wants, but matter alone does not bind us to the historical origin of human community. Human origin comes from a collective memory of a significant event, and the shared spirituality in the exchange of signs that commemorate the event. It does not come from the physical facts of a nothing-but-animals "behaviour" or nothing-but-computers "mechanism." Scientism does not grasp the sacred nature of universal human community.

A second error is an indifference to nurture and the sustaining of the creation. A human being must be nourished, instructed, raised, allowed a place in a family and given a chance to contribute to the formation of a society in historical time. That this is the way human reality operates should be beyond dispute. Consider the obstacles faced by feral children, whose humanity can never be fully developed. It is hard without sounding trite to point out that the reality we share comes from numberless exchanges of words and things we have had with parents, siblings, family, friends, co-workers and others. Exchange is a human universal (Davis 3–4; Kuper 74–78). Nevertheless, the truth of this narrative extension of any genuine human project is one for which the mad scientist has no use: for him, the success of the experiment can be measured in one quantifiable result. The moment of initial creation receives attention, while the tedious business of caring for the creature afterward is dismissed. Victor rejects his Monster the instant it takes its first breath; Moreau sends his Beast People away to the caves to be managed by Montgomery; the Rossum engineers think very little about what the robots will do or be after point of sale, and none notices when Dr. Gall adds irritability to their physiological constitution; Tyrell tricks Rachael into discovering she is a replicant only to kick her out into a world where she can be executed on sight by a blade runner. The Victor-figure builds his re-enactment of the origin of the human on the falsehood that an "objective" procedure can replace processes of caring exchange. All I need, believes the scientistic obstinate, is this body on that technological table; I can make a human out of these materials, for there is nothing but my technology to it. The difference between the supine, tortured body of the about-to-be-animated impossible human and the body of a newborn human about to be loved and cared for is one of illimitable significance. It is the difference between the staging of scientism and the scene of representation.[56]

A third error is the vain scientist's neglect of language. As we have seen in the exposition of the originary hypothesis above, we must take language to be an essential ingredient in the essence of the human, if not the most essential. To the extent that a human being does not have access to language (infants, the autistic, the mentally disabled, the comatose, those with Alzheimer's), he or she needs the protective love and care of those who do have it all the more. Now witness the language experiences of the victims of the mad scientist. Apart from hideousness, Victor's Monster must learn in observant solitude how to decode speech and read printed text; the Beast People chant their laws without understanding them; the robots talk monotonically, and express themselves as incapable of grasping key concepts relevant to adult human consciousness (pain, death, love); the adolescent inexperience of the replicants clashes with the painfully adult shocks their bodies have endured as those of mercenary soldiers, involuntary slaves, and sex-toy prostitutes. These beings have not been brought gracefully into language. How could they have been, given the contempt of their materialist creators for ordinary scenic exchange?

Language is not just a material thing. As the linguists like to say, the signifier has a material aspect: ink on paper, sound from tongue and throat and sound in ear. But the referent of the signified belongs sometimes to the material world, the signified is an idea in the mind, and the relation between them is the *immaterial relation between* sign and object and between those real creatures who exchange signs of the object. That immaterial relation is not thinkable in terms of mechanical or physical or "evolutionary" models. The deluded scientist tries to re-enact the origin of the human without any appreciation of the necessary sociality of shared representations. No wonder he fails.

When we measure the power of the vain scientist playing God the creator against that of his metahuman creatures, we discover therefore a grotesque inequality. Where knowledge is freely exchangeable, there can be no mad science. Scientism means secrecy, hoarding of information about cosmological activity, refusal to exchange findings and facts. Again, there is no resentful quarrel with institutional science in these pages. Most of what Thomas Kuhn calls "normal science" operates in a way opposite to that of the mad scientist: normal science requires the sharing of well-established means to produce "determination of fact, matching of facts with theory, and articulation of theory" (34) within an already-existing scientific discipline. John Ziman insists on the social, corporate nature of scientific inquiry (31). It is an intensely social activity, unquestionably one of the sharing and exchange of representations (models, theories, findings, hypotheses, conjectures, refutations and the like). But even so, there can be a considerable distance between scientific research communities and human communities broadly conceived.[57] That in the real world which we might think of as *mad science* violates the criteria of exchangeability, openness, and universality that a cosmopolitan faith in widespread scientific progress takes for granted. One might call to mind here the useless experiments of Hitler's minions on the victims in the camps.[58] One might remember the secretive Manhattan project, insofar as the technologies of nuclear destruction it invented are not to be universally exchanged.[59] One might consider the forces contained by the exceptional self-restriction of the genetic scientists attending the 1973 Asilomar conference.[60] The products of mad science are minimally exchangeable and maximally desacralizing. They are most spectacularly desacralizing when their object is the embodied human animal, the human being. In such cases, the analogies that wary persons draw between what real scientists occasionally prove capable of doing and the fantastic cruelties of mad science in the Frankenstein model are not unjustified. We feel the anxieties because because we have learned something of the anthropological wisdom conveyed by the Frankenstein myth.

TWO

Mary Shelley's Frankenstein (1818): Experiment and Irreversibility

Two Ways of Approaching the Book and Its Author

Witness the fame of *Frankenstein; or, the Modern Prometheus* (1818).[1] As a result of its publication, Mary Shelley in her lifetime never quite freed herself from two identifications. One labeled her "the author of *Frankenstein*." The other presented her to the British public as the mistress (or later, the second wife) of Percy Bysshe Shelley. Percy Shelley, some scholars claim, provided a real-life model for the fictional Victor Frankenstein (John Williams 181; Small 100–102, 112, 120). Although he flaunted a vaulting ambition and provoking atheism, and the propriety of his conduct toward the loyal Mary has been questioned, Shelley produced great poetry. His widow edited his work painstakingly over many years (John Williams 159–80). There is not much evidence that Mary Shelley deeply resented either identification. Moreover, the one smash success of the wife has come to overshadow, not in canonical prestige but in popular appeal, anything authored by her husband. There are many people across the globe today to whom the name Frankenstein means something, but few who have heard of Alastor or Epipsychidion. From the novel's inception, a profitable Frankenstein industry has operated without interruption. London's first theatrical version went onstage within a few years of the novel's appearance; the history of *Frankenstein* scripts and productions spans the nineteenth century. The twentieth century abounds with Frankenstein movies, comic books, imitations and spin-offs (Hitchcock; Lederer; Skal). In response to the overwhelming presence of this Frankenstein industry in popular culture, scholars of Mary Shelley, in keeping with the tendency to open up the canon of English literature, have created a Mary Shelley industry in the academy. Well-edited reprints of her post-*Frankenstein* novels have been made available as affordable paperbacks. Archivists and biographers have rescued her from confinement to the two identifications. The revisionist scholarship argues for the cultural value of the entire oeuvre of

Mary Shelley, including her travel writing, correspondence, shorter tales, and miscellaneous cultural criticism (Schor, *Cambridge Companion*).[2]

In one of the most searching studies of the novel, David Ketterer intimates there are two basic directions one can take in approaching the text of *Frankenstein*.

> But *Frankenstein* is patently open-ended. Whatever resolution may be projected lies beyond the bounds of that fictional world. The final image of a figure "lost in darkness and distance" is of a figure engulfed by the unknown. And such a figure dramatizes the reader's predicament. Unlike [Percy Bysse] Shelley, Mary [Shelley] has no philosophical answers. What she presents in *Frankenstein,* at the cost of a certain narrative incoherence, is a spectrum of possibilities . . . a philosophical appreciation of the sliding relationships between the Self and the Other. (104)[3]

Percy Shelley has "philosophical answers"; Mary Shelley has a "philosophical appreciation of sliding relationships." The "open-ended" quality of *Frankenstein* cannot be denied; it would be pure vanity for me to announce (imitating Victor) my discovery of the one key that will dispel its mysteries and dissolve its paradoxes. In these pages, no curtain will be raised on such a discovery. There are ways in which we simply cannot know Mary Shelley or Victor Frankenstein or the Monster. Blackouts, gaps, dead ends make it impossible to definitively answer fundamental questions about their relationships. And yet, adding to the judgment of Ketterer, I would suggest that the absence from the text of "philosophical answers" and the compensating presence of a "philosophical appreciation" might be better framed in the context of a struggle between anthropological or religious faith on the one hand (that which Mary Shelley appreciates); and revolutionary atheism on the other (that which Percy Shelley affirms as "answers"). For the impossible human Monster in Shelley's novel ultimately departs for a space of darkness in the distance, but Victor faces us overflowing with a mostly shameless know-it-all confidence. Mediating between the characters is a certain tacit spirit of religious humility, one I attribute to the authorial Mary Shelley. The sanity of that spirit helps guarantee the text's enduring appeal.

Ketterer's allusion to a "spectrum of possibilities" and his image of "sliding relationships" can help us find a critical position on the crowded stock market floor of commentary on the novel. The insinuation that a cautionary mad-scientist-tale interpretation of *Frankenstein* is a worn-out commonplace has itself become a commonplace. I contend nonetheless that that the power of the mad scientist figure has not yet been sufficiently explained; I take a somewhat untrendy literalizing approach, different from the psychoanalytic allegorizing approaches that have dominated interpretation of the text in recent decades. This literalizing approach may breathe new life into the mad scientist / mocked God binary, by situating it in the context of tension between the sacred and the exchangeable aspect of objects. The tension between the scientific control of Nature and the value of reciprocal exchange certainly remains relevant to the task of interpreting *Frankenstein,* for Victor's Monster is not one who can exchange things (words, yes; things, no) with others: Victor has made it impossible for him to do so.

Let us explore this suggested distinction between allegorizing and literalizing approaches to *Frankenstein.* Grant that the studious reader may well be split, to lift a phrase from Jacques Derrida, between "the two dimensions of the letter: allegory and literality" (75).[4] The allegorizing direction is practiced to its most dazzling effect mostly by academic critics of the psychoanalytic variety. Now while it would be absurd to ignore Freud's achievements in opening up public discourse about erotic experience and establishing the unconscious as a category for self-reflexive thinking, in this book, I assume that to respect Freud's achievements does not require one to subscribe to the psychoanalytic orthodoxy on "desire."[5] Let us grant the assumption that the relationship between Victor and the Monster (the Creature, the Wretch) is the central conflict in the story. The psycho-allegorical approach privileges the *doppelganger* myth, the myth of the double, as the structural bedrock of the story (Ketterer 56–65). It minimizes common-sense differences between Victor and his Creature; it prefers to frame such differences as signs of two halves of one psychic unity. With the *doppelganger* approach, the origins of the defective relationships between Victor and the family he neglects, the creature he abandons, the fiancée whose erotic attractions leave him indifferent, and the public world with which he fails to engage in productive exchange, may be located at different levels. They may be located inside the novel in the psychological dynamics of the fictional Frankenstein family, outside the novel in Mary Shelley's family dynamics, or in the interplay of the two realms. The allegorizing approach tends to lead toward discoveries about the exemplary victimization of Mary Shelley in her particular time and place; then by way of analogies made with that situated experience, it invites us to share in a resentment of her victimization (or the victimization of a group for whom she harbored sympathies). For such interpreters of *Frankenstein,* the message may be something about an exposure of a particular evil devolving from the bourgeois family, a moment in the history of capitalist patriarchy, a pattern of oppression in the British class system, or some combination of similarly under-par institutions. The diversity of grievances can stun the most intrepid explorer of the critical literature. *Frankenstein* is notorious for seeming able to mean almost anything, depending on the inflection one's subcultural resentment takes and the ideological scalpel one has sharpened.

The question of the literal reality of the Monster in *Frankenstein* is not unlike the vexed question of the "objective" reality of the ghost in Shakespeare's *Hamlet.* As Prince Hamlet's will to believe in its reality is nourished by his resentment of Claudius, who has usurped the scenic center at the Danish court, so one may argue that the Monster partakes of a dubious ontological status with respect to the maker who inspires his resentment.[6] As some *doppelganger* interpreters like to point out, we might wonder whether it is Victor who murders Ernest and Justine, Elizabeth and Clerval; we might wonder whether the murders are performed as the working out of a neurotic resentment and the Monster is a kind of fantastic externalized hallucination by means of which Victor, stuffed with hatred of his benighted family for various unowned reasons, excuses himself. At the extreme, psycho-allegorical *doppelganger* readers argue that the violence done to Victor's family does not have its genetic cause in Victor's desire to do science, but rather, the pseudo-public science is an incidental means to private (psychological, or Oedipal) ends that are more interestingly perverse.

For example, one *doppelganger* reading by Francois Flahaut, in a volume of anti-monotheistic philosophical anthropology that exploits *Frankenstein* as its discursive springboard, makes the ugliness of the monster into a legitimation for what Flahaut calls "the omnipotence of the solitary." Flahaut's thesis epitomizes the ethical open-endedness of the *doppelganger* critics in the way it ironically ends up accusing the impossible human of playing God. Because the solitary is the omnipotent, and because omnipotence is an attribute of the imperviously Divine, it is actually the Monster and not Victor who plays God, argues Flahaut; Victor suffers more than the Monster does. I do not think so. I would concede that the peculiar form of centrality the Monster claims depends on the cruelty of his circumstances: if suffering alone confers omnipotent centrality, then the Monster beats Victor in competition for the center. But Flahaut's ironization weakens our grasp of the text's dramatization of a problem following from the rise of modern science. Scientism would place no limit on the powers that experimenters such as Victor possess to desacralize cosmological objects, including the power to desacralize the physical bodies of humans (or metahumans) as cosmological objects. Scientism defends new and dangerous discourses and practices of dehumanization. Mary Shelley's text founds the myth that, when well-told, can remind us of that inexhaustibly dangerous, dangerously inexhaustible power in scientific naturalism to (pretend to) demystify human scenicity.

Rather than the allegorizing approach, I prefer a kind of literalizing attitude. But what would it mean to read "literally" the events and existents of the fictional world of *Frankenstein?* Mostly, for the investigation this chapter carries out, it means to privilege the literality of the otherness of the Monster. The Monster inside the world of the novel is real, just as in the world outside the novel the powers unleashed by the methods and findings of modern science are real. As he twists and turns, friendless and mystified, insisting that there is a painful reality in his ugliness, I uphold the testimony of the Monster as an independent, autonomous character. If he claims to know anything, it is that he is a creature or thing involuntarily separated from the chemist-anatomist who created him and repellent to the humankind to which that scientist belongs. Flahaut asserts that the Monster is omnipotent in his solitariness; if so, I reply, it is not a solitude or an omnipotence he has chosen. The more the Monster is considered somehow unreal, the less disturbing is the moral crisis created by his ugliness. It is significant that those critical analyses which pay attention to the Monster's ugliness (see Bowerbank; Dutoit; Gigante; Juengel) do not exclusively take the *doppelganger* approach that lends itself to such a lessening.

Readers unacquainted with the massive scholarship on Mary Shelley's novel may be surprised to learn there has been a debate in academia, not always a polite one, about whether it is appropriate to describe Victor as a scientist at all.[7] Most significant for us in these pages is the fact that the critics who propose limiting qualifications on our moral disapproval of Victor are those few who happen to define and to emphasize his role as a scientist. They make a heretical sect in the mass of judges who wish to send Victor with his unmitigated narcissism straightaway to the guillotine. These qualifying voices suggest that certain edges in Victor's career become more sharply visible if we concede Mary Shelley may have been imagining a promising young researcher who, somewhat tragically, loses integrity when he carries to its gruesome conclusion a daring experiment. That is the shape of Samuel Vasbinder's defense of Victor. Similarly, Jane Goodall's scholarly investigation of Victor as the victim of a Calvinist reprobate's conscience offers a defense of his career. Theodore Ziolkowski claims that the personality of Victor degenerates during the ardors of the experimental process and breaks at the moment of technological success. Such atypical defenses of Victor as moral agent attribute a significance to the radical discontinuity between his experience prior to the experimental animation of the Wretch and his experience subsequent to it. The literalizing approach I prefer takes that discontinuity into account. It is certain that one of the most necessary-to-analyze narrative transitions in Mary Shelley's text—the graphic animation of the monster, the *scene a faire* of every Frankenstein movie—is the most difficult to interpret. We can capture in a single question the most baffling puzzle the creation of the monster presents, insofar as pondering its appearance means pondering the figure of Victor as scientific agent: why does Victor create a human-like creature only to reject it at the very moment of its animation? What sense can we make of the weird way that in one and the same instant his mad science succeeds (the Monster moves) and fails (the ugliness its movement reveals is intolerable), as one and the same event?

The novel's subtitle is relevant to the issue of foresight: "the Modern Prometheus." The phrase might be taken as an oxymoron in which *modern* contradicts *Prometheus*. One might emphasize *Prometheus* so that the phrase foregrounds Victor-Prometheus at the expense of his surroundings, implying he is admirably defiant of the God of the Bible in daring to acquire forbidden knowledge. But I favor the possibility that Mary Shelley wanted us to accent the adjective and hear the phrase as "the *modern* Prometheus." Spoken thus, the subtitle emphasizes the historical frame of modernity that contains her figure of usurpation; in a modern context, we are more likely to reject the anti-democratic exceptionality of Promethean desire. Any *modern* Prometheus will be too self-importantly heroic, difficult to admire without resentment, because an egalitarian ethic dampens enthusiasm for the acceptance of would-be demigods.

The name Prometheus means "forethought" or "thinking ahead," as the theologian Ted Peters points out in his definition of "Promethean freedom," the form of freedom least compatible with religious humility. By contrast, Epimethus the brother of Prometheus (whose name means "afterthought") is in a related myth a blunderer who registers *too late* the disaster of the opening of Pandora's Box (Peters 8–10). Victor cannot count, therefore, as a genuine classical Prometheus, for his lack of foresight makes him a blundering *modern Prometheus,* all mad scientific power and no wisely humble service, all incalculable risk-taking and no regard for harm, or even for benefit. Thanks to the Prometheus of ancient Greek legend, men came to share the control of communal fire; but thanks to Victor, men share . . . nothing much, except perhaps an example they have little desire to follow.

Victor's Early Career: Discovery and Experiment

Victor's isolation at the University of Ingolstadt and the extension of the free space he has struggled to fill up at home during his childhood together condition his discovery of the "secret of life." In his oddly restrained account of that discovery of a cosmological truth, there is a contradiction between a certain humility implied by his admission that he almost happened not to discover it and the boast that he went through terrible exertions to find it. The note of humility is sounded in the sentences reporting that after "two years" of study in Ingolstadt, as he tells Walton, he made "some discoveries in the improvement of some chemical instruments, which procured [him] great esteem at the university" (1.3.78). The value of that esteem seems exhausted when his education in chemistry as such is complete:

> When I had arrived at this point, and had become as well acquainted with the theory and practice of natural philosophy as depended on the lessons of any of the professors at Ingolstadt, my residence there being no longer conducive to my improvements, I thought of returning to my friends and my native town, when an incident happened that protracted my stay. (1.3.78)

The "incident" that seems to have just "happened" is—yes—the momentous discovery of the secret of life. Before it happened, Victor was humiliated by a flattening equality with his immediate mimetic models.

We can describe that unstable equality as a point at which positive, submissive imitation of a superior model of desire is no longer possible and negative competition looms as a possibility: Victor's knowledge of "natural philosophy" has become come perilously equal to that of his professors. Consider how strangely banale and democratic it is that the mad scientist at the origin of the Frankenstein myth almost gives up. He almost goes home to his father Alphonse and his fiancée Elizabeth as one whose contribution to chemistry is only that of a technician's tinkering, that of a "'petty experimentalist'" rather than a "'man of science'" (1.2.77), to borrow a phrase from his mentor Professor Waldman. So we should notice that, in the passage just quoted, Victor has underestimated the tremendous consequences of the "incident" and not accurately measured the extent of the prolonged "stay" in his bald assertion "an incident happened that protracted my stay." His ambitious side and his irrepressible pride, those he betrays a little later in his narrative to Walton: "After days and nights of incredible labour and fatigue, I succeeded in discovering the cause of generation and life; nay, more, I became myself capable of bestowing animation upon lifeless matter" (1.3.80). It seems an "incident that happen[s]" in the passive mode of egalitarian self-effacement now (in his narrating to Walton) weighs memorably as the result of "incredible labour and fatigue" in the active mode of ambitious usurpation. Which is the truth? Did the secret just reveal itself (blame Nature for giving it up) or did Victor force its revelation (blame the scientist for violating Nature)? Whatever Mary Shelley wants us to believe, it is certain that in order to discover the "secret" in the first place he must have worked with dead animal and human flesh. He robbed graves, tortured animals, drove himself to ill-health and willfully ignored the requests of his father and sister for reciprocity in letter-writing. He cut himself off.

Victor describes the discovery of the secret of life in language reminiscent of Saul's conversion on the road to Damascus in *The Acts of the Apostles* (9:1–9).

> I paused, examining and analysing all the minutiae of causation, as exemplified in the change from life to death, and death to life, until from the midst of this darkness a sudden light broke in upon me—a light so brilliant and wondrous, yet so simple, that while I became dizzy with the prospect which it illustrated, I was surprised that among so many men of genius, who had directed their enquiries

> toward the same science, that I alone should be reserved to discover so astonishing a secret. (1.3.79–80)

Although this passage is describing the discovery of the theory preparatory to the much more spectacular practice, the animation of the Monster, it is still loaded. Consider the difference between the afigural spiritual revelation of the resurrected Christ in Saul's conversion experience on the road to Damascus,[8] and the unredeemed, non-resurrected, dead disfigured flesh on the laboratory table of Victor's scene of revelation. The quality of Victor's experience rests in a liminal space between the involuntary "election" of the tragic hero typical of the neoclassical esthetic (*I was surprised that . . . I alone should be reserved*) and the self-centralization typical of the romantic esthetic (without blinking, inside the memory of his triumph, he dubs himself one of the *men of genius,* oblivious to the vanity we see from outside).[9] Therefore, the "light" that breaks in upon Victor the apostle of atheistic modernity is the light that makes him feel unique "among so many men of genius." It makes him believe that he has been "reserved" to make the discovery. Victor becomes "dizzy with the prospect which it illustrated" not because of any feeling of awe for God's handiwork in the cosmos. He is dizzy because the secret will make *him* powerful as a creator demi-god: "A new species would bless me as its creator and source; many happy and excellent natures would owe their being to me. No father could claim the gratitude of his child so completely as I should deserve theirs" (1.3.82). The diction here is oddly vague again: what does Mary Shelley want us to imagine Victor having imagined, under the sign of "many happy and excellent natures"?

Like Moreau, like Rossum, like Tyrell, Victor operates in the conviction that (for reasons he does not specify) the human species is not good enough. To register the depth of his dissatisfaction is to begin to account for his general coolness toward, and his particular refusal to share any of the truth with, his best friend Clerval, his fiancée Elizabeth, and his father Alphonse. That he tells them nothing at all about the experiment he has performed and the Monster threatening their lives—that his narcissism seems so extreme that he cannot even conceive that the Monster might take them as prey—is the most exasperating aspect of his confessional account. Also strange, however, is that even in this recalling of the demi-god fantasy, Victor imagines only a reciprocal relationship with his meta-human creatures, without any thought of opportunities to share the results of his experiment with other people: *many happy and excellent natures would owe their being to me. No father could claim the gratitude of his child so completely as I should deserve theirs.* He is not thinking about the thanks of other humans for having found the secret of life; he is thinking about the thanks of the "gratitude" of the "many happy and excellent natures" who will "owe their being" to him exclusively. It is hinted that these children will owe him more, notice, even than he owes his human father Alphonse. Moreau will dream of proving his rival scientists wrong; Rossum and Tyrell will market their artificial humans as labour-saving products; but Victor intends only to have his creatures thank him for being created. He does not waste time wondering how others might respond to his having fathered a new race.

In Shelley's representation of what we might call the home schooling and the laboratory luck of this originary mad scientist figure, a chilling isolation emerges. There is a way in which it feels appropriate to pity Victor somewhat for the isolation that was one with his freedom. His is not a knowledge produced in noisy, energetic collaboration with others, in the give-and-take of dialogue, in the sharing of experimental procedures and results. Not his fiancee, not his father, not his best friend Clerval, not his professors even begin to get in his way. Nobody asks a pertinent question. The feeble acquiescence of his family and friends to his equally feeble explanations of the absences, depressions, illnesses, and irrationalities that are so much a part of his activity is as much one of the "horrors" of the text as his digging up corpses or torturing of unspecified animals. It is the horror of one man with too much freedom and too much power, not obliged to answer to anyone.

It may be that we can no longer appreciate the chilling quality of this part of the story, because our responses to Shelley's text have been irreversibly influenced by all the cinematic versions of the story. In those retellings, there is always at least one character—a teacher, a scientific assistant, a servant—who presents in dialogue with the recast Victor some version of the prohibition against playing God. There is no such rival character in Mary Shelley's founding text. Her Victor is condemned to freedom in a moral vacuum, given liberty by a deference mechanically granted him by everybody except, appropriately, the Monster. That weirdly empty liberty—weird in its silence, vacuity, and lack of value—that liberty possesses something of the menacing quality of the stock aristocratic villain who roams the castles and mountains of Gothic fiction, disposing of naively duped rivals and threatening virgins and moving like an insubstantial shadow in a realm seemingly devoid of human or Providential supervision. And this peculiar unreality of Victor's isolation is a feature unique to Mary Shelley's version of the myth. By the time of the modernist *Moreau* and *R.U.R.,* such a fantasy of aristocratic immunity will have become unthinkable. In modernist versions of the myth, agonistic dialogues about the moral decisions of the scientist contribute significantly to the text. Likewise in *Blade Runner,* the morality of the science that produces replicants is explicitly questioned. That Victor's Creature is condemned to an isolation equal to that of his maker, but involuntarily and without his own consent, is nightmarishly appropriate. Indeed, if I were obliged to offer my own variation on the doppelganger approach, I would do so here: Victor makes the Monster as a thing that will be totally cut off from society because he himself, while making it, has cut himself off totally from society. He expels all human ties except the *one* he has always treasured: his tie to the alchemists (Paracelsus and Cornelius Agrippa) who modeled his desire to play God, to create a human-like creature.

The passage between discovery of the secret and its practical and techno-logical implementation marks a difference roughly analogous to "pure" and "applied" research. Victor seems conscious of this difference, inasmuch as he distinguishes between the testing of the secret itself (watch for the phrase "first success" just below) and the representation of it by its incarnation in a human-like creature: "I doubted at first whether I should attempt the creation of a being like myself or one of simpler organization; but my imagination was too much exalted by my first success to permit me to doubt of my ability to give life to an animal as complex and wonderful as man" (1.3.81). Notice that Victor does not say he intends to create a human: he intends, rather, to create "an animal as complex and wonderful as man," a different kind of animal. Notice that what he doubts is not the moral propriety of giving life to an animal as complex and wonderful as man, but only his ability to give the life. He does not fear offending God, because there is to his mind no God anywhere, neither in his laboratory nor in the larger cosmos. Nor does he fear offending any moral law within himself that might generate purely anthropocentric scruples. He fears a technical failure: "Nor could I consider the magnitude and complexity of my plan as any argument against its impracticability. It was with these feelings that I began the creation of a human being" (1.3.81). Now the creature *like* a human is to *be* a human "being"; his ambition narrows, concentrates. But the "magnitude" of his plan means a measurement of design difficulties, not ethical complications; the "complexity" of the blueprint is technological, not moral.

The absence of any mention of usurping the creative position of God fits with Victor's report of youthful disbelief in the spirits and the supernatural.

> In my education my father had taken the greatest precautions that my mind should be impressed with no supernatural horrors. I do not even remember to have trembled at a tale of superstition, or to have feared the apparition of a spirit. Darkness had no effect upon my fancy; and a church-yard was to me merely the receptacle of bodies deprived of life, which, from being the seat of beauty and strength, had become food for the worm. (1.3.79)

It does not follow from a disbelief in ghosts that a cemetery should be *merely a receptacle of bodies deprived of life*. So much for the experiences of the sacred memorialized by pyramids and mausoleums, the experiences of the sacred marked by tombstones or makeshift wooden crosses. To reject superstition is one thing; to reject the sacralization of the dead, another. Victor desacralizes the human corpses he sews together. In any case, after his meeting with the vengeful Wretch on the Mer de Glace, Victor's attitude to "supernatural horrors" undergoes what should be an embarrassing change, although he never remarks upon the radical inconsistency. Once he becomes a victim of the Monster victim he has created, he calls freely and fiercely upon the spirits of his dead family to help him in vengeance. He dreams of post-mortal reunion with them. But the main point is that the more closely this originary mad scientist trying to re-enact the origin of the human approaches the moment of paradoxical animation, the moment of playing God the creator, the less is he thinking about the possible value or eventual exchangeability of the object he is creating. He is not thinking at all but emptying his mind of human thought: "Although I possessed the capacity of bestowing animation, yet to prepare a frame for the reception of it, with all its intricacies of fibres, muscles, and veins, still remained a work of inconceivable difficulty and labour" (1.3.81); "Who shall conceive the horrors of my secret toil, as I dabbled among the unhallowed damps of the grave, or tortured the living animal to animate the lifeless clay?" (1.3.82). Such emptying is hard work. For a human being, it is no easy thing systematically to pursue dehumanization.

Irreversible Experiment and the Event-Structure of Scientific Revelation

The moment of Frankenstein's animation of the creature mimics the scene of originary language use that generative anthropology hypothesizes as a minimal model for human events. The animation scene may be read as an ironic recasting of the moment in which humanity and its transcendental Other come into being together. There is a sense in which the mad scientist—as one completely alone, opposed to the human community by self-exclusion from it—comes into being together with his meta-human creation only when he creates it. Certainly, before the instant of animation Victor performs and practices as a scientist, an experimentalist, one who knows chemistry and anatomy; but his madness has not yet achieved completion. Certainly, there is some madness in a form of action so anti-social and intent on not examining unintended consequences, refusing to imagine them in the first place; but not all his madness is there yet. By suggesting that he *comes into being* as a mad scientist when the Monster comes to life, I mean to clarify that the partial success and total failure of Mary Shelley's originary vain scientist are paradoxically one. His identity as mad scientist depends upon the irrevocability of the act that will ever after define both him and the special "object" he will be bound afterward ever to care for, or else to neglect and ignore. In what the speech act theorists call the declarative speech act, the words actually bring into being the status conferred by their being spoken: *I pronounce you husband and wife* creates the status of the married; *You are to be hanged by the neck until you are dead* creates the status of the condemned; *you are under arrest* creates your status as arrested citizen. The declarative speech act does not simply report a state of affairs or just describe reality; it *changes* social reality. To perform such speech acts is quite different from stating *now they are married,* for example, or reporting *now I am doomed* or *he was arrested.* The institutional authority figure that performs the declarative speech act—- judge, priest, officer—brings into being new individual possessors of a social status (Elam 156–70; Rust 66–82). Similarly, by animating the creature Victor changes reality, but in a *special* way. He is not just changing the beach by piling stones in a mound or changing the landscape by planting a flower garden. He creates a whole new category of living thing (the impossible human) and creates in and by himself a whole new category of human overreacher, the partially successful creator ultimately bound for disastrous failure (the mad scientist trying to re-enact the origin of the human).

The baffling paradox in Victor's creation of the as-yet-unmade artificial human seems to be that, so long as it remains unmade, he cannot properly grasp its reality. For up until the successful animation of the gigantic patchwork body, the Monster is not yet human-like; the monster is *nothing but* matter. The question of why Victor does not foresee the ugliness becomes less puzzling and less pressing in proportion as we sense Mary Shelley's assumption that if we were "inside" the world of the novel, we would react to the living and animated Monster with "unfair" loathing, just as Victor and all others inside the novel do. Insofar as the loathing is a universal, instinctive, and visceral-mechanical response, we must not confuse the injustice of making the Monster in the first place with the injustice of treating the Monster badly after he has been animated. There is no doubt Shelley designs the narrative such that we are expected to feel sympathy, despite his ugliness, for the abandoned wretch. But I would suggest that she deliberately blocks and frustrates that sympathetic response and condemns it to the gesturing of a certain futile hypocrisy. She demands that we as readers sacrifice any self-elevating illusion that we could have acted other than Walton, Victor, or the cottagers if we were in the places of these characters as confronted by the animated Monster. From this we can derive a principle: the Frankenstein myth requires that impossible humans be at least as terrifying as they are pitiable, if not more terrifying than pitiable. The ugliness of the Monster is the concretization of the barrier Shelley creates to guarantee her originary metahuman's radical otherness and non-humanity. Victor does not succeed in creating a human creature. Our responses to the ugliness of his victim are not the results of mere "prejudice." We are meant to feel our horror not as the contingent effect of a "social construction."[10] George Levine observes that "the monster's isolation derives not so much from his actions as from his hideousness" (11). Indeed, there is no action the Monster can perform, nothing he can do, to remove his ugliness and replace it with unambiguous human status: that is the horror of his concocted, made-up situation. Denise Gigante, in a valuable study of ugliness as a philosophical enigma, makes a remark in line with the position I take here: "the text demands that we recognize his deformity as the precondition of the text's ethical concerns and the occasion for the narrative itself" (355).

I would even go so far as to argue that the descriptions that Mary Shelley does provide of the monster's physical appearance are one of the "flaws" of the text. They are flaws because, ironically, they do not make the Monster look frightening enough: they do not suffice as explanations of the terrified reactions to the Monster that she has her characters perform. More strangely, it might be best to concede that perhaps no verbal account at all could achieve the effect that a perfect impression of such *mythic* ugliness would require. Reading the textual descriptions, we may well think to ourselves, *well that does not seem too bad; I could look on a creature like that and not run away in terror. Why does everyone in the book run away in helpless terror or turn away in disgusted repugnance?* For those of us schooled in decades of Frankenstein movies, it is even more likely that the Monster-as-described-in-words may seem less than horrible. Those who love the 1931 James Whale film love Boris Karloff's "humanized" monster. However, Shelley's founding verbal text is not the movie. I contend that, as an index of Mary Shelley's intentions concerning the attitude she wishes readers to take toward the Monster's ugliness, the evidence provided by the explicit active responses of the characters when they see the creature should be weighed more heavily than the "evidence" of the features listed in the portrait-sketches. I suspect that Shelley did not wish her readers complacently to congratulate themselves *if I were in the novel, I would be nicer to the Monster than Victor or anybody!* She hoped, on the contrary, we would wrestle with the frightening spectacle and terrible fact of a human-like creature who could not by any means be assimilated into the human community. The ugliness of the creature is the externalization of that impossibility of assimilation. We are to "side with" the humans in the novel to the extent that we are to identify with their disgusted fear of the Monster, assuming that such fear and disgust would be an inescapable and inevitable response to his presence were we in their positions.

Victor uses the word "ugly" to describe the Monster's appearance prior to its animation. However, Victor assumes that Walton will take for granted, as he does, the before/after dichotomy of the event-structure of what he has experienced as a scientific revelation. He assumes that Walton will accept the implication that it qualifies the experimenter's vulnerability to moral blame:

> Oh! No mortal could support the horror of that countenance. A mummy again endued with animation could not be so hideous as that wretch. I had gazed on him when unfinished; he was ugly then; *but when those muscles and joints were rendered capable of motion,* it became a thing such as even Dante could not have conceived. (1.4.86) [emphasis added]

The creature was ugly before animation; however, the post-animation ugliness is of a different order. It is the *animation,* the event of muscles and joints being made capable of motion, which makes the irreversible historical difference. Later, when the now-independent Creature confronts him and Victor tries his futile best to do his moral duty and respect its desire for a female companion, he structures the verbal account of his mixed feelings in exactly the same terms of before/after, static as opposed to moving. His moral sense is overcome by instinctive repulsion at the creature's anti-erotic power, because the *motion* of the Monster blocks the power of its verbal claim to moral reciprocity: "His words had a strange effect upon me. I compassionated him, and sometimes felt a wish to console him; but when I looked upon him, when I saw the filthy mass that moved and talked, my heart sickened, and my feelings were altered to those of horror and hatred. I tried to stifle these sensations . . ." (2.9.171). Quietly to believe that we are better than Victor, and that we would have embraced the Monster with pity and gentleness, is to enjoy the benefit of a moralizing victimary fantasy at the cost of appreciating the amoral horror of a tragic disaster.

Regarding this puzzle, I suggest that Mary Shelley, in creating the moment of this irrevocable error, was thinking about a fundamental shift in the relationship between human beings and the cosmos, or, as I would refer to it in the vocabulary of originary analysis, the scenic object experienced as cosmological. Almost two centuries later, we are now familiar with many images that both represent the power of technology and raise perplexing moral questions about human interference in Nature: the finger on the switch, the key in the ignition, the pliers about to cut the wire that will defuse or ignite the bomb, the palm full of pills or the syringe prepared to inject the chemical cocktail into the flesh. These stock figures of a decision to choose to implement an available technology represent the power of modern science to intervene (God-like) in the cosmos. In certain contexts, they signal the burden of responsibility the human bears to exercise powers where the intended result risks unintended consequences of huge magnitude. Modern science means there is no need to call upon Mephistopheles to appear or to dialogue with the genie who has come out of the bottle. Mephistopheles and the genie, like God, have apparently vanished; we have appropriated their powers for ourselves. God seems no longer to intervene in cosmological history; but we certainly do. Therefore, it is within the power of the experimental scientist or the world-altering engineer to have Nature "speak" to him more or less directly, via his experimental tools. The experiment becomes a new form of revelation whereby Nature, the only source of "evidence" there now is, shows itself or reveals itself. Such evidence has something of the power of God.[11]

Let us step back and recapture the novelty of this context of the newly mediated relationship with Nature before we pass judgment on Victor's "going ahead" with the decision to hit Enter and animate the creature, on Victor's failure to see in advance its animated ugliness. Consider how many switches we pull every day of our lives in a world where technological wizardries unimaginable for Mary Shelley fill up our households. Consider, instead, a world where such switches are *awe-inspiring novelties.* That is the world of Victor Frankenstein performing his experiment. In such a clash of contexts, honesty might make us murmur some hesitations before we join in a rush of scandalized denunciation to condemn Victor for his failure to foresee. Before the instant of animation, the experiment itself is unsuccessful and thus incomplete; after animation, the experiment is complete. It is alive.

My claim is not that before the animation, Victor tries to imagine what the creature's life will be like but goes ahead anyway and so should be blamed for self-consciously reasoning *this creature will suffer horribly but I do not care.* The descendants of Victor may exhibit precisely such conscious indifference to the suffering of others (Moreau and Tyrell both do), but Victor here at the romantic founding of the Frankenstein myth does not. He shows himself capable of such anticipatory imagination later on, when he refuses to complete the construction of the Monster's female companion. But the first time is unique. Only by positing his imagining-almost-nothing in the first place can we make sense of his failure to foresee, and make sense to some extent of the reticence in Shelley's text (the Creature's laments aside) concerning his blameworthiness. Here is Mary Shelley's almost colorless, outline vision of the scene that tends to reappear in almost every visual re-telling of the Victor story.

> It was already one in the morning; the rain pattered dismally against the panes, and my candle was nearly burnt out, when, by the glimmer of the half-extinguished light, I saw the dull yellow eye of the creature open; it breathed hard, and a convulsive motion agitated its limbs.
>
> How can I describe my emotions at this catastrophe, or how delineate the wretch whom with such infinite pains and care I had endeavoured to form? His limbs were in proportion, and I had selected his features as beautiful. Beautiful!—Great God! His yellow skin scarcely covered the work of muscles and arteries beneath; his hair was of a lustrous black, and glowing; his teeth of a pearly whiteness; but these luxuriances only formed a more horrid contrast with his watery eyes, that seemed almost of the same colour as the dun white sockets in which they were set, his shriveled complexion, and straight black lips.
>
> The different accidents of life are not so changeable as the feelings of human nature. I had worked hard for nearly two years, for the sole purpose of infusing life into an inanimate body. For this I had deprived myself of rest and health. I had desired it with an ardour that far exceeded moderation; but now that I had finished, the beauty of the dream vanished, and breathless disgust and horror filled my heart. Unable to endure the aspect of the being I had created, I rushed out of the room, and continued a long time traversing my bed-chamber, unable to compose my mind to sleep. (1.4.84—85)

Unanimated, the creature is only a test case. The new weapon has not yet been exploded, the new medicine not yet been ingested, the DNA not yet been recombined for the first time ever. There is a certain injustice in demanding that Frankenstein fear the monster *before* the monster has been animated because the unanimated monster is not, strictly speaking, to be feared. That feature of the cosmological object which is revealed in the success of the experiment appears always to have existed, but prior to the revelation of its existence, there is no way of *knowing* that it has existed.

Roslynn Haynes comes close to a similar conclusion in her list of the ways in which Mary Shelley's novel "pre-empted many of the philosophical and psychological considerations that have subsequently been recognized and inextricably linked to scientific research." The second "consideration" in her list is this:

> [Mary Shelley's] chief purpose . . . is to explore the ethical consequences of the *success* of Frankenstein's experiment. In scientific terms, the creation of the Monster is a brilliant achievement; yet Frankenstein's horror begins at the precise moment when the creature opens its eye, the moment when for the first time Frankenstein himself is no longer in control of his experiment. His creation is autonomous and cannot be uncreated any more than the results of scientific research can be unlearned, or the contents of Pandora's Box be recaptured. Hence, paradoxically, it is at the moment of his anticipated triumph that Frankenstein qua scientist first realizes his inadequacy. (*From Faust* 97)[12]

The creature cannot be uncreated and the results of scientific research cannot be unlearned: there is the irreversibility of what I am naming as the *event-structure* of the scientific "revelation" Victor experiences. In the animation scene, Mary Shelley shows us the disturbing link between the event structure of scientific revelation and religious revelation. Unlike religious revelation, scientific revelation does not require or work on a whole community. The impossible human of scientism is a scene-destroying (or God-expelling) monstrosity because its presence as the result of a scientific "creativity" founded on the willful forgetting of all ideas of God and moral equality between humans does something irreversibly violent. The presence of the artificial human obliterates the structure in which the historical human community (ours) and the irreducibly-to-be-shared sacred emerged together. Instead of a sacred object in the center that we unite in resenting even as we refrain from appropriating, we have in the center a to-be-tortured object—the object of a maximally desacralizing, minimally exchangeable gesture; and on the periphery, we have a vain human alone, deluded by scientism. The configuration is that of the unmediated subject-object model object of reductionist materialist metaphysics, from which all mediation by human others or by a scenic central Other has been expelled. The operating human seems to be alone with cosmological objects, with demystified and immediate "Nature" out there, existing on a horizontal plane where shared transcendence is not in question.

Victor fails after the success of the animation to learn anything from his mistakes. Ironically, he seems to lose his contempt for the supernatural and he gradually regresses into an inchoate, curse-making spiritualism

(3.7.224; 3.7.226; 3.232). His regression hints at possibilities unleashed by modern science: any sacralizing representation of the God-indifferent-to-humanity of ritual religion may now (in modernity) meet its match in a scientific representation of the human-indifferent-to-God: the scientistic usurper. It is a scandal for religion that "God" lets Victor get away with it. In playing God with some minimal success (the animation works), Victor proves that in the universe he inhabits, God is either dead or so passive that that he might as well be dead. George Levine's apt observation regarding the atheism in the novel's dramatic itinerary fits here: "The whole narrative of Frankenstein is, indeed, acted out in the absence of God" ("Ambiguous" 7). The Wretch in Mary Shelley's founding text suffers all too terribly from the disfiguration, misbegotten as erotic and social potential, abandoned and expelled, useless and alone. Frankenstein is to the monster as Satan is to Job. At his most candid, in his death-bed speech, Victor aptly compares himself to Milton's Satan.

> All my speculations and hopes are as nothing; and, like the archangel who aspired to omnipotence, I am chained in eternal hell. My imagination was vivid, yet my powers of analysis and application were intense; by the union of these qualities I conceived the idea, and executed the creation of a man. Even now I cannot recollect, without passion, my reveries while the work was incomplete. I trod heaven in my thoughts, now exulting in my powers, now burning with the idea of their effects. (3.7.233)

Victor in one breath claims to suffer torment ("I am chained"). In the next, he recalls what to him were the halcyon days of being free to play God, not yet disabused of his dream or encumbered by consequences such as the agency of the progeny—"while the work was yet incomplete." There is no mention or sign in his discourse to Walton of lasting remorse for the useless suffering of the Monster. Victor repents of nothing. If he regrets errors, he regrets them not out of sustained compassion for another being but out of a steely interest in his own ontological state, which he believes is pitiable. The mad scientist playing God the creator imitates Satan the accuser and tormentor. The mad scientist oblivious to the model of God incarnate as the Crucified, deliberately tortures a body other than his own and traps the being of that body in a hellish destiny. I would go so far as to suggest that the way in Shelley's founding text that no God intervenes to stop Victor in his tormenting of the creature is a prophetic intuition of the de-humanizing possibilities of the worst of scientism—atheistic, reductive, obstinately materialist, "hard" science that expels from its considerations that level of being we have identified as the "anthropological." The mad scientist is a real evil to feared, not just a fictional one to imagine. For if the vain scientist wishes to prove beyond a doubt that *God is dead,* there is no faster, more efficient way to make the point and prove the central scenic death than to despise, disfigure, and annihilate human beings—those creatures alone in the cosmos whose first memorable idea was the idea later to be formulated as the name-of-God, the idea of a shareable transcendent scenic center that *is.* If one can perform almost the same violent disfiguration on a mock-human synthetic creature, so much the better for the scientistic obstinate.

On the Expulsion of the Monster

I turn now to the career of the Monster, specifically to two mock-creation scenes analogous to the scene of the animation of the Monster analyzed in the previous section. My claim is that we should resist a victimary approach to the Monster: while we might wish to sympathize with him in his misery, we do better to resist the temptation to make of him mostly a *human* victim. To try to humanize the Monster ironically dehumanizes him even more than Victor already has. The interactions of Victor and the Monster present a demonic parody of the Divine-human interaction in the moment of human origin as hypothesized by generative anthropology. Instead of God opening up the human to the historical future by being misapprehended as the victimary object of scenic resentment and then later recalled as the sacred first Person of communal memory, here in *Frankenstein* a human scientist playing God obliterates the possibility of any economically satisfying future for the metahuman he has created. Scientism intent on remaking the human is toying with the one "object" exceptional in the cosmos for its belonging to the scene of representation. Thus the violence scientism implemented would do to humans as cosmological objects is violence of an order entirely other than that to which belongs the violence it might do to gases and liquids, fruit flies or army ants.

The way all humans in the novel instinctively flee the Monster externally verifies his not-being-human. If we want to experience Mary Shelley's novel in its fullness, those of us weaned on the many cinematic

Frankenstein monsters, from Karloff's inarticulate giant to Robert DeNiro's scarred outcast (in the film directed by Branagh), must do our best to expel such images from our heads. We need especially to erase from our minds the square-headed, grrrr-ing giant with bolts in its neck. For Mary Shelley's Monster, the original, is neither a sad gentle giant nor a fully humanized figure. He is neither laughable nor embraceable. The corpse-stitched, mummy-skinned, opaque-eyed figure that Mary Shelley proposes for our reflection, if we take the time to linger over him as she describes him, and to listen to his tale, proves terrifyingly different from us. He has the resentment, desire, guilt, linguistic capacity, esthetic taste, religious yearning, erotic appetite, and productive capacity expected in a healthy human person. But Mary Shelley has, for our difficulty, made the Monster manifestly not to be one of us. If we are not afraid, we are not paying attention to the fact of his unimaginably total solitude. We look unwillingly upon a Monster who seems human in all respects *except* the face that prohibits our interior conviction of equality with him. Our fright depends upon the proximity of the Monster to us: except for his ugliness, he might pass for human.

Let us take a quick inventory of the reports of the unprovoked human hostilities that confound him. Having been abandoned by his maker, having stumbled into the forest and gradually gotten control of his sense perceptions, long before he has even learned to make fire or to speak, the Monster enters a small hut where an old man is getting his breakfast: "He turned on hearing a noise; and, perceiving me, shrieked loudly, and, quitting the hut, ran across the fields with a speed of which his debilitated form hardly appeared capable" (2.3.132). His first meeting with a human consists almost wholly of the flight of the other. It is not a *meeting* at all. Next, the untutored creature is driven from a panic-stricken country village: "I had hardly placed my foot within the door, before the children shrieked, and one of the women fainted. The whole village was roused; some fled, some attacked me, until, grievously bruised by stones and many other kind of missile weapons, I escaped to the open country" (2.3.132). When the Monster first perceives his disfigured face in a mirror of still water, he reports a sensation diametrically opposed to that contained in the equivalent moment of the Narcissus myth.

> . . . but how was I terrified . . . ! At first, I started back, unable to believe that it was indeed I who was reflected in the mirror; and

when I became fully convinced that I was in reality the Monster that I am, I was filled with the bitterest sensations of despondence and mortification. Alas! I did not yet entirely know the fatal effects of this miserable deformity. (2.4.139)

He hopes against all probability that the ugliness will not destroy his chances as he awaits a verbal interview with the senior member of the DeLacey family: "I cherished hope, it is true; but it vanished, when I beheld my person reflected in water, or my shadow in the moon-shine, even as that frail image and inconstant shade" (2.7.156). After his first murder, that of the boy Ernest Frankenstein, the Monster contemplates a miniature portrait of Victor's late mother that Ernest had been wearing around his neck. At this juncture—he has just impulsively sacrificed his moral innocence to the indulgence of a romantic delusion about the satisfactions of revenge—his resentment rapidly matures. The Monster explicitly identifies the ground of that resentment as an intuition of eternal erotic deprivation: "I remembered that I was forever deprived of the delights that such beautiful creatures [as this human woman] could bestow; and that she whose resemblance I contemplated would, *in regarding me,* have changed that air of divine benignity to one expressive of disgust and affright" (2.8.167) [emphasis added]. I intend this inventory to help us resist a temptation to which much contemporary criticism, with a few exceptions (e.g., Lipking; van Oort *End* 148–169) has succumbed. It is the temptation to ignore the universality of the human response to the Monster's ugliness.[13]

The corroborating report by Robert Walton of his confrontation with the Monster following the death of Victor contains the most extended description of the creature's face. Elsewhere in the text we have already learned that the Monster has white teeth, thin black lips, and the opaque, clouded eyes of a corpse; we know that the blood vessels beneath his skin are visible; eight feet tall, he was sewn together from various body parts, animal flesh conjoined to human body pieces. Frankenstein made him huge because from a technical point of view it was easier to work with big rather than small anatomical parts. Walton has told us that Frankenstein died only hours ago; now it is midnight on Walton's ice-bound ship; he has been writing to his sister; he hears a cry and returns to Victor's cabin.

> Over him hung a form which I cannot find words to describe; gigantic in stature, yet uncouth and distorted in its proportions. As he hung over the coffin, his face was concealed by long locks of ragged hair; but one vast hand was extended, in colour and apparent texture like that of a mummy. When he heard the sound of my approach, he ceased to utter exclamations of grief and horror, and sprung towards the window. Never did I behold a vision so horrible as his face, of such loathsome, yet appalling hideousness. I shut my eyes involuntarily, and endeavoured to recollect what were my duties with regard to this destroyer. I called on him to stay.
>
> He paused, looking on me with wonder; and, again turning towards the lifeless form of his creator, he seemed to forget my presence, and every feature seemed instigated by the wildest rage of some uncontrollable passion. (240)

Walton is unique in the world of the novel, as the only human *without* violent intent to call on the Monster to "stay": that uniqueness of response explains the Monster's responding to his gesture of reciprocity by "looking on [Walton] with wonder." He can hardly believe a sighted human other than Victor is willing to ask him to stop and talk. Although showing courage in a way nobody else has, Walton must struggle with the feeling of repulsion at the "horrible" face, its "loathsome . . . appalling hideousness." The struggle between involuntary, instinctive repulsion—"I shut my eyes involuntarily"—and calculated, rational "duties" is a struggle similar to that Victor experiences when forcing himself to consider with justice the Monster's demand that he make him a female companion (2.9.171). There is something fabulously excessive in the ugliness of the Monster, as if the fearsomeness of the originary sparagmos somehow has been concretized bodily in the un-presentable face of this one mythic figure, a face on which more than once "every feature seem[s] instigated by the wildest rage of some uncontrollable passion."[14] According to Alan Richardson, the Monster's barely contained "rage" of "uncontrollable passion" may indicate a taint of animality transferred by the beast flesh that contributes to his organic being. [15] Given the temptation to condemn Victor and believe ourselves morally superior to him, it can perhaps never be emphasized enough that long before the Monster confronts his maker and tells his tale, long before Victor himself is called to account, the Monster is rejected by humankind. He is rejected by people who know nothing of Victor and have nothing to do with the family of Frankenstein. I believe Shelley intended us to grasp that he is rejected by people like us.

The Mock-Creation Scene of Failed Integration

We are justified in calling the Monster's interactions with the DeLacey family a "mock-creation scene" in rivalry with to Victor's animation scene because, as many critics have noticed, Mary Shelley is carrying out a hypothetical imaginary narrative, a "thought experiment," in which a human-like creature is brought forward to the boundary of social interaction without having been nurtured by any of the elementary forms of human care. The Creature has no "primary caregiver," neither an ape-parent as Tarzan has nor a wolf-parent as the feral child has. Timothy Morton remarks that Shelley was "fascinated by one of anthropology's founding myths, that of 'first contact' between humans from different cultures taking place upon a supposedly neutral ground" (264). The remark fits with our idea that in his effort to enter human society, the Monster prepares and executes his own creation scene—a fiasco of failed integration.

When we re-trace his solitary efforts in learning mobility, food gathering, self-preservation, and language-learning, and when we reflect on the gentle benevolence of his bringing firewood and clearing snow to lighten Felix DeLacey's workload, we ought to perceive the melancholy opposition between the Monster's hopeful desire to belong to human society and Victor's bizarre determination to take himself quite out of it. The Monster's attempts to belong, however, only confirm the impossibility of his ever succeeding: one who deserves to be regarded as if human is doomed instead to a fate the vain scientist set for him in the first place. His non-humanity is a matter not of economic deprivation in the simple sense of physiological survival, but of being wholly deprived of a position of participation in economic exchange.

Let us recall in Gans' originary hypothesis the completion of language events in the sparagmos or the exchange of things, the sharing of portions of the divided-up sacred food object: without some exchange of valuable things, there is no human scene. The Monster is damned to experiencing the sociality of language and the use of the sign without ever wholly consuming a portion of the valuable object to which the sign might refer, precisely because as an *impossible human,* his participation in community is ruled out. The critical literature has perhaps not sufficiently registered the significance attributed to the *economic* dimension of human activity in the vision of the human that Shelley propounds in the Frankenstein myth. The prohibition against the Monster's ever exercising the freedom to exchange *things* in reciprocity with "real" humans structures his deprivation. He can look but he cannot touch. He may watch through the window from outside a human family eating dinner, but he will never be invited inside. He has the bodily power to survive as nothing but an organism on his own, but he is not satisfied to be nothing but an organism. The delusion that his ugliness might somehow be overcome nourishes his desire to integrate with the DeLacey family. His delusion is the same one that his victimary defenders to this day hang onto. His artificial origins, despite the victimary impulse to embrace him, must always be a liability.[16] Furthermore, the Monster's interactions with the kind Old Man DeLacey (kind only because blind) reach their catastrophic end *not* at the point alone when the young man Felix intervenes brutally, but at the point when the Monster is also asked who he is—when he is about to be forced to tell the whole horrible story of his being made out of corpses and shocks.

The Monster's elaborate, patient, nervously self-conscious preparations leading up to his conversation with the blind man are preparations far more suspenseful than those contained in Victor's account of the tests and preparations leading up to the animation. The apparatus of the electrocution into life of the supine body that we know from the movies outdoes in visual spectacle the humble image of two figures sitting in chairs beside a fire, conversing. But unlike the moviegoer, the reader engages more intimately with the Monster's slowly gradual approach toward his interview with DeLacey (not with Victor's design of the animation equipment). We find the victim's preparations excruciating to witness because we know his creative project will fail.

One of Shelley's models for her description of the Creature's psychological development may be found in Etienne de Condillac's *Treatise on Sensations.* As the Monster moves about the forest, having been abandoned by Victor, he learns to distinguish shapes and sounds, forms and perceptions. He recalls that earliest stage of conscious life as an experience of being something like an abandoned infant grotesquely trapped in a big adult body: "I was a poor, helpless, miserable wretch; I knew, and could distinguish, nothing; but, feeling pain invade me on all sides, I sat down and wept" (2.3.129). One difference between Shelley's account of the Creature's acquisition of language acquisition and that imagined by Condillac, one of her sources, is that her Creature is *not paired* with another of his kind. Gans points out in his analysis of Condillac's hypothetical scene of language origin that the motif of the "passion" shared by Condillac's two isolated children who begin to interpret each other's signifying gestures betrays an intuition that mimesis is at work in specifically human competition and sharing. The presence of the motif suggests Condillac's grasp of the necessity of mimetic paradox as the precondition of the first human ostensive sign.[17] Frankenstein's Monster, by contrast, occupies a scene bereft of companionate interaction. Like a feral child, the Monster engages in nurturing reciprocal exchange with no human others. The situation in which he acquires human language has him observe *from a secret distance* other creatures who speak: he eavesdrops on the DeLacey family through a hole in their cottage wall, slowly but gradually associating words with things (2.4.137–38).

He learns language much more quickly when the Arabian Safie, the lover of Felix DeLacey, joins the family and herself gets taught English as a second language: "Presently I found, by the frequent recurrence of one sound which the stranger repeated after them, that she was endeavouring to learn their language; and the idea instantly occurred to me, that I should make use of the same instructions to the same end" (2.5.143). The Creature remembers himself as such a keen student that he felt the vanity of mimetic competition, regardless that his competitor was unaware of his very existence: "My days were spent in close attention, that I might more speedily master the language; and I may boast that I improved more rapidly than the Arabian, who understood very little, and conversed in broken accents, whilst I comprehended and could imitate almost every word that was spoken" (2.5.144). Next, Shelley places an abandoned sack containing three books in the Creature's path, which he takes up and—here the threadbare plot contrivance collapses into total incredibility—proceeds to read without any assistance. The texts are Plutarch's *Lives,* Milton's *Paradise Lost* and Goethe's *The Sorrows of Young Werther.* The Creature's sense of interiority is produced more powerfully by the experience of eavesdropping on conversation combined with reading, than by eavesdropping alone.

But why does the Monster want language? He wishes to belong to a community: he wishes the exchange of signs to be *completed by the exchange of things.* That she gives her metahuman character such a wish indicates Shelley's intuition of the non-instrumental, mimetic nature of symbolic exchange. Recall from chapter One that according to generative anthropology, a human sign intends any scenic object of appetite as something that can (eventually) be shared in the exchange of things. The human sign intends the object as a significant thing with an economic value. The Monster does not want just to survive; he wants the kind of exchange that will permit him a place on a communal, collective scene. He hungers for the transcendental effect of sociality, which can never be "consumed" as an object but only experienced as an effect of human exchange.

The Monster does become a secret economic producer, but his gifts to Felix and the Delacey family will never be reciprocated. In secret, he works for the DeLacey family, gathering firewood and doing nighttime work in their fields: "I afterward found that these labors, performed by an invisible hand, greatly astonished them; and once or twice I heard them, on these occasions, utter the words, *good spirit, wonderful;* but I did not then understand the signification of those terms" (2.4.140). While acting as an "invisible hand," the Monster slowly acquires language and increasingly desires economic participation. Victor during his experiments must work with great intensity to de-humanize his intellect, forcibly to expel scenic intuition; the Monster by contrast must work to humanize his raw mind, must struggle with determined patience to try to enter the scene of representation. We may elaborate on this contrast by saying that Victor is so deeply mistaken as to despise the human in such a way that he believes in the desirability of creating a substitute for it, whereas the not-yet-disillusioned Monster is sufficiently innocent to worship the human in such a way that he believes in the possibility of belonging to it, even against all odds. Here he describes the humble DeLacey family:

> I looked upon them as superior beings, who would be the arbiters of my future destiny. I formed in my imagination a thousand pictures of presenting myself to them, and their reception of me. I imagined that they would be disgusted, until, by my gentle demeanor and conciliatory words, I should first win their favour, and afterwards their love. (2.4.140)

Retrospectively, the Monster is mocking the naïve vanity—*I formed in my imagination . . . I imagined*—of his belief that his ugliness could have been effaced by his acts of generosity or his use of language (his moral equality as one capable of exchanging signs). The scenic configuration of language, foundational to the human, comes up against the ontological barrier of the ugliness that separates him as a "marked" artificial being.

The Monster puts his faith in the *minimal* form of human representation. By contrast, the mad scientist puts his faith in a *maximal* form of representation: in laboratory science with its equipment, devices, instruments, literature; its heavy, expensive cultural overhead. After the Monster is animated, the reversal signaled by Victor's calling on supernatural agents is augmented by his feeling a new hatred for those "chemical instruments" that embody the aspirations of scientific practice. Victor hates packing up and unpacking them, concrete reminders of his delusion: "Ever since the fatal night, the end of my labours, and the beginning of my misfortunes, I had conceived a violent antipathy even to the name of natural science. When I was otherwise quite restored to health, the sight of a chemical instrument would renew all the agony of my nervous symptoms" (1.5.95). Our sense of the Monster's moral innocence is seconded by the primitive simplicity of the cultural resources he employs when negotiating with the DeLacey family. It is interesting to notice that in proportion as the Monster becomes more pathologically obsessed by his revenge against Frankenstein, he becomes more reliant on technological equipment (and by association less primitive, less innocent).[18]

Readers sympathetic to the Monster's plight feel a growing dread as he approaches the crucial interview: "The more I saw of them, the greater became my desire to claim their protection and kindness; my heart yearned to be loved and known by these amiable creatures; to see their sweet looks turned towards me with affection, was the utmost limit of my ambition. I dared not think that they would turn from me with disdain and horror" (2.7.156–57). His ineradicable ugliness—the absence of esthetic value, the presence of anti-erotic disfiguration—is the reality that will destroy his fantasy. His delusion persists: *I dared not think*. A variation of the initial scene of his creation is enacted: as Frankenstein spurned him at the moment of his "birth," so will the DeLacey family spurn and abandon him.

The Creature's interview with the grandfather is charged with a sense of finality. The dialogue does contain the kindest words that will ever be directed his way. The peaceful old man offers help: "'If you will unreservedly confide to me the particulars of your tale, I perhaps may be of use in undeceiving them. I am blind, and cannot judge of your countenance, but there is something in your words which persuades me that you are sincere. I am poor, and in exile; but it will afford me true pleasure to be in any way serviceable to a human creature'" (2.7.159). But the Wretch already expects that he cannot pass as a human creature. The Old Man asks the question that brings the verbal exchange to the point of ostensive crisis: "'May I know the names and residence of those friends?'" If the Monster is to let old DeLacey know the names and residence, then he will have to refer to "you and your family" and he will have to name the residence as "right here, where you and I are now." The Monster's hypothetical discourse about an absent fictional scene must somehow be converted into an indicative report of a present real exchange. And the ostensive presence will be tested for its immediate verifiability. Are *you* my friend? *Here* and *now?* The mistiming of Felix's return to the scene does not cause the disaster. Rather, the Monster fails before Felix returns. Words fail him: he cannot tell about his origin.

> I struggled vainly for firmness sufficient to answer him, but the effort destroyed all my remaining strength; I sank on the chair, and sobbed aloud. At that moment I heard the steps of my younger protectors. I had not a moment to lose; but, seizing the hand of the Old Man, I cried, "Now is the time!—save and protect me! You and your family are the friends whom I seek. Do not desert me in the hour of trial!"
>
> "Great God!' exclaimed the Old Man, "who are you?"
>
> At that instant the cottage door opened, and Felix, Safie, and Agatha entered. Who can describe their horror and consternation on beholding me? (2.7.160)

Although the blind grandfather cannot see the Wretch's ugliness, the sobs—in their pre-verbal violence, their bottomless depth—form, Mary Shelley hints, an auditory substitute for that visual horror. The Monster's ugliness is to human sight what his inconsolable sobbing is to human hearing: Old DeLacey is expressing shock when he cries "'Great God!'" Furthermore, to ask *who are you?* is to ask a question that the Creature cannot answer (*the particulars of your tale*) without revealing the secret of his origin in Victor's laboratory. The Creature has already asked this most painful question of himself many times: "And what was I? Of my creator and creation I was absolutely ignorant; but I knew that I possessed no money, no friends, no kind of property. I was, besides, endowed with a figure hideously deformed and loathsome; I was *not even of the same nature as man*" (2.5.145) [emphasis added]. Again: "From my earliest remembrance I had been as I then was in height and proportion. I had never yet seen a being resembling me, or who claimed any intercourse with me. What was I? The question again recurred, to be answered only with groans" (2.5.147). And again: "My person was hideous, and my stature gigantic: what did this mean? Who was I? What was I? Whence did I come? What was my destination? These questions continually recurred, but I was unable to solve them" (2.7.153). The Monster has no useful answers to these questions; when he reads Victor's laboratory journals (2.7.155), he hears a story that fills him with baffled rage. The Creature's conversation with his human fantasy friend must end at old DeLacey's *Great God! Who are you?* He senses the vast improbability of his social integration even before he suffers blows from Felix without returning that violence (2.7.160).

It is a short distance to the Monster's despairing vow to take revenge on his human creator and to force an alternative, to demand the creation of a mate, a second member of his "species." The DeLacey family leaves the neighborhood, never to return; the Monster burns down the cottage in a fury of insane despair (2.8.163). He decides to seek out Victor Frankenstein, whose scientific journal just happened to be in the pocket of the coat he put on when he was stumbling away from the laboratory as a giant-sized mental infant (another threadbare plot device connected to the motif of reading material) (2.7.155). The Monster decides to put to his creator an ultimatum: since he cannot join human society, Victor must create a female companion for him: "I am alone, and miserable; man will not associate with me; but one as deformed and horrible as myself would not deny herself to me. My companion must be of the same species, and have the same defects. This being you must create'" (2.8.168). That ultimatum leads to the third and final creation scene of the novel.

The Vain Scientist as Pseudo-Savior

The difference between Victor's second experiment, the making of the female companion, and his first, the making of the Wretch to whose story we listen, is a source of some of the more bitterly ironic pleasures in the text. We witness Victor doing what he did voluntarily then, under duress now: "It was indeed a filthy process in which I was engaged. During my first experiment, a kind of enthusiastic frenzy had blinded me to the horror of my employment; my mind was intently fixed on the sequel of my labour, and my eyes were shut to the horror of my proceedings. But now I went to it in cold blood, and my heart often sickened at the work of my hands" (3.2.189; see also 3.2.183, 3.2.187). When the Monster confronts him on the Mer de Glace and forces him to sit and listen, following the murder of Ernest and the framing of Justine, during a family holiday in which Victor's baffled father and fiancée hope in vain to lift him from his pit of gloom, Victor chafes, resists, complains and moans. We ought to place the cause of mimetic desire first: it is the habit of Victor as one of a temperament inclusive of the desire to play God, to exude an Olympian violence that never can bear a position of inferiority to anyone, human or metahuman. The mad scientist refuses to confess that he acted in madness (see also 3.2.187; 3.238). That he cannot conceive of a difference between being humbled and resenting humiliation is part of his madness. In the apt description of Pamela Clemit, his "self-justifying confessional narrative collapses into unwitting self-condemnation" (34). By force of will, he ignores all the evidence that might lead him to share a little of the blame for the extensive destruction (Ernest killed, Justine framed and hanged, Clerval murdered, Elizabeth murdered, his father destroyed by grief, his health ruined).

By the time that the Creature has made his ultimatum and demanded that Victor construct for him a female companion, he has proven his physiological and moral superiority to the chemist. For the reader, his story and situation are more engaging than Victor's. Who among us does not find the middle part of the book the most intriguing, stimulating, enjoyable? Even in the misery of his solitude, the outcast is more "interesting" than the fuming, self-excusing Victor. Henceforth, Victor writhes, curses, faints, hates, swears revenge, calls on spirits, gets depressed, mystifies his family, tells them nothing, curses, weeps, stares at the floor, lies in the boat in the middle of lake like Rousseau, staring at the sky, self-centralizing with energies drawn from a bottomless pit of resentment. But "everybody knows," to quote a phrase from a song by Leonard Cohen: everybody knows by this point that Victor Frankenstein has lost the fight, that Victor Frankenstein of Geneva and the University of Ingolstadt, to use an unkind epithet popular among the hardened, market-wise young people of North America these days, is a *loser.*[19]

The central dialogue between mad scientist and impossible human calls Victor to account: the "object" in the desacralized center comes alive as its own speaking, subjective center with its own voice. The once-dead central object intended by the scientistic imagination, now fully animated, reveals itself to Victor as, unexpectedly, a thing outside and beyond the control of the powers of reductionist science: an impossible human, but very like a real one. Whatever result Frankenstein intended, the unintended consequence obtrudes itself upon him; the consequence cannot be escaped. The ethical as embodied by the non-human Monster reveals itself as a mode of understanding more powerful than the materialistic mode that Victor believed sufficient at the moment of animation. The moral self-assertion the Monster performs as a human-like being overpowers for a moment the fact of his ugliness. It even overpowers, briefly, Victor: "I was moved. I shuddered when I thought of the possible consequences of my consent; but I felt that there was some justice in his argument. His tale, and the feelings he now expressed, proved him to be a creature of fine sensations; and did I not, as maker, owe him all the portion of happiness that it was in my power to bestow?" (2.9.170). Reluctant and soon forgotten, these moves are the clearest Victor ever makes in the direction of pitying the Creature. Glancing at the consequences of Victor's failure to acknowledge the category errors in his project, we become conscious of the brutal way in which there is far more to the "secret of life" than the technique by which one might engineer the initial animation of an unanimated corpse. To become human life, raw bulk undifferentiated "life" requires social interactivity and communal belonging: it requires great involved activity on what generative anthropology calls the scene of representation. In keeping with his status as the originary modern mad scientist trying to re-enact the origin of the human, Victor sits on the side with those thinkers in our time who nod toward the residual mystery attached to the question of the cosmological origin of natural life only with the most grudging resentment.[20]

In the Monster's preparations for his mock-creation interview scene with Old DeLacey, as we have seen, there is suspense. In the dramatic reversals Shelley stages in the preparation for the second Monster-construction scene, there is a similar thoroughness in the exploration of a narrative irony. In a desolate outpost on the Orkney Islands, Victor manages to stitch together the anatomical parts for the female creature. During his journey to Scotland and the preparatory procedures, he has been tormented. Under the Monster's extortionist thumb, he has been forced to be disgusted by the very things at which, in the first case, he forced himself to feel *no* disgust. All the thinking Victor did not do in the first case, emptying his mind of moral considerations, he has done in this second case. The original experiment was lubricated by enthusiasm, propelled by a desire to know the technological outcome whatever the cost; now it is clogged, obstructed by a hatred of scientific practice itself. Victor in the Orkneys resembles the ancient Prometheus chained to a rock by Zeus as punishment, his liver gnawed daily by the eagle of disgusted self-awareness.

Frankenstein pauses before he animates the companion female thing. He reports to Walton a series of five ethical considerations, a "train of reflection" (3.3.190). First, he claims he knew the risk incurred by not knowing in advance the female Creature's innate psychology: "I was now about to form another being, of whose dispositions I was alike ignorant; she might become ten thousand times more malignant than her mate" (3.3.190). We may well accuse Victor of self-serving memory alteration inasmuch as "alike ignorant" falsely suggests he considered the possible "dispositions" of the first monster in advance, when he did not report any such planning. Regardless, the risk is real: the female creature will be her own self. Second, in a closely related move, while meditating on the mate as purely *potential* being, Frankenstein is doing what David Heyd calls "genethics" and is confronting "pure genesis" problems, asking the kinds of questions that the Biblical God in the book of Genesis would have had to ask before creating humankind. If one chooses to make a person of a certain description under certain conditions, what consequences follow from the choice, given that definition and those conditions? The condition here is that of the female companion not necessarily feeling bound by a promise that she herself did not make, bound by a contract she did not sign: "He had sworn to quit the neighborhood of man, and hide himself in deserts; but she had not; and she, who in all probability was to become a thinking and reasonable animal, might refuse to comply with a compact made before her creation" (3.3.190). Those bioethicists who worry about the conditions for the first human clone (for example) take up such considerations. The clause *but she had not* reminds us of the bioethical proposal that we should worry that the first human clone might wish that others had decided not to clone her, because clone-being might seem burdened with an excess of un-freedom. Leon Kass articulates such a proposal, one skeptical of the scientism informing defenses of human cloning: "First . . . any attempt to clone a human being would constitute an unethical experiment upon the resulting child-to-be. As the animal experiments (frog and sheep) indicate, there are grave risks of mishaps and deformities. Moreover, because of what cloning means, one cannot presume a future cloned child's consent to be a clone, even a healthy one" (155–56). Justine Burley and John Harris present a counter-argument: "We maintain that unless it is shown convincingly that 'living in the shadow' [of the cloners' design] is somehow both horrendous and more autonomy-compromising than the plethora of other widely accepted and permitted upbringings a child might be 'forced' to undergo, the liberal principle of freedom in matters relating to procreation overrides the concern about autonomy-related welfare deficits that will be suffered by clones" (244–45). I cannot resist some anachronistic transfers here. Victor as vain scientist has exercised *the liberal principle of freedom in matters relating to procreation.* And perhaps the existential deprivation of the Wretch may be described as something like an *autonomy-related welfare deficit,* but intensified—yes, intensified.

Third, Victor paints a picture opposite to the Monster's imaginary vision of mutual affection: "They might even hate each other; the Creature who already lived loathed his own deformity, and might he not conceive an abhorrence when it came before his eyes in the female form?" (3.3.190). We must give Victor credit again for imagining, this time around, the *freedom* of the metahumans to feel hate. A fourth consideration spells out the possible consequences of a separation between the monsters: "She might also turn with disgust from him to the superior beauty of man; she might quit him, and he be again alone, exasperated by the fresh provocation of being deserted by one of his own species" (3.3.190). If the free female companion abandons him, he will then have motivation anyway to wreak vengeance on Victor's family, rendering futile the humiliated scientist's attempt to pacify the blackmailer by submitting. Notice the systematic quality of these ethical considerations. Victor is performing cost-benefit calculations, thinking in the fashion of William Godwin, the coolly rational philosopher father of Mary Shelley; indeed, he is thinking in a way not unlike that of the bioethicist in our time.

Victor's fifth and final ethical consideration concerns an enduring theme of the Frankenstein myth: *the apocalyptic threat of the rival species.* It tips the scale of deliberation against the Monster; in it, species identity is one with ethical identity.

> Even if they were to leave Europe and inhabit the deserts of the new world, yet one of the first results of those sympathies for which the daemon thirsted, would be children, and a race of devils would be propagated on the earth, who might make the very existence of the species of man a condition precarious and full of terror. Had I a right, for my own benefit, to inflict this curse upon everlasting generations? I had before been moved by the sophisms of the being I had created; I had been struck senseless by his fiendish threats; but now, for the first time, the wickedness of my promise burst upon me; I shuddered to think that future ages might curse me as their pest, whose selfishness had not hesitated to buy its own peace at the price perhaps of the existence of the whole human race. (3.3.190)

One may contrast Frankenstein's original dream of the "species" that would "bless" him as its "creator" (1.3.81–82). Now, taking into account later references to this projection, the perverse irony is that Victor will position himself as the pre-emptive self-sacrificial savior of the whole human race, when the need for such saving never would have arisen had he not, alone in the first place, completed the experiment that engendered the crisis.

Victor twice repeats his references to the apocalyptic threat of the rival species. The repetition supports my claim that it is the one among the five utilitarian calculations that tips the scale. Alphonse has managed the release of his difficult son from legal confinement (the Monster having managed slyly to frame Victor for the murder of Clerval). Victor has been lying in bed deliriously raving about being "guilty" for the deaths of Ernest, Justine, and Henry: "'A thousand times would I have shed my own blood, drop by drop, to have saved their lives; but I could not, my father, indeed I could not sacrifice the whole human race'" (3.5.209). With the figure of "shed . . . blood"; with the sacramental overtones of "drop by drop"; with the choice verb "sacrifice" in the context of saving "the whole human race"; with these clues, we may infer a substitution of Victor's godlike self for the figure of the Crucified. Shelley has Victor follow the report with a deflating fact: "The conclusion of this speech convinced my father that my ideas were deranged, and he instantly changed the subject of our conversation, and endeavoured to alter the course of my thoughts" (3.5.209). Is Victor implying that his ideas were *not* deranged? I believe Shelley wants us to notice that his deliriously claiming messianic centrality (not unlike "playing God") has gone so far that even his normally unflappable father is moved to change the topic.

The other repetition occurs when Victor is making his last death-bed statements to the enthralled young Robert Walton, having decided his conduct has not been blameworthy: "In a fit of enthusiastic madness, I created a rational creature, and was bound towards him, to assure, as far as was in my power, his happiness and well-being" (238). In a stunning but typical ellipsis, he says nothing about the fact that he did not in the first place do *what was in his power* to assure the nurturing into prosperity of the "rational creature." He did nothing. He abandoned the helpless thing. Therefore, the pronoun *This* which opens the upcoming passage is utterly without referent. It refers to a duty he did not even begin to fulfill, although he insinuates otherwise.

> This was my duty; but there was another still paramount to that. My duties to my fellow-creatures had greater claims on my attention, because they included a greater proportion of happiness or misery. Urged by this view, I refused, and did right in refusing, to create a companion for the first creature. (238)

The cruelty of the abandonment and neglect of the Monster (sins of passive omission) are ignored; the purported saintliness of the deliberate omission in the refusal to gamble on the Monster's companionate dream is spotlighted. With the calculating phrase *greater proportion of happiness or misery* we hear the Godwinian utilitarian calculus at work once again. The irreconcilability of the duties—to the human species, to his abandoned "rational creature"—may be for Victor a source of death-bed moral self-approval. However, for the Monster, that irreconcilability is the cause of his unbearable oneness. Victor's self-approval is vacuous: his playing God the re-enactor of human origins in the first place was the cause of the entity who is now condemned to the irreconcilability.

It would not be hostile to the novel's play with ethical considerations to note that suicide could have solved the problem of the Monster's blackmail. The motif of suicide runs through the text's extravagant victimary rhetoric. Victor and the Monster routinely claim to be damned, tormented, driven to despair; they profess powerful wishes for self-annihilation. Suicide would shut down their dance of death. Absent Victor, the Monster would have no reason to make other humans suffer:

the Monster takes his vengeful pleasure most sweetly in the pain of Victor alone, the pain of the one who made him ugly. A pre-emptive suicide to protect Elizabeth and Henry and others never occurs to Victor because his desire for self-annihilation is bogus. Charles Schug (in what remains, over thirty years later, one of the best analyses of the novel) has perceived the damage done to Victor's "savior" defense by his never owning up to the unglamorously limited scope of the Monster's targets.

> . . . for all practical purposes, Victor has failed his second duty as well as his first: once the Monster has eliminated all of the people Frankenstein held dear he is no longer a threat to "beings of [Frankenstein's] own species." The Monster chooses from the very beginning to wreak revenge on his creator, not humankind at large. Frankenstein's real duty was to William and Clerval and Elizabeth, not to his 'species' in general. (616)

In other words, Schug hints, Victor's imaginary substitution of the whole *species* as the target of the Monster's vengeance is another megalomaniac move, elevating Victor to the god-like (to which he never should have aspired) rather than confining him to performances of the ordinary and domestic (which sphere he has studiously neglected).

The abortive destruction of the female prototype makes the second most graphically horrific moment in the novel. The hallucinatory tendency of the text's surface demonstrates itself when the Monster pops up voyeur-like again: his face comes into view in the window of the makeshift Orkney Island laboratory, just after the mental exercise that has resulted in Victor's grandiose notion that he might save humankind from the apocalyptic threat of a rival species.

> As I looked on him . . . I thought with a sensation of madness on my promise of creating another like to him, and, trembling with passion, tore to pieces the thing on which I was engaged. The wretch now saw me destroy the Creature on whose future existence he depended for happiness, and, with a howl of devilish despair and revenge, withdrew. (3.3.191)

In another macabre scene staged that same night, Frankenstein rows a boat out onto the water and drops overboard the dismembered parts of the aborted female companion thing. Again we see in the isolated concentration on the figure of Victor as one alone, that which should be a shared activity—the sparagmos, the taking apart of the central scenic object. Victor's is a non-communal rendering of the scenic object. The grudgingly sober deliberation in his disposal of the companion's unanimated body parts (3.3.197) contrasts with the wildly feverish, fearful all-night pacing in the courtyard after the irreversible animation (1.4.86). The frustrating formal barrier that was first created by the event-structure of the animation and the universality of the fear of the Wretch's ugliness becomes a barrier even more palpably frustrating. What seems to us a rational, considered decision must also and equally seem a horribly cruel decision. Victor's tearing-apart of the allegedly threatening female thing condemns the Monster to eternal solitude. Victor is like an ethicist who would compel us to justify torture to dispel a threat to humankind, a threat which the ethicist himself has brought into the world. Our pity for the Wretch is aroused. However, our reluctance to see the companion creature animated rises out of something more than speciesism. Shelley shows a certain ironic wisdom in making fear of the Monster's sexually reproductive capacity the long-term historical-propositional form of our ostensive fear of his ugliness. Pragmatic considerations force us to acknowledge that sharing the earth with "a race . . . who might make the existence of the species of man a condition precarious and full of terror" would be at best an enormous risk. We fear that risk in proportion as we value our belonging to the "species" labeled human.

And yet, as generative anthropology contends, the human is not just a species. There is no way to predict with certainty that the rival species would in fact be hostile to the human, so there is no way to guarantee the accuracy of the judgment that weighs prudence heavier than the Monster's need for a companion. What we see here is not the (anthropological) value of human community being preserved so much as the (cosmological) fact of human biology being respected. Therefore, a conflicting suspicion murmurs that the destruction of the companion is a total moral disaster, because the Monster, apart from his ugliness, is so very much like us, so very close to being human. On the other hand, as Charles Schug suggests, "While the Monster gains sympathy, he does not establish unimpeachable credibility" (618). We cannot justify the Monster's murder of Ernest and his framing of Justine, those acts which, committed before the ultimatum, cast a heat over it that flames easily into Victor's resentment. The Monster's implicit request that he be trusted would have been much more believable if he had not exercised such violence: "no amount of misery can justify his murdering [Ernest] and his framing of Justine, a totally innocent bystander" (Schug 610). The Monster gains nothing and loses much moral credibility in causing those deaths. The necessary evil of violent consumption as human survival demands it is one thing; the luxury of the violence performed to satisfy revenge is another. Although it is easy to understand the passion of revenge, it ought to be difficult to justify its exercise. Victimary interpreters of *Frankenstein,* who side with the Monster in opposition to Victor, can do so only at the expense of legitimating an ethic of victimary revenge. They must expel from consideration that core of properly scientific courage that gives Victor a certain modern identity in which we ought to claim a share if we wish to avoid hypocrisy. Science is not reducible to scientism; Victor has proven his scientific genius in the same irreversible event that proved his scientistic idiocy. Meanwhile, the vengeance is shared in their world-inside-the-world. The most tedious aspect of the textual surface of *Frankenstein* is the recurring, overblown rhetoric of revenge, a rhetoric Monster and mad scientist together deploy relentlessly.

I will not linger over the revenges that follow the abortive destruction of the mate, including the murders of Clerval and Elizabeth. Nor will I analyze Victor's brushes with the ineffectual law: the trial of Victor for murder (he is framed as Justine was framed) and his absurdly belated attempt, after the murder of Elizabeth, to convince a Genevan magistrate of the Monster's "objective" existence and thus rouse a human hunting party to track down and kill it (3.7.221–23). Nor will I reflect at length on the passages leading the Monster and Victor up to the North Pole. It is perhaps sufficient for us to register that in the reflections Victor performs just prior to the tearing apart of the never-to-be-animated female monster, we witness the birth of a discourse something like that of utilitarian bioethics a century and a half before the discipline proper is born—one more testimony to the prescient insight of Mary Shelley's text. Bioethics is the branch of philosophy that tries to limit the subtle inequalities and gross terrors threatened by reproductive technology and genetic engineering, among other scientific practices that now give us the opportunity to play God in brave new ways. It tries, but as the authoritative study by John Hyde Evans establishes, bioethics has at best limited success. All the bioethics commissions so far convened have proposed (for example) only *danger* as the obstacle to human cloning; it is a thin debate, not a thick one. The dominant value in bioethical discourse is *autonomy,* and it has crowded out the other considerations the commissions attempted to include. If we wish, we may choose scientism. No one, no God is necessarily there between us to stop us.

A Concluding Retrospective

I have acknowledged more than once that the abject Monster deserves our sympathy. To remove from the text the moral ground of its victimary rhetoric is a surgery too drastic, in my view, to be worth the benefit of any analytical or historical distance thus gained. Feminist criticism has found the Monster's story full of possibilities for comparison with the arbitrarily imposed sufferings of women, more particularly those of Mary Shelley herself and the women of her time (see Mellor; Jacobus; Moers). *Frankenstein* has been put to service in disability studies: "therapists, social workers and educators have used film version of *Frankenstein* to stimulate discussion among the disabled" (Hoeveler 59). Other studies have shown the ways that the Monster has been appropriated by conservative forces to satirize mob-sized democratic resentment and to mobilize fear of such putative entities as the monstrously rebellious Irish and the masses of the monstrously angry poor (Sterrenburg). These various appropriations of the figure of the Monster, from different political wings and limbs, may all be to the good. Shelley had the insight to represent the power to dehumanize of that which we now call scientism; she also had the prescience to imagine the extremity of the pain suffered by the victims of scientism. And there have been too many such real-world victims since the time of Mary Shelley.

Meanwhile, there is a potential for harm in exchanges of resentful accusations, exchanges such as those which the Monster and his vain scientist maker dizzyingly engage. In the contemporary context of debates over the authority of scientific institutions and those difficult questions where *science* can be exploited by either side in a political debate, the economies of accusation and apology, outcry and reform, are seldom free from destructive hostility and self-righteousness. The politicization of victimary claims can be harmful if a minimum of civility and the exercise of reason is not respected (Elshtain 63–91; Good, *Humanism*). As Charles Frankel suggests in his useful exposition and critique of irrationalism, "'Reason' . . . from a sociological point of view . . . is simply the name for forms of behavior by which individuals' beliefs are concerted without recourse to force or authority. It codifies the elementary principles of courtesy without which the maintenance of a liberal civilization is impossible" (929). The literary analyst who converts the rhetoric of the Monster into a metaphorical victimary politics may well not be respecting the conspicuous deletions of narrowly specified content by which Mary Shelley, in my opinion, sought to steer us away from such politicizing conversions. The lack of any explicit connection between the Monster and one specific human party, cause, camp, or agenda might give us pause. We might try anew to let the absence of such restrictive content itself guide us toward granting the Monster his uniqueness, his aloneness, his being the odd, anomalous, illogical thing he is: *one of a kind.*

Esther Schor is correct, in my view, to name him "the loneliest character in the English novel" (*Cambridge* 1) and to describe Shelley's achievement in fleshing out the figure of the self-justifying Victor as "the century's most blistering critique of Romantic egotism" (2). To Walton at the end of the novel, the Wretch declares: "Once I falsely hoped to meet with beings, who, pardoning my outward form, would love me" (242). The Monster is never embraced by another affectionate body. Not once. The Monster is correct when he says he is worse off than Satan in *Paradise Lost,* that Satan had friends.[21] I am inviting us to try out the idea that *the Monster's resentment is uniquely legitimate inasmuch as it is uniquely non-human.* We may grasp that the Monster is not fully human—that he is an *impossible human*—even as we concede that his not once having been tenderly embraced is a horror. His is more than a romantic tragedy. It is desolation.[22]

The textual source of the modern myth Mary Shelley founded has its most forceful effect when approached as an experiment in anthropological hypothesizing. The novel dramatizes with bold abandon the clash between two all-too-human fantasies. At one pole, there is an ill-advised scientific atheism that attempts to evacuate the imagination of moral considerations, reducing the human to chemical, mechanical, or instrumental categories (Victor). At the other pole (in the voices of the Monster's contemporary defenders more than in the voice of the Monster himself, in my view), there is the possibility of a victimary romanticism that loses legitimacy in proportion as it submits to an appetite for non-negotiable revenge, raging with stone-hard fists from a terrible depth in resentment of the seeming historical passivity of the God of Biblical revelation. In the modern world, science proves itself by having a technological impact on cosmological bodies. Victor the originary vain scientist proves himself by having a technological impact on the mass of stitched-and-scarred flesh that becomes the animated Wretch. But at the same time Victor proves nothing. He proves nothing significant about the natural world and everything about the significance of his anti-anthropological delusions. He symbolizes the confusions of a fictional (but not just fictional) world from which experiences of the human object as sacred, as esthetic, and as economic have been expelled. A pure scientism fully implemented produces nothing of value. That is why the Monster has nowhere to go in the end except into a solitary bonfire of his own making.

> Do not think that I shall be slow to perform this sacrifice. I shall quit your vessel on the ice-raft which brought me hither, and shall seek the most northern extremity of the globe; I shall collect my funeral pile, and consume to ashes this miserable frame, that its remains may afford no light to any curious and unhallowed wretch, who would create such another as I have been. I shall die. (3.243)

The central object into which the Wretch plans to throw himself is not a burning bush from which a God who might be believed by some to love humankind might speak, a God who might suffer rather than do nothing but cause suffering. This one for the Wretch is a bonfire where the originary accuser of humankind sits in waiting to burn us up, telling us we are *nothing but cosmological objects.* But we are not cosmological objects alone. We are humans always in the presence of one another, and our fate, thank God—or thank "God or whatever means the Good"[23]—our fate need not be the desolation of Mary Shelley's originary impossible human.

THREE

Allegories of Playing God in *The Island of Dr. Moreau*

H. G. Wells and Biological Thinking

The abiding shock value of *The Island of Dr. Moreau* (1896) testifies to the danger in raising questions of human origin. H. G. Wells stands tall among literary artists as one who brought to public awareness the anthropological implications of Darwinian science. Not without consequences for his intellectual career, however, he imagined the problem of the origin and end of humankind as a *biological* rather than an anthropological problem; that is, a problem likely to be solved by "scientific" rationality applied to humans rather than the study of humans as moral and historical actors. The young Wells studied biology under no less an authority on natural selection than Thomas Henry Huxley, Darwin's close ally and public promoter.[1] According to Roslynn Haynes, that one year of study formed the ". . . turning point in Wells' life. In intensity its effect was nothing less than that of a religious conversion" (*Discoverer* 12). The concept of natural selection liberated young Wells from the painful restrictions of his low church Fundamentalist upbringing. Ever after, he would reflect on the human in terms of the biological category of the species: "No other concept ever made an equivalent impact on Wells—rather, the criteria of biology became his yardstick to measure the claims of all other disciplines—astronomy, physics, sociology, politics, even theology and art" (*Discoverer* 16; cf. Brome 44). Darwinian science expelled the Biblical God of a punctual creation from the cosmos (Mayr 96; 99). While certainly liberating for many in its effects, the expulsion did not lead Wells into complacency. He never became, for example, a Social Darwinist in the mould of Herbert Spencer, letting natural selection underwrite laissez-faire economics and social inequality (Philmus and Hughes 179–80). On the contrary, Wells "hated things as he knew them"; he campaigned for a more rational and "planned" society, tirelessly imagining better worlds: "What Wells despised above everything was what we might call the domesticated imagination—the human mind closed to speculation" (Reed, *Natural History* 3; 86). He displayed irrepressible energy as a storyteller, educator, utopianist, polemicist and controversialist. *The Outline of History* (1923), a best seller, inspired hundreds of thousands of average readers in Europe and North America to think outside the box of authoritarian Christian orthodoxy. Wells lived out his prolific writer's life as a formidable example of that now rare creature, the public literary intellectual. A century later, most educated people know of the nature-nurture debate. Leon Stover credits Wells with setting the debate "into motion," calling Wells " . . . the first biologist to discriminate between nature and nurture, or what in 1896 he called a distinction between the 'hereditary factor' and 'the acquired factor'" ("Introduction" 2–3).

However, Wells' adherence to Darwinian science was mixed with signs of a personal spiritual dislocation. He oscillated in his responses between the twin poles of apocalypse (dread of human extinction) and utopia (dreams of guaranteed survival). The oscillation is dramatized in his famous "scientific romances" of the 1890s: *The Time Machine* (1895), *The Island of Dr. Moreau* (1896), *The Invisible Man* (1897), *The War of the Worlds* (1898), *The First Men in the Moon* (1898) and *When the Sleeper Wakes* (1899). To this day, these remain his most critically revered works.[2] In the apocalyptic mode, Wells confronted the Darwinian cosmological decentering of the human with fantasies in which humankind either deploys the power of scientific inquiry triumphantly, or survives the humiliations of near-annihilation. Humankind is nearly exterminated by advanced inhuman creatures from the planet Mars in *The War of the Worlds;* it is washed from the surface of the planet Earth, traceless, erased by 800,000 years of evolutionary time, as if it had never happened to exist, in *The Time Machine.* In *The Island of Dr. Moreau,* the human is represented most thoroughly by one survivor, the narrator Prendick, whose sense of belonging to the species is all but extinguished in the course of privately disastrous events. At the other pole, in his utopian mode, Wells proposed a variety of futuristic social worlds from which ignorance, poverty, and violence have been eradicated—societies shaped by aggressively statist policies and managed by scientific elites, societies that justified their revolutionary purges of the "weak" and "unfit" by a violent opposition to the threat of species extinction. The troubled oscillation between the polar obsessions of extinction or apocalypse (on the one hand), or utopia and violent species-purification (on the other), has been frequently observed in the critical literature on Wells (Wagar, *Traversing* 9; Vernier 78–79; Hughes, "H. G. Wells: Toward"; Farrell 146).

The Island of Dr. Moreau reconstructs the scene of the human-animal separation in the context of Darwin's impact on evolutionary thinking, to particularly distressing effect. (Readers unfamiliar with the story might now wish to review the concise version offered at the opening of chapter One.) Roger Bowen accurately notes that "in *Moreau* speculation about evolution is at the heart of the narrative as it is not in the remaining novels" (319) (the "remaining" being all the novels after *The Time Machine* and *Moreau*). Wells in *Moreau* has given his imagination over to the apocalyptic polarity of his responses to evolution. His gesture toward utopia in the text offers only the strangest glimmer of hope in a gloom of despair: Moreau's unrealized fantasy of a species tortured into being-beyond-pain. Critics mostly agree that the story is unrelievedly bleak (Fried 101; McConnell 89; Sherman 869; Bergonzi 112). Experts on Wells' whole massive career output praise *Moreau* as one of his best three scientific romances, the others being *The Time Machine* and *The War of the Worlds* (see, for example, Ruddick, "Annotations" 319). Manuscript evidence and two scholarly editions have established the re-worked, polished quality of *The Island of Dr. Moreau* (Philmus, "Introducing" xviii; Reed, *Natural* 198). It has been distinguished as "arguably the most intensely physical and visual of Wells' works" (Steffen-Fluhr 433), and as his "starkest and most sustained treatment of man's dual nature" (Haynes, *Discoverer* 24).

Despite its bleakness, critics praise the riddled density of its literary overcoding. As Roger Bozzetto quips, "Contemporary criticism of *Moreau* has been at once fascinated and disconcerted by it. The only thing that all accounts in effect agree on is the difficulty of apprehending the author's intentions" (34). Bozzetto points out "a richness of interpretive possibilities which reveal the presence of a figurative meaning that resists any reduction to univocal or unambiguous allegory" (38). John Hammond, founder of the H. G. Wells Society, contends that in *Moreau* as in all of Wells' novels, we sense a "didactic intention . . . there from the outset" but become aware also of "layer upon layer of allegorical and symbolic meanings: literary, religious, philosophical, social" (30). Elaine Showalter argues likewise: "The psychological, literary, social and intellectual sources of *The Island of Dr. Moreau* are enormously complex, and this story deserves to be read in the company of Kafka, Conrad,

Camus and Coetzee" (77). However, the praise grows faint when we hear certain hints that Wells may have been simply confused and inconsistent about what position to take respecting the question of the human-animal boundary. Darko Suvin hints at a certain unsatisfactory indecision in the text: "Yet such virtuosity cannot mask the fundamental ambiguity that constitutes both the richness and the weakness of Wells. Is he horrified or grimly elated by the high price of evolution?" ("Wells" 29). Suvin's "high price" may be represented in the chance, waste and pain that Prendick witnesses on the island; Suvin's "horrified" would identify the apocalyptic pole, his "grimly elated" the utopian. In a fine structuralist analysis, John Huntington complains: "*The Island of Dr. Moreau,* while it may be Wells' most systematic study of the evolutionary dilemma, arrives at no conclusions. The prevailing mood is of a confused flux . . . The darkness of the book comes not from a real pessimism, but from a kind of muddiness that the inconclusive yet systematic separations and crossings create" (68). A "real pessimism," Huntington implies, would have made it possible for us to identify a philosophical position consistently occupied by Wells regarding the "evolutionary dilemma." I agree with these hesitant voices. I believe that during the experience of studying *The Island of Dr.* Moreau, one's resentment is given an excess of targets. The targets might seem to overlap and co-exist (if one feels comfortable with internal contradictions), but they also threaten to cancel each other out, erasing meaning with the steam of incoherence.

At the time of writing *Moreau,* H. G. Wells had been provoked by August Weismann's displacement of Lamarck. He was despairing in his new conviction of the irreparable animality of the human (Philmus and Hughes 179–85). Here are Philmus and Hughes, describing Wells roughly at the time of his composing *Morea*u: "He was reacting [in 'Bye-products in Evolution' (1895)] to 'Weismannism,' which, by dealing the death-blow to the doctrine of the inheritance of acquired characteristics, left open no avenue of evolutionary innovation except through chance variations selected for 'fitness' or extinction by the environment. 'Bio-Optimism' (1895) marks the further retreat of Darwinian orthodoxy to its last line of defense, an iron Calvinism of struggle and the election of the 'fittest'" (183). Later, Philmus and Hughes remark on the dark mood that was to infuse *Moreau.* Their information provides an especially helpful context for the exchange of speculations about the psychogenesis of this tormented narrative. I quote from their work at length.

> Already he had spelled out the diabolism of the theme of the sculpting God in *The Island of Dr. Moreau,* the "theological grotesque" then awaiting a publisher. Faced with the dilemma of harmonizing his need to believe in some kind of Lamarckian inheritance with the scientific disproof of Lamarck by Weismann, Wells *suffered the pains of* writing Moreau, but then effected a characteristic "disentanglement" from the impasse by the expedient of accepting Weismann's verdict while simultaneously denying that the secular advancement of humanity has depended upon biological evolution at all. That is, he ceased to speculate in biological terms of how man became man or will become any other entity, and turned instead to cultural evolution, which he labeled "The Acquired Factor." (184) [emphasis added]

The key point here is that when writing *Moreau,* Wells was speculating in biological terms of how humankind might become another entity. That is, Wells had not yet discovered or invented a "culturalist" solution to Weismann's finding. Weismann in his experiments on frogs had proved that acquired characteristics could not be transmitted by inheritance (it is worth remembering that Darwin himself had not rejected the Lamarckian option of intergenerational inheritance of acquired characteristics). When writing *Moreau,* Wells had not yet formulated an alternative to imagining the survival of the human in the painful biological processes that are personified by the character of his mad scientist. If Wells had already found a culturalist solution—as he would later, with the celebration of the possibilities for species renewal in education in the essay "Human Evolution, An Artificial Process"—then the story we might enjoy despite its perplexing horrors would never have been written.

In this chapter, I argue that although *The Island of Dr. Moreau* appears to insist on the assimilation of humanity to animality, and seems to endorse the shocking modernist truth that the human is demeaned by Darwinian science's revelation of a cosmological temporality indifferent to the human, it really does more; it contains its modernism but is not contained by it. *Cosmological temporality indifferent to the human* is perhaps an obscure phrase. With it, I wish to suggest ideas brought into the view of the British public most forcefully by Darwin and Huxley. The following by Robert Philmus, from the "Introduction" to the now-standard scholarly edition of *Moreau,* summarizes them:

> Darwinian evolution at once exacted and engendered an unprecedented broadening of humankind's temporal perspective. Furthermore, the universe it posited was not only far vaster in temporal scope than the human had hitherto supposed it; it was also continuously changing so as to make the tenure of *Homo sapiens* in it precarious. Indeed, the very survival of the human species depended [in Wells' view] on its capacity to adjust . . . (xiv)

Despite the conflicting narrative ambitions, didactic intentions, and educational aims informing this one text, Wells ultimately represents the narrator Edward Prendick as a moral authority; and Wells suggests by way of Prendick's tragic fate that the scene of human language generates a temporality irreducible to cosmological or biological quantification. Wells does so with hints, clues, indirections. We can be sure, however, he knew as a storyteller better than to accept the despair into which the implications of Weismann's doctrine threatened to plunge him. His sane knowledge is expressed in his leading us not only to identify with Prendick throughout most of the narrative, but also to break with Prendick's self-understanding in the final pages of the novel. I am arguing that *The Island of Dr. Moreau* as an esthetic experience reminds us that the human is not primarily a biological identity at all.

Put more crudely, my thesis is that Dr. Moreau the character is boring in an important way, and our interest in the text ought to be mostly an interest in Prendick. Moreau's monologue, the routinely foregrounded and endlessly analyzed chapter 14, "Dr. Moreau Explains," contains shocking ideas, yes. But it is one segment of discourse in an extended narrative of many remembered actions and dialogic exchanges. Any interest we might take in Dr. Moreau is surely exhaustible in the way that an interest we might take in the psychology of torturers is: we should have the courage to exercise the human intuition that cruelty, amorality, and indifference to the pain of others, whether cloaked in stern subservience to cosmic or "historical" law or dressed up in militant submission to a utopian future, will finally prove boring. Seeking the causes of such cruelty is one thing; attributing a poetic profundity to the cruel agency's justification for its volition is another. Goodness fascinates more than evil and love fascinates more than hate; creation is prior to and contains destruction. Moreau justifies all his cruelty by assimilating human temporality to cosmological temporality; once we register that mistake, his vaporous aura of mystifying intelligence falls away. Once having seen his mistake, we find him unmasked: he is a monomaniacal hypocrite, excellent as a skilled vivisectionist but little else. Moreau is no more interesting than Hitler, Stalin, or Pol Pot, whom we study surely not as fascinating models to be scrupulously imitated but as deluded men, whose delusions are ideas to be rejected. That which we reject can only be so interesting. The torturer's ideas must ultimately be set aside as valueless—valueless in the esthetic mode of the tawdry and valueless in the economic mode of the non-exchangeable, the useless. I will argue that Wells is interested in the psychology and ideas of Edward Prendick far more than those of Moreau. A respect for Prendick—no more than an insistence on our putting ourselves in his place and asking ourselves what we would have done differently, and considering what emerges in turn from our answering such self-reflective questions—will require me to disagree with those analyses that put him down. True, Prendick fails to present a counter-text as self-composed and self-assured as Moreau's explanation (or purported explanation) of his amorality.[3] However, it would be perverse of us not to own Prendick as our ally, however imperfect, in resistance to Moreau's amorality.

To support these claims, I analyze three different "ways of playing God," layers of meaning and targets of satire or speculation, in the text. We will first examine Moreau playing the punctual God of orthodox creationism; second, Moreau playing the God of the liberal theology that hopes to compromise with evolutionary gradualism; last, Moreau playing the nightmare "God" of the utopian species-shaper. A fourth segment will defend Prendick. Overall, the conservative element in Wells' creative genius is more valuable than his polemical advertisement of the Darwinian doctrine in which he felt trapped by Weismann's thought. Our experience as readers of Prendick's account of his victimization should be an occasion for affirmative mourning rather than scandalized despair.

Moreau Playing the God of Punctualist Creation Theology

One layer of my analysis proposes that Wells in *Moreau* seems to assert the assimilation of humanity to animality. Wells' invention of the subject-to-degeneration Beast People forms the center of a satire on humankind. Here, we have a Wells for whom we are saved from brute animality by a "thin veneer" of civilization: "Ostensibly a tale about animals changed into humanoids, it really dramatizes the constant threat of humankind's melting once more into bestiality"; "men are beasts and will remain so" (Reed, *Natural* 36, 96). Just as the Beast People have been prevented from dissolving back into animality only by the imposition of the hypnotic authority of Moreau, so are humans prevented from owning their animality only by religious delusion. The insult directed at humanity is identical to one directed at religion: humankind deluded by creation theology is despicable or grotesque (or both). Prendick is shocked (as the reader is meant to be) by the spectacle of the Beast People massed in their dark cave chanting an absurd Law they do not understand. In this episode, one of the most powerful in the movies based liberally on the novel, we sense the directorial hand of a resentfully "scientific" H. G. Wells who wishes to argue that religious worship is nothing but collective delusion.

> "Not to go on all-Fours: *that* is the Law. Are we not men?
> "Not to suck up Drink: *that* is the Law. Are we not men?
> "Not to eat Flesh or Fish: *that* is the Law. Are we not men?
> "Not to claw Bark of Trees: *that* is the Law. Are we not men?
> "Not to chase other Men: *that* is the Law. Are we not men?
> "Not to go on all-Fours: *that* is the Law. Are we not men?"
> (12.57)

The laws of the pseudo-religion of the Beast People are based on fear of pain—the pain inflicted by the mad scientist.

> "His is the House of Pain.
> His is the Hand that makes.
> His is the Hand that wounds.
> His is the Hand that heals."
> (12.57)

The involuntarily bewildered Prendick, in his sane self-distancing from these chants, speaks for Wells when he regards the Law of the Beast People as nonsense ("deep down within me laughter and disgust struggled together" [12.57]). Whatever else Moreau has done, he has not made creatures capable of participating in a genuinely human ritual scene. They do not know what they are doing.

At this level of the text's satire on humans-as-beasts, the division between the three humans and the many Beast People on the forsaken

Pacific island is an analogue of the division between what we might call the educators in English society, the class of priestly and political leaders, and the benighted mass of exploited people who accept the ideas given them. This is a Wells who posits religion as the Enlightenment boogeyman of an institutional reign of punishing oppression and gawking superstition. Moreau uses hypnosis to induce "fixed ideas" in his tortured-into-humanness subjects. His assistant Montgomery (the man who first rescues the castaway Prendick and negotiates his passage onto the island) continues the education, assisted for a while by some Kanaka natives. But the hypnosis and the lessons in rote memorization function as Wells' analogues of the indoctrination of untutored people into politico-religious subservience. The power of the elites in late Victorian England, we are to understand, is no more legitimate than the power of Moreau. Once we grasp the innocence of the Beast People as victims of torture and hypnosis, the target of the satire shifts somewhat. Wells' point seems not to perform a sneering reduction of all humans to gullible idiots, but rather to propose, more narrowly, an indictment of those who promulgate doctrines that deny the animality of the human; for the Beast People in being "humanized" are prohibited from indulging their natural drives to walk on all fours, drink from pools, feed carnivorously and the like. In the light of this shift, the degeneration of the Beast People is not the vehicle of Wells' satire; on the contrary, Moreau's "humanization" of the Beast People with surgery and hypnosis symbolizes the violence and fraud of religion. The eventual killing of Moreau and the overthrow of the Moreau-Montgomery-Prendick regime on the island, with its whips and revolvers, become analogues of the much-to-be-desired "death of God" and abolition of religious institutions. Moreau represents the cause of religious delusion more than the Beast People represent its effect. As Moreau deserves to die, the crude God of fundamentalist superstition deserves to die: their reigns have alike been based exclusively on fear of punishment.

At this level, we are to celebrate the reversion of the beast folk back into animality, after the deaths of Moreau and Montgomery as their release from captivity. John Reed has accurately pointed out the cruelty in Moreau's engendering in the beasts prohibitions against the very appetites that come naturally to them: "There is nothing wrong with an animal running on all fours or lapping water, but Moreau has proscribed these natural instincts to his creations, because he does not want them to *be* animals. By resisting nature, [Moreau] has called into being the need for a law; in doing so, he has set one law [his own] against another [natural appetite]" ("Vanity" 141). Prendick, although he perpetuates the false religion after Moreau's death out of a desire for individual self-preservation, correctly sees Moreau's work on the Beast People as a wrenching of the creatures from their proper ecological sphere into a doomed artificial hybridity. Humans are really animals; creation theology cruelly suppresses their natural animality. At this level, we assume the centrality of what Robert Philmus calls "the satiric idea that these creatures [the beast folk] are travesties of humankind" (233); we place weight upon "the continuity between animals and humans" that Philmus posits as "Wells' premise" (216). The most significant analogy between the "society" of Beast People and the victims of religious orthodoxy in English society is to be found in the innocent ignorance they share: "This, [Wells] asserts, is the truth of the human condition; recognize man for what he is—a beast thinly disguised as a rational creature. Prendick, like Gulliver before him, is the solitary preacher asserting the reality of man's nature" (Hammond 37). Prendick's concluding thoughts, on this reading, derive from a painful but bracing education into truth about the animality of humankind: our humiliating biological ancestry proves the supposedly demeaning continuity between human and nonhuman life-forms.

The critical literature on *Moreau* includes several valuable variations on this interpretation. At Roger Bozzeto spells it out, "[t]he island becomes . . . a laboratory version of Genesis, the text for its own purposes recapitulating in a burlesque manner the Creationist hypothesis, with Moreau in the role of God" (39). We notice, in the "Dr. Moreau Explains" chapter, his speaking to Prendick of "days" of his creations in ways that echo the first chapter of Genesis; we notice the echo of the serpent in the Garden in Moreau's creation of the lethal wriggling creature that escapes from him and kills some of the Kanaka assistants; we notice the mock-ascension of Moreau announced when Prendick tells the Beast People that Moreau is not dead, but gone to the sky, where he continues to watch them from above (Bozzetto 42–43; see also Hammond 36; Beauchamp 413). In a second variation, John Hammond defends the "satire on humankind" interpretation by emphasizing connections between *Moreau* and the fourth book of Jonathan Swift's *Gulliver's Travels.* Hammond privileges the similarities between the Beast People and Swift's Yahoos, between Gulliver's disgusted resolution to avoid contact with humans after his stay among the Houyhnhnms and Prendick's inability upon his return to England to abide people, his studious isolation. For Hammond, Wells' didactic intention foregrounds "the notion that man is and remains inherently animal and that his instinctive nature [will] continually reassert itself despite a façade of law and decorum" (34).

A third variation is Gorman Beauchamp's elucidation of Wells' later decision to classify the text as a "theological grotesque." Beauchamp points out that the phrase "'theological grotesque'" is usually "taken to describe" the "aspiration endemic to literary scientists from Victor Frankenstein" onward to "usurp the divine power of creation, to play God" (408). However, Beauchamp wishes to get beyond that focus on Faustian hubris: "Its theological grotesquerie stems, not from Moreau's playing God, but rather . . . from God's playing Moreau" (408). That is, the performance of Moreau as scientist is objectionable in the way that the performance of the God of Genesis is: "Put baldly, the pain that Moreau inflicts upon the creatures of his little island reenacts microcosmically the macrocosmic process of evolutionary creation through suffering. God is thus Moreau writ large . . ." (408). With the phrase "process of evolutionary creation," itself a little oxymoronic, we cross the boundary into our second level of allegory—Wells attacking the God of liberal theologians. For now, however, note Beauchamp's crucial move, made at the end of his argument: an exposure of biographical information relating to Wells' fundamentalist upbringing. As a child, Wells suffered a bad dream that showed him "'Our Father in a particularly malignant phase, busy basting a poor sinner rotating slowly over a fire built under the wheel'" (quoted in Beauchamp 412). Beauchamp ventures the following psychogenesis: "Wells has artistically resurrected his childhood spectre of a diabolical deity and superimposed him, in a particularly malignant phase, on the figure of Dr. Moreau . . . The hatred of his childhood God, at whatever level of consciousness, remained intense for Wells, at least through the writing of *Moreau,* and animates the grimly inverted theodicy of the fable" (412). The ultimate card that Beauchamp plays is a claim about a deeply intense resentment of fundamentalist religion at the center of the Wellsian imagination.

Taking the text as a satirical attack on religious faith and an argument for the need to recognize the fact of human animality, as Darwin had "proved" it, opens up much of its power. It remains, unsurprisingly, a limiting interpretation. One drawback is that it attributes to Wells an extraordinarily crude reduction of the role of religion as such in human society. Thus, I have insisted on *narrowing* the target of the satire at this level from "man" to the priestly elites responsible for the authoritarian promulgation of orthodoxy. It is best to say that the novel as "satire" at this level does not attack humankind; rather, it attacks a fundamentalist Christian dogma inclined to bibliolatry. Furthermore, if we limit our decoding to this allegorical level, we restrict the novel's brilliantly mixed suggestiveness to a resentfully self-cancelling thesis: for the more weight we attribute to the intentionality of Wells' satire against the naïve creationist God, the heavier the implication that Wells remains bound in belief despite himself to the very object of his resentment.

A contemporary reviewer in *The Guardian* (in June 1897) observed: "Sometimes one is inclined to think the intention of the author has been to satirize and rebuke the presumption of science; at other times his object seems to be to parody the work of the Creator of the human race, and cast contempt upon the dealings of God with his creatures" (Parrinder, *Critical Heritage* 53). Wells is on record speaking favorably of that review alone.[4] The *Guardian* reviewer also argued calmly: "The inevitable reversion of these creatures to bestiality is very well described; but it ought to have been shown that they revert inevitably because they are only man-made creatures" (53). The gentleman believer is hinting at a weakness in Wells' attack on Christian faith. Certainly, the Beast People "believe in" Moreau, after a fashion; but Moreau, the gentleman hints, is not the God of Christian revelation. The *Guardian* reviewer has not, let us notice, been scandalized by a text which we might believe Wells had hoped would scandalize. *They revert inevitably because they are only man-made creatures* affirms the massive differences between the God in whom an ordinary Christian believes and the figure of Moreau; given such affirmation, the analogy is easily dismissed. Any analogy between Moreau and the Biblical God must be an invidiously selective one that throws all its weight on the cruelty of a certain cartoon-version of the Old Testament God at his most punishingly capricious.[5] The broken body of Christ on the cross is hardly an analogue for the body of Moreau holding the scalpel that tortures the puma. Nor is the prize puma, except at the most crudely reductive level of flesh made to suffer, a persuasively appropriate analogue for the Word made flesh.

The more seriously Wells takes the "God" in whom he wishes to destroy others' belief, the more that wish itself betrays his own ironically residual belief in the effectiveness—the indestructibility—of that God.

The severity of Wells' attack betrays a pathological rivalry with the very idea that a self-contained atheist should have dropped as something beneath his notice. There is no escaping the paradox that the atheist may always try to be one for whom the concept of God is empty, but must remain condemned to be one for whom the concept has certainly operated powerfully in human history (Gans, *Originary* 42–43). Gans elaborates pithily on this point: "The persistence of irony is proof that *resentment of the divinity outlasts faith in it;* the ironist is an atheist who condemns God for his failure to exist" (*Signs* 69) [emphasis added]. Wishing to move up to the next allegorical level, I will aim now to humanize a little, by literalizing, the evil Dr. Moreau. I will discount his function as this mock-God, while I add value instead to his role as a fallible human scientist who personifies faith in the gradualist cosmic process of evolution; and I will propose that his desacralizing experimental work symbolizes the Darwinian cosmic process itself.

Moreau Playing the Gradualist "God" of Liberal Theology

Liberal theology, more actively in dialogue with science than fundamentalist faith, is able to account in greater detail for more aspects of Moreau's behavior. A preoccupation with blasphemous moral hubris gives way at the next allegorical level to an examination of science's epistemological limits. Here, Moreau the experimental scientist replaces Moreau the self-protecting law-enforcer. Metaphorical embodiments of the *process of evolution* and cosmic necessity displace metaphors of the act of divine creation. Since he is a human scientist who personifies evolutionary process, Moreau plays the God assumed to be responsible for evolution. The readers targeted do not deny human animality as do our first class of believers. The readers here make a different mistake: they believe that Darwinian evolution can be reconciled with theism; they believe they can reconcile faith in the goodness of evolution with faith in the goodness of God. They are wrong. Wells forces into the view of such compromisers the "chance, waste, and pain" of the cosmic process. The cosmic process cannot possibly be the responsibility of a deity deserving of worship: evolution is *impersonal, indifferent, cruel and wasteful* just as Dr. Moreau is all these things: "Moreau is no loving God; he personifies the insentient, mechanistic process which Tyndall and Huxley had so ruthlessly explored" (Haynes, *Discoverer* 31). As Frank McConnell puts it:

"how ludicrous, in the light of the evolutionary facts, appear our attempts to locate or describe a god—how silly, weighted against the immensity and incomprehensibility of universal process, are the mythologies we invent to render that process human" (97). Here the attack is directed against faith in science and scientific "progress," for scientists themselves are refusing to advertise the anthropological puzzle of an impersonal, God-abandoned natural order. Even deism should be out. Here at this level, we see the expression of Wells' "powerful intuition of the invalidity of the synthesis between Darwinism and progress" (Ruddick, *Ultimate* 74). It is the level at which Wells teaches, in Darko Suvin's words, that "science is a hard master. Like Moreau, [science] is indifferent to human suffering; like the Martians [in *The War of the Worlds*], it explodes the nineteenth-century optimistic pretensions, liberal or socialist, of lording it over the universe" (Suvin, "Wells" 24; see also Philmus, "Introducing" xxviii; Haynes, "Unholy" 16).

This allegorical register invites us to recognize the way Doctor Moreau squeezes the eons-long temporality of evolution into a visible series of discreet experimental acts. Bergonzi's groundbreaking early analysis of the novel recognized this: "He appears . . . as God: not the traditional God of Christian theology, but the sort of arbitrary and impersonal power that might be conceived of as lying behind the evolutionary process" (104—105). In the words of Nicolai Vallorani, "Moreau's experiments . . . spring from a scientific impulse partially supposed by the theoretical frame offered by Darwin's and Huxley's theses. The scientist starts from the assumption that it is possible to recreate the evolutionary situation by compressing the required timespan through a series of surgical operations" (249–50). The squeezed evolution metaphor is not itself ethically charged. There is nothing inherently cruel in a process that requires long stretches of time to complete itself; such a process might be neutral. The ethical enters with the waste and pain involved in Moreau's experimentation. The auditory image of the screaming in pain of the puma that Moreau is vivisecting and re-sewing, screaming to which Prendick listens for hours, is a condensation of the pain suffered by sentient creatures in all of evolutionary time.

Personified by Moreau the vivisectionist, the evolutionary process is sadistic. And the ugliness of the Beast People at their genesis takes on new meaning, recalling the ugliness of Frankenstein's Monster. From his very first nightmare impressions when he is recovering on board ship, having seen the Beast Person we later know as M'ling (Montgomery's servant), Prendick is disturbed and haunted by the bodily awkwardness and unrightness of the creatures he only later learns to name the Beast People. From the beginning, Prendick fears something is amiss with the experiments Moreau is performing on his "biological station." The raw material represented by the innocent animal flesh is wasted, in that the experimental goals of the properly human form and mind are not achieved. Bernard Bergonzi recognizes this horrifying aspect of the allegory, claiming that Moreau is "Frankenstein . . . in a post-Darwinian guise" (108). Bergonzi points toward "the alarming implications of what [Moreau] symbolizes: he stands for both science—unhindered by the ethical considerations which had concerned Huxley—and evolutionary nature, in all its *violence and arbitrariness:* 'The study of Nature makes man at last as remorseless as Nature'" (207—108) [emphasis added]. Robert Philmus puts it in a nutshell: "Moreau the experimenter with living tissue is thus the Shaping God of Evolution" ("Introduction" xxviii). The essential difference between this level and the first is that Moreau playing the God of liberal theology represents not the impossibility of removing animality from the human but the impossibility of adding human scenic intuition to the animal. Moreau's failure to successfully endow animal materials with the "additional" features of human form and mind symbolizes the impossibility of genuinely progressive evolutionary process.

I have touched on "waste" and "pain." Let us give "chance" or the "randomness" of evolution its due. Clearly, Moreau is not interested in the limits of individual plasticity alone as an abstraction, for the plasticity of frogs or rats would not satisfy him. When Prendick asks him *why* he has chosen (of all living things to try to create) the human animal, Moreau claims he did so by "chance." Some critics have suggested that Wells, by means of the Moreau-Montgomery discourse on "chance," wished to teach his readers about the role of randomness in evolution, thereby endorsing its anti-theological implications. But as John Reed has observed, the exercise of free will was always a central value in Wells' ethic. Leon Stover and Robert Philmus, in their scholarly editions of the text, each consider the paradoxical nature of Moreau's claim (for Stover's considerations, see *Island* 137n102). Philmus comments in some detail:

> Here, however, the "chance" by which the doctor seizes upon "the human form as [his] model" has a specifically Darwinian connotation:

> it identifies Moreau with a Nature which works blindly through Natural selection and which consequently has no intentions or design whatever, benevolent or otherwise. There is, of course, a paradox in this, insofar as Moreau functions as God over his Creation; and that paradox is contained in the seeming self-contradiction of the statement "he had *chosen* that form by *chance*." (96n50)

An irony emerges if we reflect on Philmus' observations: we have *not* escaped the problem of the Punctual Creator at this level of the allegory after all; the "chance" of gradual process at this higher level collapses into the "choice" of punctual intention at the lower level. For inasmuch as we wish to worry over the amoral purposelessness of the much-feared *cosmic process,* we are expecting in advance the meaningfulness of a question that the process, correctly apprehended, should prevent us from asking in the first place.

Why doesn't the cosmos care? asks the body surprised by the Darwinian randomness of the human lot in the universe. We should ask what such a question could mean—how it could be answered. It could be answered only by personifying the cosmos as a Being that might care—a Moreau, or a God. Wells is asking us to be scandalized by the indifference of a cosmological temporality indifferent to the human; paradoxically, however, a proper submission to evolutionary doctrine should abolish the scandal altogether by assimilating the human to the cosmological. On the one side of the debate, we may claim that the human species is unique in the universe because it alone possesses a scenic moral sense, in which case the human difference from the cosmological would dissolve the scandal of the ethical indifference of the cosmic process. We (humans) are to be moral for each other. On the other side, we may accept the idea that the human must be assimilated to (a part of) the cosmological, and face the scandal of being reduced to something unworthy of moral consideration. But if the moral is absorbed into the mechanical-physical domain of the cosmological, then we must cease expecting the cosmic process to respond to our cries about the insignificance of the pain of sentient creatures.[6] H. G. Wells, no fool, in the calm depth of his imaginative exploration of this debate, sensed the dizzyingly irresolvable quality of the paradox. The blurred muddiness of Moreau the character, the daunting overdetermination of his conflicting symbolic aspects, is, from this perspective, the effect of an artistic success rather than an ideological confusion.

We need briefly to consider the new symbolic roles this level generates for the Beast People in the plot. In our first reading, the killing of Moreau marked the crucial reversal by which the Beast People were freed to begin their return into natural "innocent" animality. Here, the reversal that liberates the Beast People occurs when each creature begins to degenerate naturally—at which moment Moreau gives up on each, sending it into the wilderness, scapegoat-like. The only such event that Moreau recounts in detail is that of the initial reversion of the first gorilla man whom he teaches to think and to speak human language, but whom he later discovers swinging from a tree and speaking gibberish (14:76). This event—Moreau's disappointment at the failure to educate the Ape-man into language—should interest us for two reasons (at least).

First, it provokes Moreau to admit the all-too-human motivation of resentment for his allegedly dispassionate scientific work. He had dreamed of returning to England with a fully humanized beast person. It has not been sufficiently underscored in the critical literature that Moreau's motivation is not scientifically detached; it is mimetic, conditioned by resentment and rivalry, by competition for prestige. He wants one day to be able to take revenge for his sacrificial expulsion from England by creating a "rational creature of [his] own" (14:78). He dreams that this rational creature of his own will become so undeniably human that, when he returns in triumph, nobody will be able to deny that he had been right all along (14:76–77). The account of the ape-man's initial reversion forces Moreau to confess his failure—specifically, his failure to give his creatures human minds. We touch the boundary of our next allegorical level here, where Moreau's insane ideas about the super-humanizing effects of excessive pain take over. The Beast People's crude social life mimics the human hope of species self-improvement. The loss of their crude "culture" when they lose the power of speech therefore takes on the coloration, however minimally, of a small-scale tragedy—rather than representing a happy return to natural animality. Their "'upward striving'" (14.79) symbolizes the incompleteness of human evolution, so that the killing of Moreau is here a regrettable overflow of aggression that the Law, by means however clumsy and grotesque, had contained. It is not a release from captivity we are to celebrate.

Second, in the context of his disappointment, Moreau confesses that the minds of all the Beast People generally are the real obstacle (14.78–79). The ape-man's betrayal is only one such failure. Perhaps because Wells, through Prendick, is more forthcoming with concrete details about the physiological similarities between the Beast People and English people rather than their shared mental deficiencies, criticism on the text has not, in my view, properly emphasized the fact that the human thing Moreau openly confesses his failure to "create" is the human mind. Vivisection is only part of the process of man-making. The under-described hypnosis and endowment of "fixed ideas" is a treatment each creature undergoes in Moreau's laboratory and afterward, through his employees (Montgomery and some Kanaka natives they train). Each time, without fail, the education does not "take"; each time, when the failure occurs, Moreau gives up the beast person and lets it go out from his enclosure to join the other rejects in the huts. Victor Frankenstein's single abandonment of the Monster, Moreau repeats anew with each failed creature. But where Victor felt immediately something had gone wrong, Moreau feels no self-doubt and keeps going: more animals, more vivisection, more surgery, pain and waste; more hypnosis, more terror of return to the House of Pain, more chanting of the absurd Law. Such is his allegiance to "scientific" method. Moreau is a God who has decided to re-enact the mythical fall of man as many times as it might take to produce a "rational creature of his own."

And how does Edward Prendick fit at this level? A formula such as "Gulliver is to the Yahoos as Prendick is to the Beast People" no longer holds. The self-examining quality in Prendick's account of his own mental illness should open us to sympathy: Moreau does not question himself, but Prendick does. At the ending of the narrative, he knows that he is mistaken, even though he cannot change his mind. His anti-social discomfort with English society is at this level not a revelation of the truth of the animality of the human, but the result of an extended series of revelations (in keeping with the gradualist, serial quality of the horrors he has experienced). The ultimate object of the revelation is diffuse: its content is the incommensurability of cosmological time and human (historical) time. Certainly, the people in England remind him of the Beast People. Prendick may well have absorbed the idea of the pitilessness of the cosmos before having arrived on the island; thanks to his experiences there, he is newly traumatized into believing humans are nothing but animals.

At this level of Moreau as the personification of (squeezed) evolutionary process, Moreau himself must ironically be "contained" by the very cosmic process that he attempts to control inside his fictional world, the process he is meant to personify for readers outside the fictional world. Prendick himself will include Moreau as one among other victims of cosmic indifference. That he includes Moreau as a fellow being subject to "the vast, pitiless mechanism" that he begins to see as the governing structure of the universe should not be ignored: "A blind fate, a vast, pitiless mechanism, seemed to cut and shape the fabric of existence, and I, Moreau by his passion for research, Montgomery by his passion for drink, the Beast People, with their instincts and mental restrictions, were torn and crushed, ruthlessly, inevitably, amid the infinite complexity of its incessant wheels" (16:96). From Prendick's gesture of inclusion, it follows we might observe that Moreau is only (literally) human. Moreau may be a *metaphor* of the "cosmic process" for the reader, but he is no such thing for Prendick. The Beast People kill the scientist and Prendick does not mourn him: the human Moreau of this level is a small-scale tragic victim of his own delusions. The delusions, however, may be seen to grow to larger and more dangerous proportions; then they become the highlight at a next, more disturbing level of the text's allegorical activity.

Moreau as One Who Believes in Scientific Species-making (The Atheist Plays God)

I return to the "conservative" clause in my argument: H. G. Wells in *Moreau* aimed to show that the human as a biological species is demeaned by evolutionary science's revelation of a cosmological temporality indifferent to the human. For religion and myth, the human has an origin in a sacred event; for religion and myth, the human and the cosmos interact, intersect, exchange—as if the cosmos is alive or endowed with being. The theologian John F. Haught reminds us that with the advent of modern science, it was life that came to need explaining; before, it was death that had called for explanation. Before modern science, the idea that humans could anticipate an otherworldly afterlife was "rational" because it was the occurrence of living, moving things ceasing to move that was felt to be irregular, inexplicable. The idea of matter being *naturally* dead or inanimate was then incomprehensible. But after Newton, we were forced to begin imagining the universe as dead matter at its foundation; after Newton, all forms of life from the microscopic to the gigantic, including the human, began to appear anomalous in a mechanical cosmos composed of inanimate parts. Life became the "new" that needed to explain its presence. Darwinian science, among its other effects, completed the project begun by Newton: it pushed into the foreground of secular cosmology the exceptionality of "life." We were obliged to imagine a cosmic process which contained nothing but the eons-long activity of a lifeless landscape—volcanoes, vapors, inorganic chemistry—before the emergence of life. *Cosmological temporality indifferent to the human* means in part the ground of this disorientation: neither living things generally nor humans particularly may feel at home in a dead cosmos of matter detached from creative animation, abandoned by invisible but living agencies (deities, ghosts, or souls).

As is well known, in damaging the intellectual respectability of faith in the Judeo-Christian narrative of Divine cosmic redemption, Darwinian science seemed to some to expel from the cosmos the creator-being that Western theology had named "God" (e.g. Mayr 96; Dupre, *Darwin's* 60). Let us pause to mark this evacuation of "God" from the human universe. God has been evacuated from the fictional world represented in *The Island of Dr. Moreau*. God has been evacuated from the real world of those tough-minded late Victorian readers for whom Wells intended his last, most morally unsettling, most politically radical layer of meaning. The targeted readers here recognize without resistance that the text of *The Island of Dr. Moreau* is built upon a confident, serene atheism: confident in science, serenely indifferent to religion. For these readers, God does not exist. No argument is necessary. To assume the absence of God is bolder than to argue it; the assumption transcends the paradox inherent in your resenting that which you say does not exist. For such readers, even Immanuel Kant's program of grounding human moral intuition in formal reason must be rejected because of a residue of theistic universalism. Instead, we enter the territory marked out by a work such as Kai Nielsen's *Ethics Without God* (1990): the assumption is that we remain systematically hostile to religion only because we may not yet, for political reasons, enjoy the concrete circumstances that would permit our being indifferent to it. Too many unliberated cowards and untutored fools still believe.

The utopian vision of a new human species, the unique vision of Dr. Moreau as heroic atheist-scientist, enters now. At this allegorical level, Moreau is neither incorrect in his scientific methods nor questionable in his cruelty to animals. On the contrary, he may be the hero both for tough-minded readers and, according to Leon Stover, for the author H. G. Wells. To linger over the screaming in pain of the puma and the fear suffered by the Beast People and the indignities visited upon Prendick is to get trapped by petty scruples. Any "bungling" with the Beast People as results of experiments gone wrong is justified by the utopian end the scientist has in mind, an end to which the Beast People are nothing but a means. No longer pointing vertically at the absence of God, the figure of Dr. Moreau is pointing horizontally at an empty future to be filled by a *new species.* Certainly, Wells and Moreau concede, it upsets conventional morality to listen to the pain of the puma or to witness the ugliness of the Beast People and remain calm, detached, rational. But now the highest value is just such calm, detached, rational inquiry. Nothing may be exchanged for it. Our intuitions of moral reciprocity are set aside, because "humanity" is set aside. Stover puts it this way: "As a master physiologist, Moreau makes men out of beasts, but his true vocation is learning by experiment how to carve over-men out of half-hewn humanity" ("Editor's" 20). Or again, "It is not the humanizing of animals, but the deanimalizing of man with which Moreau is concerned" ("Editor's" 14). The first and second allegorical forays are now rejected for the mistake they make of presupposing that for H. G. Wells "the human" *as we know it* is to be valued.

Even more intriguingly, this utopian vision is founded on the collapse of the paradoxes governed by the tension between God-as-punctual-actor and God-as-process-presence. Those paradoxes are solved by the deification of the man-as-god scientist. The dichotomy between punctual God as creator who intervenes and process God as withdrawn witness or immanent accompaniment to eons-long evolution dissolves. It dissolves because neither idea of God counts or matters. Into the vacant scenic center steps, strolls, and strides the scientist-leader, the leading scientist. The scientist legitimately replaces God in directing human history. Elitist, revolutionary, radical utopian science replaces democratic, co-operative, timid liberal science. The attack on *liberal* science has been exhausted already; it was complete at the second level of allegory. It no longer makes sense at this level to speak of Moreau as an "embodiment of the cosmic process"; he is an embodiment of nothing. He is (literally) a scientific rationalist who chooses to try to make new people of a certain type. The cosmic process itself now becomes something *other than* Dr. Moreau himself, because it is the thing external to himself that he alters, a thing he directs and manipulates. Yes, he is working on "the other," without hesitation or compunction: he is subject working on object. It makes all the difference that Moreau is *not the one feeling the pain* when he tortures his subjects, using pain to "humanize" them into becoming the new species-beyond-pain of which he dreams. In this difference between torturer and tortured, we have Wells' evocation of the person who grasps scientific truth absolutely free to do with that truth, and in the name of that truth, whatever he believes must be done for the improvement of humankind. Wells is setting Moreau's atheism against the less radical, more humane atheism of other scientists. Bergonzi points out that "the romanticizing of Moreau, and his specific identification with the arbitrariness and indifference to suffering of what Huxley called the 'cosmic process' would have offended scientifically inclined readers" (112). Wells' point now is that atheists should not fear changing the human species themselves; they should not fear biological engineering. Since it is presupposed that theological thought has already been exploded into irrelevance, this allegory is purely political. Self-confident, self-knowing scientists in a secular age should not shrink from taking drastic measures in social engineering.

The critical literature on *The Island of Dr. Moreau* has not described the relative autonomy of this third allegorical level by means of a focus on "ways of playing God," but it has come close to doing so by displaying an anxiety with the disturbing centralization of a political thought that crystallizes only when the reader begins to grasp the autonomy of this "atheistic" level. Nicholas Ruddick compares Moreau to Kurtz in Conrad's *Heart of Darkness:* "If Kurtz, the emblematic, hollow rhetorician utterly freed of ethical constraints, anticipates the Hitlerian demagogue, then Moreau foreshadows the concentration camp doctors, equally unconstrained, whose experiments on their 'subhuman' victims only affirm the inhumanity of pure reason" (*Ultimate* 67; see also McConnell 92, Hinckley 89, Sherman 870–871). Jean-Pierre Vernier condemns many of Wells' heroes for their excessively cerebral detachment from the feelings that bind ordinary people together: "The intellect is constantly shown as leading to inhuman behavior: Griffin [from *The Invisible Man*] and Moreau are both endowed with more than ordinary intellectual faculties; the Martians [invaders in *The War of the Worlds*] and Selenites [from *The First Men in the Moon*] are beings entirely governed by their brains and devoid of all feelings. The paradoxical situation seems to be that evolution leads to a dehumanization of man because the brain will tend to supersede the body" (77–78).

The important readings of the novel that make the most of this connection between the fictional Moreau and totalitarian science are those by Frank McConnell, Leon Stover, and Elana Gomel. In one of the most important studies of *Moreau,* Elana Gomel posits an "ideological continuity between Nazism and the fin-de-siecle bio-ideologies reflected in Wells' great novel" (396). She defines fascism as "an anthropological revolution, aiming at the creation of a new type of human being" and a political force animated by the "utopian project of remaking humanity" (394). Richard Shorten in *Modernism and Totalitarianism* has also located the concept of the post-Darwinian New Man in the central currents of totalitarian thought. The desire for a "New Man," emblem of the utopian future society toward which fascism believes it is moving, is one response to what I have been calling a *cosmological temporality indifferent to the human.* Gomel contextualizes its appearance: "The New Man's subjectivity incorporates Darwin's denial of essential humanness, while at the same time neutralizing [that denial's] potentially anarchic emphasis on randomness, heterogeneity, and accident" (398). The new man is the leader, the commander, the scientist (or other fascistic personality) who plays God. I will quote Gomel's work at some length here, asking that we keep in mind the figure of Moreau torturing the puma-creature with the end of making a "rational creature of [his] own."

> The victim's subjectivity is shattered, but the violator's is reborn in the image of the sublime, irresistible, indivisible might. By identifying with the unspeakable, the New Man positions himself beyond the judgment of words. He is now outside the reach of the Symbolic as the Real itself, the Thing, the unutterable, alpha and omega, God. (406)
>
> [. . . .]
>
> Dr. Moreau is one of the first portrayals of the New Man of eugenics who later evolves into the New Man of fascism. Both the island and the camp are factories of pain, and neither can be understood outside the ideological rationale for its existence. The cruelty of Moreau and Mengele is not the cause but the effect of their involvement with "a peculiarly modern institution" of ideologically motivated torture. (412)

One of the virtues of Gomel's interpretation is that it reminds us of the invisible but unbreakable link between politics and anthropology: the link between utopianism and fascism as political structures at one pole, and at the other pole the self-defeating "anthropological" project of theorizing the human as if the moral reciprocity modeled by the exchange of signs could somehow be stamped out while we yet remain human.

Gomel's admirably searching argument meets its limitation in a fetishizing of "the body." She wishes to posit totalitarian cruelty as the effect of a political institution. She betrays a hope to contain the anthropological in the political: "cruelty" in her argument signifies a unique form of universal human bodily experience, so that she takes up belief in *the body* as a response to the horrors of totalitarian amorality, in the good trust that others will join the political fight against cruelty. I can sympathize with her impatience over notions of a context-free human cruelty that emerges *ex nihilo* and her impatience with a purely abstract formal morality. But to take (by contrast) the reciprocal exchange of signs as the fundamental model of human interaction, as generative anthropology does, is not to take up something purely abstract. The scenic model attaches itself to a concrete hypothesis of the origin of the human in historical time and space. We had bodies at the origin of human language, of course, but when we began to be aware of exchanging signs, we began to become aware of each other's intentions: human mind began, with "the little bang of language" (Gans), to emerge. It is impossible for any version of physicalism, however sophisticated, to grasp the human. The basic model of human interaction in generative anthropology is the reciprocal exchange of signs; it is not just "the body" that exchanges human signs. I would suggest that discourses of "the body" alone will not be able to protect us from the likes of Moreau. It is rather the case that we need both memory and the exchange of signs that signify and preserve memories, before we can protect ourselves from totalitarian science. This is to agree with Gans' hypothesis that the anomalous "cruelty" peculiar to the horror of the Nazi death camps is their scenelessness (*Signs* 164). Where human culture is scenic, victims are meant to be remembered. It is no coincidence that the horror of Moreau's island may also be verified by its peculiar scenelessness: on the island, no traces are left to corroborate Prendick's confessional account of the evil that he witnessed there.

Leon Stover's account of Moreau playing the utopian species maker is unusual, in that Stover assimilates the utopianism of the character Dr. Moreau to the utopianism of the historical person H. G. Wells, indicting not just the character Moreau but Wells the author for bad science, bad politics, and anti-human thinking. As Michael Sherborne points out in a review of Stover's annotated edition of the novel, "Stover is perhaps the only reader to conclude that the mad scientist is not just someone for whom we are meant to feel some reluctant respect, but the out and out hero of the book" (52; see also Wagar 73–74). I claimed above that to read Moreau as playing God in the way of the utopian species-shaper is to move the allegory from a theological to a political level: the true scientist is unafraid to perform social engineering in pursuit of utopia. Stover is unique among Wells scholars because he persists with an aggressive, unsparing critique of Wells' political utopianism, linking it to the horrors of fascism, Stalinism, and other totalitarian ideologies of the twentieth century: "Like every ideologist in the business, he [Wells] offers first a total critique of the existing order, and then a prophesy of ineffable things to come, with which commonplace reality is invidiously compared" (Stover, *Island* 205n190). It is beyond the scope of this study to offer a decisive verdict on the question of Wells' anti-semitism or Wells' seeming endorsements of fascism, although the studies of Brian Cheyette and Philip Coupland offer at least some backing for Stover's un-ease.[7] Meanwhile, the reader might well ask, where is the utopianism in *The Island of Dr. Moreau,* a novel which may surely be said primarily to envision (on the contrary) a dystopia?

Dr. Moreau aims to create the prototype of a new species. The new species to grow out of that prototype will be a species beyond pain; its creator will be himself immune to the moral intuition that might flinch at inflicting pain on others. In his explanations to Prendick, Moreau speaks favorably of the puma on which he is currently working, with its high pain tolerance, as a promising specimen for which he has high hopes (14.79). For Moreau, pain is only a transitory feature of transitory species, a "cosmic irrelevance" in Frank McConnell's phrase: "And this mere epiphenomenon in the history of consciousness will disappear, like all useless organs, when the perfectly rational man is attained" (91—92). For Stover, the pain functions as a symbol of the *necessary evil* that statist regimes, in their pursuit of the impossibly utopian society, justify as the "means to the end" of that very utopian society.

> Everything Moreau does . . . is based on the author's premise that "[t]he social man is a manufactured product of which the natural man is the raw nucleus." [Wells:] "In the past, the making of a man

> has been . . . a planless process." With Moreau, the product is a consciously made artifact, but his method of mastering nature is nature's own painful and lethal method. He at once embodies the action of applied natural history and the pitiless actions of nature's impersonal, self-running mechanism. Pain, intelligently directed, humanizes. ("Introduction" 21)

Stover gives the name "inverted homeopathy" to the strategy of deliberately inflicting pain to produce a creature that is immune to pain: "The pain to end pain is admittedly cruel, but is a saving cruelty. Doctor Moreau, however, attempts to justify the excruciations of inverted homeopathy by depreciating pain itself" (45). The relative autonomy of the allegorical level of Moreau playing "God" as political scientist pursuing the new utopian society should now be visible. The Beast People are experimental trash to be exploited, just as human victims of genocide are obstacles to be exterminated.

It is not easy to imagine a species without pain, as Moreau asks Prendick to do when he "explains" himself. I would suggest, however, that to push our imagination further in that direction leads us to the error at the center of Wellsian anthropology. That error is his overt biologism paradoxically masking a radical culturalism, his totally fanciful utopianism masquerading as hard-headed scientism. Once we focus on the fascination with the idea of the species in this particular version of Wells, we recognize the icy fact that Wells here seems not particularly attached to the morphology of the human as we know it. Our attachment to our human shape is the source of the shock in Wells' description of the Martians in *The War of the Worlds.* It is the source of the shock in his imagining of the crab species 800,000 years in the future in *The Time Machine,* and of the strange non-human Selenites in *The First Men in the Moon.* The most disorienting visions in Wells' scientific romances are those by which he forces the reader to recognize the arbitrariness of our species as a biological entity. By this, I mean that Wells leads us to a realization that, yes, as a mere organism the ontological status of which is determined by its being either alive or extinct, persisting in or erased from cosmic time, the "human" can only be dwarfed by cosmological temporality. Who could disagree? But I would return to a point made in chapter One: the historical human community is not assimilable to the biological human species.

The force of the scandal inherent in Moreau playing the species-maker depends on the binary opposition between existence and annihilation. The most valuable feature of the human, from this perspective, is its aliveness: we are valuable only as long as we can continue to exist, to maximize reproductive fitness, as opposed to other species. The perspective is unfounded: such a model of human ontology rests on an impoverished quantitative opposition between annihilation and extinction. Stover may well be correct that, in thrall to this model, some part of Wells was risking the suggestion that there is nothing unique about us as humans because our primary goal must be to continue just to exist. If the social engineers must inflict pain and perform purges to save the species, so be it. Avoiding extinction is all; preserving "the human" is nothing. Language, culture, religion, symbolic exchange, even the bodily forms we currently occupy—none of these matter, given the Darwinian revelation of a cosmological temporality indifferent to the human. Moreau proposes to Prendick that pain is just a biological mechanism designed to warn the organism about dangerous forces in the environment, most obviously predators or competitors in physical contact, that the organism must fight against (14.73–75). A species without such a primitive warning system would be, by implication, be a species with an infinitely more refined warning-system—presumably a perceptual system equipped with auditory and visual mechanisms that could provide warnings well in advance of physical contact or proximity of the enemy organism. The super-species would have triumphed over all the big competition: the very fact of its possessing a super-intelligent warning system would verify its "advanced" occupation of a secure biological niche. The super-species would be "out of" danger in the curiously literal sense of being "out of" a sphere of existence bearing any resemblance to that of any current mammalian species on planet earth, given that mammals have nervous systems capable of feeling pain.

If we ask whether *a species beyond pain* such as Moreau imagines would be so different from us that it would already not be identifiable as the human, Moreau would answer that species identity itself must take cosmological survival as a value higher than species morphological continuity: that is, better survival in a form unrecognizable to current members of the species than annihilation. This contempt for *what is* resembles the reductionism of ultra-Darwinian genetic determinism, according to which the (biological) persistence of genes (selfish genes, lumbering robots) appears to be more valuable and real than the (anthropological) persistence of individuals, families, societies or cultures. At this level, the anthropology of the character Dr. Moreau is anti-human in the literal sense that "the human" is no more than that abstract biological blank-space in which that part of us which survives in the face of the indifference of the cosmos is valued. This anti-human tendency should impel us to ask whether the properly human dimension latent in the text of *The Island of Dr. Moreau* might be discovered not through a fixation on its fascinating vain scientist, but through considerate attention to its confessional narrator, Prendick.

The Island of Mr. Prendick

The main clause of my argument is that *The Island of Dr. Moreau* can lead us to the anthropological truth that the human is not primarily a biological identity. The credibility of such a suggestion about the novel's thematic tendency depends on our giving priority to the suspenseful narrative action and our readerly interactions with Edward Prendick. To give priority to Prendick's confessional account of his experiences does not require that we judge him a moral paragon. In a number of otherwise sophisticated scholarly interpretations of the novel, Prendick has been subjected to unsparing moral censure: he is more unfeeling than Montgomery, he is a collaborator in Moreau's cruelties, he is a prig, he is a do-nothing, he is a coward, and the like.[8] In the context of such a tendency to narrator-bashing, we might remind ourselves that nothing is easier to resent than the center (and nothing is easier than to resent the center). Edward Prendick, not Moreau, is the center of this text. He is with us before Moreau and after Moreau, before and after the Beast People, and he transmits to us everything we know about them. If we identify with Prendick's feelings and beliefs about the Beast People, we lose nothing of the "shock" value of Wells' experiment in modernist storytelling. Meanwhile, we gain something in access to an anthropological wisdom that it seems reasonable to attribute to a more calmly thoughtful and a less fiercely resentful Wells: Wells the storytelling artist as distinct from Wells the provoking ideologue. We might try pulling back from our fascination with the character Moreau—try to resist being distracted too much by the "Dr. Moreau Explains" chapter—and account for our experience of the whole story.

The reflection Prendick transcribes after writing down his memory of the mercy-killing by which he prevented what would have been a massapproved imprisonment and torture of the Leopard Man contains the most succinct statement of his ethic.

> Before they had been beasts, their instincts fitly adapted to their surroundings, and happy as living things may be. Now they stumbled in the shackles of humanity, lived in a fear that never died, fretted by a law they could not understand; their mock-human existence began in an agony, was one long internal struggle—and for what? It was the wantonness that stirred me. (16:95)

Prendick judges that the Beast People are not human but "mock-human." When trying to survive on the island, he believed they deserved his pity, but the feeling was bounded by a recognition of their dangerous otherness and a refusal to assimilate them to the human. My claim in this closing section is simply that Prendick represents a small residue of hope, hope that remains after the apocalyptic dystopia ruled by the horrible Moreau has wreaked its havoc on its victims. If we are to find in the text an affirmation of human being, we will find it in and through Prendick.[9]

This thesis requires that we take sides *against* Montgomery and Dr. Moreau insofar as we hold them accountable for their victimization of Prendick. Wells structures the ethical difference between characters by overcoding it with an epistemological difference: the secrets kept from Prendick by the managers of the "biological station" concern a morally indefensible science. The multiple evasions of Montgomery and Moreau indicate the moral indefensibility of their project; their secret-keeping links them to the secret-keeping of all mad scientists. And they keep many secrets (2:11–12; 4:19; 5:24; 7:32–35; 8:37; 10:49). Their evasiveness also makes our uneasy disorientation as readers one with the disorientation of Prendick the character. In telling us his tale, Prendick is a skillful narrator, determined to make us feel what he felt when first experiencing what he is now recalling and transcribing. It is no small matter that he believes the indignities he suffered, despite the painfulness of their being revealed, ought to be shared: there is nothing accidental about the literary quality of the text. Its esthetic quality is intended by Prendick. I cannot emphasize enough that Prendick writes with the purpose of therapeutic self-opening, in a reaching out for contact. This text is unusual among Wells' scientific romances in the decidedly tragic tinge added to its other qualities. It is a personal record of a private tragedy. Everything builds toward the powerfully subdued ending, reminiscent of but very different from the end of *Gulliver's Travels.* Prendick's trauma, furthermore, is inseparable from his ethical sensibility. Montgomery managed with the flippant decadence of an addict; Moreau carried on with the steely indifference of the utopian experimentalist. Moreau and Montgomery would have kept secret the very things Prendick is exposing to us, the things his nephew as heir and editor has decided to reveal. If Prendick had been as unfeeling as they, he would have felt no compulsion to write. Wells therefore frames the text we have in our hands as a testament to Prendick's struggle to try to make sense of seemingly senseless horrors. It is with a note of reluctant humiliation, not the trumpeting of heroic triumph, that Prendick on the first page describes his tale as one that is "as horrible, and certainly far stranger" than the "published story" of the shipwreck of the *Lady Vain.*

Beyond these differences in personality, there is a more powerful difference resulting from Prendick's having lived alone among the disfigured hybrids, without weapons, without human companions, and without the stabilized regime of man-animal difference that had structured Moreau and Montgomery's interactions with them. We miss everything if we miss the difference it makes that only Prendick can claim to have become "one among" the Beast People, in a way that Montgomery (far more the colonial hypocrite than Prendick) could never have claimed to be one among them, in a way that Moreau (with his delusions of self-sufficiency) would never have wished to be one among them.

Once Montgomery and Moreau have been killed, Prendick suffers a wary co-existence with these impossible humans. That experience gives him a peculiar authority. He begins his second-to-last chapter with a startling preparatory entrance: "In this way I became one among the Beast People in the Island of Dr. Moreau" (21:118). Not one *of* them, but one among them: as close as he could be to the impossible human without ceasing to be human. The Dog-man who becomes Prendick's one loyal companion among the Beast People during their slow reversion into animality reports to his secret master that the other Beast People see Prendick, in a way they never saw Montgomery or Moreau, as one of them: "'Even now they talk together beyond there. They say, 'The Master is dead; the Other with the Whip is dead. That Other who walked in the Sea is—as we are. We have no Master, no Whips, no House of Pain any more'"(21:119). Prendick is *that Other who is . . . as we are.* Neither Montgomery nor Moreau endures such a situation of forced intimacy with the Beast People. Prendick confesses that during the "ten months" that he struggled for existence "as an intimate of these half-humanized brutes," many things happened which he finds painful to be forced to remember: "I prefer to make no chronicle for that gap of time . . . There is much that sticks in my memory that I could write, things that I would cheerfully give my right hand to forget" (21:121). The conditions of solitude and involuntariness shaping his experiential expertise should give us more confidence to make *Prendick* rather than Moreau or Montgomery the most reliable authority on the Beast People.

On the Mercy-killing of the Leopard Man

For the sake of argument, let us concede as much as possible to those critics who find Prendick's conduct blameworthy or priggish, and zero in on an episode for detailed analysis that shows him at his least defensible. The episode begins when Prendick gives in to Moreau after the mad scientist's pseudo-explanations: "By way of answer to his second question, I handed him a revolver with either hand" (14:79). Thus ends the dramatic tension in the Arabian-nights discourse of the "Doctor Moreau Explains" chapter. The first question had been "'What do you think?'" Moreau's "second question" had been: "'Are you in fear of me still?'" (14: 79). By returning the revolvers, Prendick is telling Moreau he is no longer "in fear" of him—but we ought to construe this narrowly: Prendick no longer fears, specifically, that Moreau is engaged in the vivisection of human beings. Certainly he continues to fear Moreau. He tells us he did: "I will confess that then, and indeed always, I distrusted and dreaded Moreau" (13:68). In deciding not to persist with an ultimatum, Prendick grants tacit approval to Moreau's continuing experimentation on the puma. Satisfied by the explanation that the victims are nonhuman animals rather than humans, he is unglamorously guilty of what we might now call speciesism. It is impossible to rescue him from the indictment of thus condoning, even if only temporarily, cruelty to animals.

But the condoning is in fact only temporary, for Prendick next acts decisively in defense of a hybrid beast-human. The narrative segment begins with the discovery of a bloody, half-eaten rabbit, which arouses Montgomery's fears; Prendick in turn reports having seen a mangled rabbit his very first day on the island, identifying the Leopard Man as the culprit in that earlier instance (16:87). Montgomery's fears grow: "'You see,' he said, almost in a whisper, 'they [the Beast People] are all supposed to have a fixed idea against eating anything that runs on land. If some brute has by accident tasted blood. . . .'" (16:88). Montgomery reports these facts to Dr. Moreau: "Moreau took the matter even more seriously than Montgomery, and I need scarcely say I was infected by their evident consternation. 'We must make an example,' said Moreau. I've no doubt in my own mind that the Leopard Man was the sinner'" (16:88). If the Beast People begin to hunger for meat and take each other as prey, the humans on the island will be in danger of becoming prey. The Leopard Man's fall into bloodthirsty animality, indicated by his killing of the rabbit, is his fall from the impossible-to-maintain mock-humanity Moreau has forced upon him.

What follows the discovery of the bloody rabbit, therefore, is a hunting party, one dramatically different from an earlier one in which Prendick was prey, the one that ended with his listening to the "Dr. Moreau Explains" discourse. As Prendick was earlier hunted by all when convinced that Moreau was torturing men (11.52–13.69), so now the Leopard Man is hunted by all. Before the Leopard Man flees in terror, a ritual scene is set up: in front of the gathered beast folk, Dr. Moreau plays God, deliberately cracking his whip and speaking in "a voice of thunder" (16:91) and interrogating the mass, demanding that the culprit identify himself. Prendick recalls Moreau enjoying the authority: "It seemed to me there was a touch of exultation in his voice" (16:91). Prendick attends to the evidence sufficiently to notice that the hyena-swine might also be guilty of the recent rabbit killings. By contrast, Moreau has his mind made up that the Leopard Man alone is guilty: "Moreau looked into the eyes of Leopard Man, and seemed to be dragging the soul out of the creature" (16:88). Prendick feels some pity for the Leopard Man. He is troubled by the Ape-man's gleefully malicious yelping "'Back to the House of Pain, the House of Pain, the House of Pain!'" (16:93). Prendick lets go of his self-preserving hostility: "When I heard that I forgave the poor wretch all the fear he had inspired in me" (16:93). Moreau still wants the Leopard Man taken alive so that he might make a tortured example of him and thus restore the Beast Folk's fear of meat-eating. However, that desire to engineer a successful sacrificial fury is frustrated. All those in the hunting party compete to corner the Leopard-Man, all against one. After he wins the race, Prendick defiantly chooses to kill the creature rather than let Moreau take him back to be tortured.

> It may seem a strange contradiction in me—I cannot explain the fact—but now, seeing the creature there in a perfectly animal attitude, with the light gleaming in its eyes, and its imperfectly human face distorted with terror, I realized again the fact of its humanity. In another moment others of its pursuers would see it, and it would be overpowered and captured, to experience once more the horrible tortures of the enclosure. Abruptly I slipped out my revolver, aimed between its terror-struck eyes and fired. (16:94)

This seems to me a case of deliberate mercy-killing. However, one of the more puzzling victimary-minded attacks on Prendick blames him for acting "spontaneously" here. I would counter that the text supports a claim that the quick decision is also a thoughtful one, based on a reasoned consideration of alternatives; there is nothing reckless, underthought, or casual in it.[10] Prendick is apologizing for the killing only as a possibly incomprehensible reversal of attitude, not for the ethical stance it presupposes: that is, he confesses (to us readers) a self-awareness that at the beginning of the ritual interrogation of the mass of beast folk, he had felt differently. At that time, he reports, he had felt stupid for ever having believed in the human-ness of the Beast People: "I looked round at their strange faces. When I saw their wincing attitudes and the furtive dread in their bright eyes, I wondered that I had ever believed them to be men" (16:90). Their fearfulness—suggested by the body language of "wincing" and by the "furtive dread" in their eyes—indicated the fearful cowering of animal creatures overpowered by Moreau. However, here at the end of the chase, Prendick realizes in the individual Leopard Man what he calls "the fact of its humanity." But the "strange contradiction in me" for which he apologizes is not really so strange. Given the earlier pronouncement, one need not believe Prendick has changed his mind: the Leopard Man remains more leopard than man, more animal than human. The quality that Prendick confers on the cornered, helpless creature with the attribution of *humanity* seems close, rather, to something like *vulnerability* or *the pitiable.*

That is to say, the humanity of the Leopard Man for Prendick at that instant is its seeming to merit moral reciprocity: Prendick will do for it what he would wish it to do for him if their positions and capacities were reversed. He believes the Leopard Man is better dead as an object of his memory, than alive as an object of the purposeless torture inflicted by

Moreau in the House of Pain (16:91). Wells ironically has Prendick here "play God" instead of Moreau: Prendick makes the life-or-death decision favoring euthanasia, a decision that usurps the absolute power to decide that Moreau would otherwise enjoy (a usurpation of which Moreau predictably in turn expresses his resentment). Prendick believes his intentions are good. It would be impossible to argue that the intentions of Moreau and his horde of persecutors are good. Although we have seen Prendick return the weapons to Moreau at the end of his self-explanations, here and now the newcomer to the island has differentiated himself from Moreau as one who rejects unjustifiable cruelty to animals.

This death, a death of one of the Beast People, has a far greater emotional impact on Prendick and moves him to far more searching reflections than any reflections ever reported by Moreau or Montgomery. The power of the mercy-killing to provoke reflection is explicable not simply as a result of Prendick's having been the one who did the killing, nor as a result of the facts that Moreau dies "offstage," his corpse discovered after the violent incident (18: 105) and after Montgomery receives his fatal wounds without Prendick present (19:110). Prendick recalls with clarity the death and burial of the Leopard Man because it is the only incident during his involuntary sojourn on the island when all—Moreau, Montgomery, the beast folk, and Prendick himself—all enact a performance of sacrificial violence. It is the one episode when the artificial mock-society of the island, its caricature of a culture, organizes into a purposeful mob that includes him. There is perhaps no clearer example of Prendick's ethical independence in the narrative: he differs from the whole community when he saves the Leopard Man from certain torture.

Now in the critical literature on Moreau, Prendick's reflections on the death and burial-in-water of the Leopard Man are frequently analyzed. There is the passage concerning "the realization of the unspeakable aimlessness of things on the island" (16: 95). There is the "Poor Brutes!" passage that I (too) have taken as exemplary of Prendick's ethic, contrasting their bestial life with their "humanized" existence trapped in the "shackles of humanity" (16:95). There is the complaint about Moreau not having any "intelligible object" in his experiments (16:95) and the passage about a "blind fate, a vast, pitiless mechanism" (16:96) seeming to Prendick to "cut and shape the fabric of existence" (16:96). There is a passage perhaps more frequently quoted than any other, however. Prendick is up high on the headland, looking down on the Beast People and their two human masters. The Bull Men have carried the dead Leopard Man's body out to sea.

> They [the Beast People] were all intensely excited, and all overflowing with noisy expressions of their loyalty to the Law. Yet I felt an absolute assurance in my own mind that the Hyena-Swine was implicated in the rabbit-killing. A strange persuasion came upon me that, save for the grossness of the line, the grotesqueness of the forms, I had here before me the whole balance of human life in miniature, the whole interplay of instinct, reason, and fate in its simplest form. The Leopard Man had happened to go under. That was all the difference. (16:95)

Why should this scene of excitement at a burial-in-water represent for Prendick "the whole balance of human life in miniature" and seem to him more typical of "human life" than the chanting of the Law in the caves or the interrogation of the amassed Beast People by Moreau? The passage speaks of an *interplay of instinct, reason, and fate.* The *instinct* is an instinct to survive, to pride oneself on one's survival, to believe oneself invulnerable. The *reason* is indistinguishable from submission to the "Law" that holds the pseudo-community together: in other words, the intense excitement that the Beast People are feeling is an undignified version of "reason." They have been excited by the death of the Leopard Man (he has died, but they live), and they reasonably believe the difference between him and them is owing to the "Law." But Prendick believes that inasmuch as the Hyena-Swine is as guilty as the Leopard Man, the Law in this case has done no justice at all: the Leopard Man's fate is that of an arbitrary sacrificial substitute. Therefore, the intense excitement ("instinct") of the Beast People, their "noisy expressions" of "loyalty to the Law" ("reason"), and the *fate* of the Leopard Man as one whose death represents only a tragic absurdity—together, those add up to a "miniature" of "human life." The interpretation of the scene as one in which the central victim is subject to fate approaches tragic despair, in keeping with Prendick's other near-hopeless reflections at this point in the text. He despairs at the irrational mob cruelty seemingly generative of the human-like social order.

And yet there is a significant residual hope, in tension with that despair: the "human life in miniature" here for Prendick is *not a biological phenomenon.* "Instinct" is there on the scene of culture; reason and fate are there on the scene of culture; but Prendick, unlike Dr. Moreau, does not reduce "human life" to species survival. Justice has failed here, but *it is the aspiration to avoid injustice that Prendick sees being mocked in the scene.* Therefore, for Prendick, the scenicity of the human is a moral reality; the mock-human here is a sad caricature of moral reality. Dr. Moreau is not troubled by the botching of his experiments because the creatures are not "perfect" humans. Prendick, however, will be haunted by his memories because he recognizes in the Beast People some artificially infused rudiments of the capacity for scenic exchange even as he senses the non-humanity of the Beast People. Even here, with the Beast People at their most sadly deluded, their attribution of the wrong meaning to the violent death of one of their own still counts as an act of giving meaning. Their striving for justice will always reveal nothing but the impossibility (they are impossible humans) of their having a communal existence independent and free from the impositions of Moreau, their scientistic designer who plays God the creator.

Hypnotism and the Unnatural Language of the Beast People

My defense of Prendick depends on our assent to the claim that the Beast People are not human as we are, but impossible humans. To distinguish Moreau's failure from that of Shelley's Victor is helpful here. Victor creates only one creature, but that one impresses us with "human" qualities far more persuasively than do all Moreau's Beast People put together. The monstrous Wretch possesses all that would make him human—an *innate* capacity for language, human desire, human resentment of the centers from which he is excluded, human esthetic intuition, need for communal belonging and economic participation—except that his overpowering ugliness forms an ineradicable barrier that excludes him, sacrificially, from participation in a human economy of exchangeable value. The Beast People, by contrast, have no *innate* capacity for language; they have no innate scenic desire or resentment apart from the brutal imposition-by-hypnosis of Moreau's "Law." They have no esthetic sense. They have no religious awe, no need for communal belonging mediated by a sacred center, no inclination toward economic exchange. Most decisively, the "language" of the Beast People has emerged only as the unnatural by-product of physiological scars inflicted by Moreau, scars later capable of naturally healing—unlike the spiritual wounds inflicted on Prendick, which do not fully heal. The effortless reversion of the Beast People to a speechless, lawless animality after the death of Moreau is evidence that their human qualifications are missing.

Consistently preoccupied with the fragile linguistic capacity of the Beast People, Prendick remembers his reflections on his first day on the island: "I wondered what language they spoke. They had all seemed remarkably taciturn, and when they did speak, endowed with very uncanny voices" (7:33). During his flight from Moreau after the change-to-human sound in the puma's cries, Prendick relies on the Ape-man for help, and seems willing to consider this creature human solely on the basis of his capacity for speech: "'You,' he said, 'in the boat.' He was a man then—at least, as much a man as Montgomery's attendant—for he could talk" (11:54–55). But the attempt at a meaningful conversation with the Ape-man fails: "I tried him with some other questions, but his chattering prompt responses were, as often as not, at cross-purposes with my question. Some few were appropriate, others were quite parrot-like" (11:56). Led toward the huts by the Ape-man, Prendick remembers, he had been hopeful; later, he would be undeluded: "I guessed . . . I might perhaps find them friendly, find some handle in their mind to take hold of. I did not know yet *how far they were from the human heritage I ascribed to them*" (11:55) [emphasis added]. When they fail to obey Moreau's command quickly enough to prevent him from escaping, Prendick describes their "bestial minds" as "happily slow" (12:62). When Moreau has handed the revolvers over and he, Montgomery and Prendick are heading back to the enclosure, Prendick studies further the bodily signs of mental activity in the Beast People: "They seemed, *as I fancied then, to* be trying to understand me, to remember something of their human past" (13:67) [emphasis added]. His fancy then was mistaken. The Beast People possess, Prendick will learn from Moreau's explanations, neither a human heritage nor a human past. The supposed "language" of the Beast People is not a form of voluntary exchange of symbols; it is, rather, an artificial effect of the experimental violence of their maker's hypnotism.

The Beast People's non-human status does not follow from a description of their "language" alone. Other components of the human in the company of originary language—resentment, desire, guilt, experience of the sacred, moral equality—are equally artificial impositions on their animality. Prendick witnesses the slow erosion of such components during the "reversion" of the creatures, experiencing the reversion as a relief rather than a disappointment. What do the Beast People desire? Whom do they resent? All of their communally mediated desires may be reduced to one: the desire not to return to the "House of Pain." Their behaviorist resentment of Moreau fades away once he is dead and absent. They do not naturally resent Moreau, for Moreau is not naturally their center of attention; naturally, animals have no centers of scenic ritual attention. Nor do the Beast People compete to take Moreau's place as humans compete for scenic centrality. Left to themselves, the Beast People just do not share in the center-periphery structure of human cultural experience. Their rebellion is no more meaningful than the mechanical war that Capek's robots will engineer to wipe out humans. The Beast People, just as they are happier without the "language" Moreau injects into them, are happier without the mechanical fear-of-pain "religion" that Moreau and his adjuncts impose on them.

The anthropology that we are trying out in this book is one founded on a minimal model of language events. Therefore, it is of particular relevance to our thesis that the failure of the mock-humanity of Wells' impossible humans is one with the falseness of Moreau's conception of symbolic exchange. Here is Moreau defending to Prendick his biologistic delusion that "the great difference between man and monkey is in the larynx" (14:73).

> In our growing science of hypnotism we find the promise of a possibility of replacing old inherent instincts by new suggestions, grafted upon or replacing the inherited fixed ideas. Very much indeed of what we call moral education is such an artificial modification and perversion of instinct; pugnacity is trained into courageous self-sacrifice, and suppressed sexuality into religious emotion. And the great difference between man and monkey is in the larynx, he said, in the incapacity to frame delicately different sound-symbols by which thought could be sustained. In this I failed to agree with him, but with a certain incivility, he declined to notice my objection. He repeated that the thing was so, and continued his account of his work. (14:73)

According to Moreau's description here, the purpose of human language is to have a certain effect on "instinct": somehow language lets its programmers or creators replace "old inherent instincts" with "new suggestions";

somehow language enables the "modification" or "perversion" of instinct. One form of instinct is "pugnacity," another is "sexuality." This structural opposition between language and "instinct" is already a clue to Moreau's biologistic fallacy: Moreau fails to differentiate between animal appetite (goal-seeking behaviors not mediated by the exchanged sign) and human desire (behaviors mediated by the scenic center around which reciprocal symbolic exchange is organized). This positing of a pre-linguistic, pre-cultural desire (prelinguistic desire itself an oxymoron) is wholly typical of the modernist esthetic, as Eric Gans has explicated its limits.[11] Dr. Moreau reduces the mystery of human language to a spontaneous side-effect of the purely physiological capacity in the larynx "to frame delicately different sound-symbols by which thought could be sustained." All mammals could speak like humans, Moreau implies, if only they had the organs in the throat capable of producing human sounds. Human language, on this account, is simply an epiphenomenon of instinct: it is different from barking, meowing, chirping, mooing, hissing, chattering and the like only because of humans' organic capacity to produce a greater variety of phonemes, humans' greater control over the physiological production of sound waves. It is difficult to believe that Wells took this idea seriously; Wells must have given some thought to the difference between human and animal minds. No wonder Prendick expresses skepticism, and no wonder Moreau can only "repeat," without giving reasons for, the idea.[12] Certainly, Moreau proves that the physiological capacity of animals to produce sounds can be surgically altered; and Moreau proves that animals can be hypnotized into a religion of fear-and-pain. But his reduction of human language ability to a physiological capacity corresponds perfectly to his mad scientistic reduction of the anthropological to the biological.

Prendick is horrified at the degeneration of the beast folk's speech—but not because the Beast People are *humans* degenerating and he wishes they would stay the same. On the contrary, he feels horror because it is *language roughly like the language of humans* that he witnesses degenerating.

> It was about May when I first distinctly perceived a growing difference in their speech and carriage, a growing coarseness of articulation, a growing disinclination to talk. My Ape Man's jabber multiplied in volume, but grew less and less comprehensible, more and more simian. Some of the others seemed altogether slipping their hold upon speech, though they still understood what I said to

> them at that time. Can you imagine language, once clear-cut and exact, softening and guttering, losing shape and import, becoming mere lumps of sound again? (22:122)

Notice the reduction of symbolic exchange to sound-production here: the "softening and guttering" and the "becoming mere lumps of sound" represent failures of surgically-created capacity "to frame delicately different sound-symbols by which thought could be sustained" (14:73). The physiological fallacy rests in the implication that "thought" follows automatically from the plastic-surgery creation of organic capabilities in the larynx, as opposed to "thought" being created communally in the scenic structures of symbolic exchange. We can almost hear the wounds and scars from Moreau's torturous surgery healing themselves naturally, and in that reassertion of "Nature," the Beast People are freed from the terrible necessity of fearing the pain of Moreau and being compelled to chant the Law.

Prendick's Farewell

From the fact that Prendick is a unique victim of traumatic experiences shared neither by Montgomery nor by Moreau, it does not follow that we must agree with his negative evaluation of the humans to whom he returns in England. The comparisons between Edward Prendick and Lemuel Gulliver that we are irresistibly led to make, while legitimately based in external evidence about Wells' general admiration for Swift and his particular debt to the model of *Gulliver's Travels,* mislead when not carefully qualified. One might wish that the critical tradition on the novel were a little more forthright in owning up to the problems involved in assessing Prendick's final state-of-mind. It should no more be easy to assimilate Prendick's final ideas to those of H. G. Wells, than it should be easy to identify the home-returned Gulliver with Jonathan Swift.

I believe we ought to emphasize the manifest differences between Gulliver and Prendick. Whereas Gulliver is blissfully ignorant of and oblivious to the ethical implications of the comically shocking disgust he feels for human bodies and beings, a disgust accompanied by his ludicrous love for the plain horses that he confuses with houyhnhnms, Prendick is aware of the moral problem in his involuntary disgust. Gulliver is imperviously proud of his rational contempt for Yahoos; Prendick is nervously ashamed of his visceral dislike of Englishmen and women. Gulliver enjoys the self-assurance of the "elected" neoclassical protagonist, but Prendick is burdened with his own insignificance.

> I could not persuade myself that the men and women I met were not also another, still passably human, Beast People, animals half-wrought into the outward image of human souls; and that they would presently begin to revert, to show first this bestial mark and then that. But I have confided my case to a strangely able man, a man who had known Moreau and seemed half to credit my story, a mental specialist—and he has helped me mightily. (22:130)

Prendick openly acknowledges the unreasonableness of his "fear" that humans will degenerate into Beast People. He confesses an inability to "persuade" himself; because self-persuasion fails, he seeks help from a "mental specialist" who proves unable to cure him of his delusions but helps him muddle through. Gulliver never even begins to believe he might need help. Ultimately, every passage in the novel's magnificent ending where Prendick reports his disassociation from human beings is a passage infused with qualifications. These qualifications of regret, embarrassment, apology and self-alienation differentiate his case from that of Lemuel Gulliver. Prendick in his "strange" condition—"strange" is one of his favorite qualifiers (21:124; 22:129; 22:130)—wishes he could feel otherwise.

The involuntariness of his final condition therefore prompts us to ask what we are to make of his final statements. First, we might judge that Prendick is to a significant extent the victim of Moreau, as a number of critics have argued of late (Showalter 82; Ruddick, *Ultimate* 66; Jackson 24). Oddly, it is a judgment neither universally passed not uniformly emphasized. Certainly, Prendick is an imperfect character, neither an Achilles in war nor a Christ in peaceful self-giving. He approves of vivisection under certain circumstances in a way that some readers will find unforgivably repugnant—but he openly confesses that (7:35; 14:73; 17:97–98). No wielder of a sword of self-righteous justice, he lacks a certain brand of chivalric courage—he confesses that too (8:38; 20:117). Just as certainly, Prendick is punished disproportionately for any imperfections he has: to condemn him from outside as a convert to Moreau's doctrine (as Stover does) or a willing collaborator with Moreau (as Gomel does) is shooting fish in a barrel.[13] Prendick's reluctant, disturbed, and self-critical submission to the intimidations of Moreau is morally blameworthy given a certain attitude; but only given an attitude determined to resent his narrating centrality. The only moment at which identification with Prendick should become impossible, I believe, is the moment at the very conclusion, when Prendick removes himself *involuntarily* to an impossible distance from us. That withdrawal is the source of the novel's tragic depth; its tragic depth is the only thing that moderates the otherwise scathing anti-human tenor of its contents. We need not accept his judgment that humans resemble the Beast People, for we have witnessed his sufferings from the outside. His judgment is a symptom of a psychological disorder; we need not accept the judgment to pity the disorder. Prendick is wrong. His wrongness is a side effect of the more terrible wrongness of Moreau's scientism.

In his farewell to the reader, Prendick claims that his mental health is improving and that the "mood" in which he felt like "an animal tormented with some strange disorder in its brain"—a concise description of the Beast People—affects him "more rarely" (22:131). He has not lost faith in human language and culture; on the contrary, he refers to "wise books" as "bright windows in this life of ours, lit by the shining souls of men" (22:131). His solitude is not, we do well to notice, total solitude: "I see few strangers, and have but a small household." He does see people; he does share a household. In the end, however, his "experiments in chemistry" take second place to his "study of astronomy."

> There is, though I do not know how there is or why there is, a sense of infinite peace and protection in the glittering hosts of heaven. There it must be, I think, in the vast and eternal laws of matter, and not in the daily cares and sins and troubles of men, that whatever is more than animal in us must find its solace and its hope. I hope, or I could not live. And so, in hope and solitude, my story ends. (22:131)

I believe it is a mistake to interpret this attachment to *the vast and eternal laws of matter* as Wells' sarcastic way of suggesting that Prendick has been seduced into submission to Moreauvian doctrine. On the contrary, Prendick's openly conceding the ineffability of the "how" and "why" of the "sense of infinite peace and protection" differentiates his attitude from the scientism of Moreau. The motive and purpose of his studies, their giving him "hope"—a word repeated thrice in the passage, hope without which he "could not live"—differs from the species-altering, obsessive utopian motivation that drives the research program of Moreau. Certainly, it matters that, as John Reed has demonstrated, the stars were one of Wells' most favored images of serene faith in the human future, an image by which he expressed trust in the cosmos as being a fit habitation for humankind despite its vastness. It is in "laws" that Prendick believes he might find hope—"laws of matter" which certainly suggest astronomy and physics rather than biology. There we find the move Prendick has made: he does physics now, but not biology. He does not move away from delusions about the biological *human species* toward truths about the anthropological *human being;* he moves instead away from the biological altogether, toward the chemical and physical.

The stars with their demonstration of *the vast and eternal laws of matter,* however beautiful those stars or laws might be, are a million miles away, literally and allegorically, from the mundane human exchange of signs and goods. It takes no genius to see the improbability, if not the absurdity, of translating such "laws" into something that might give us "hope" in human affairs. Many cases of scientistic obstinacy demonstrate an obliviousness to that improbability. Prendick's farewell does not recall the intrusively reductionist materialism of Moreau. But it does mark his sad collapse into a certain evasion of human reality. In Wells' text at this point, we should feel the invitation to to sacrifice our identification with Prendick to our sense that we as readers cannot take him, any more than we can take Moreau, as a defender of human community. In letting Prendick go, we reaffirm something of our humanity as readers: we reject his belief in the resources of solace that astronomical matter presents.

Moreau's most telling victim (pun intended) is the human person who has most intimately witnessed the horrible effect of his experiments. This person, this Edward Prendick, has been traumatized to such an extent that he remains unable to re-join fully the very human community that his narrative does so much not only to shock and scandalize, but also to reassure. Prendick's testimony may reassure us, because of his self-differentiation from Moreau, that the One from whom humankind differs at its ontological origin has nothing whatever to do with the Dr. Moreau who tries to play God. Prendick's testimony may reassure us as well that, despite the initial shock and despite the trauma he has suffered, we humans need not fear that we are nothing but something like hypnotized Beast People. But the price of this reassurance is our loss of contact with him: for to be able to suggest this, he has had to misrecognize his own truth. We are invited to recognize the anthropological truth that the story of Prendick represents more clearly than he himself recognizes it. Humankind is not the ultimate object of scandal in the text. Rather, the strange violence done to the resisting, tragic Prendick is the true scandal of Wells' instantiation of the Frankenstein myth. It is a fiction of a scientist who plays God by trying to torture into dissolution the difference between animality and the human.

FOUR

Karel Capek's *R.U.R.* (*Rossum's Universal Robots*) Mechanical Not Erotic

R.U.R. as Rebellious Heir to *Frankenstein* and *Moreau*

After enduring the horrors of *Frankenstein; or, the Modern Prometheus* and *The Island of Dr. Moreau,* to move through the ninety-odd pages of Karel Capek's dramatic script for *R.U.R.* (*Rossum's Universal Robots*), a play first performed in January 1921 in the National Theater in Prague to resounding applause (Bradbrook, *Karel*) and still occasionally performed today, is to meet a writer whose lightness of touch offers a pleasant change. The knitted-brow melancholy of Mary Shelley and the fist-making pugnacity of H.G. Wells give way to a witty cheerfulness. Karel Capek (1880–1938) was the most beloved Czech literary artist of the first half of the twentieth century, "a great artist who has to be reckoned with as one the major figures of contemporary [1936] literature" (Wellek 46). According to Ivan Klima, "Through his remarkable creative output, Capek became the leading representative of democratic Czech culture" ("Introduction" x). Darko Suvin recognizes Capek as a major figure in the history of science fiction as well: "Capek is—together with Eugenij Zamyatin—the most signficant SF writer between the World Wars"(*Metamorphoses* 280). Compact but powerful, *R.U.R. (Rossum's Universal Robots)* re-works the Frankenstein myth so as to get us *laughing* at scientism's lack of Promethean foresight and its Pandora-like power to unleash the incalculable. Meanwhile, Capek carries out the serious task of rearranging the relative weights assigned to the essential components of the human in the myth. There is nothing in *R.U.R.* as irreversibly cruel as the unforeseen ugliness of the Wretch in *Frankenstein.* And in place of Wells' implied historical past where humankind is helplessly cast back into a jungle, Capek substitutes a vision of humankind cast forward into a swirling vortex of consumerist self-destruction. In this experimental drama mixing Ibsenite drawing-room dialogue with expressionist techniques,[1] perhaps the most inventively fanciful of Capek's variations is the swift efficiency of the disaster itself. There is no laborious digging in graveyards here, no graphic series of bloody trials and errors. The species extinction that is merely a distant threat in Shelley and Wells gets completed in the few hours it takes to perform the script.

Far more spectacularly than Shelley or Wells, Capek puts on stage the apocalyptic usurpation of a rival species. He also gets us wondering with an uneasy curiosity about the eschatological force of technological inventions wedded to market freedom. He imagines the world ending, decades before the detonation that started the nuclear age and decades before apocalyptic discourses of ecological collapse entered the public sphere. Victor feared the monster and his mate would reproduce and spawn a whole race to wage war on humankind; recall the belated cost-benefit analysis that legitimated Victor's taking on the role of mock-saviour. One of Capek's parodic moves in *R.U.R.* is this: the businessmen who run the Rossum's factory do many such cost-benefit analyses, but their accounting fails to save the human from annihilation because it does not include a thoughtful, let alone correct, analysis of *what the human is and does.* The tragic uselessness of Shelley's Wretch gets converted into the over-usefulness of a bazillion inexpensive mechanical domestic servants. Capek gives us a hint: maybe we should fear vain businessmen as much as vain scientists. What is a mad scientist stuck alone in a laboratory or on an island, deprived of distribution, and consumer demand for his irresistibly unnatural products? It is one thing for Victor Frankenstein's pale, privileged European family to be selectively executed, another thing for humankind to be annihilated because of bad business planning and crazy consumer behaviour.

Consider Victor's fear that monster and mate might possibly have children. Capek goes ahead with that dreaded multiplication of the impossible human. Perhaps he has grasped that to concede the absurd doubling of the origin of humankind by technological device is already to concede the triviality of humankind. We might as well get cancelled out anyway, if our origin is reducible to a tawdry chemical-mechanical process. If the true beginning of world history can be reproduced experimentally, then history is absurd; if history is absurd, then it is not so unbearable that it should end. The lengthy middle section of the play after the Comic Prologue and first intermission, Acts One and Two, shows us the members of the Rossum's Universal Robots management team gathered in the sitting room of Helena (maiden name "Glory"), for ten years now the wife of their boss Harry Domin. The people gradually accept their coming deaths and exchange opinions on the end of civilization. They take the blame (most of them), another reason for the text's lightness: the audience wastes none of the energy that readers of Shelley and Wells expend on moral exasperation, on resenting the evasions of Victor and Moreau. Even though the men on the Rossum team have played God, they are somehow understandable; they make an effort for each other to explain their situation. At the end of Act Two, some robots crawl through the window at the back of the stage and stab to death—stab in the back—Dr. Hallemeier, head of the Institute for robot education, a large red-haired man who is caught off-guard while piling furniture up against the door. The other humans have already exited by other doors. Robots onstage soon report those others are "done"—killed (Capek, *R.U.R.* 2: 69–70).[2] In Act Three, one human being, the bricklayer Alquist, remains on stage, economically useless like Victor's monster, unable to perpetuate either his own species (no women survive) or the robot "species" (the manuscript containing the secret formula for robot creation has been burnt in the fireplace).

Recall that Dr. Moreau dreamed of a utopian species beyond pain. One of Capek's jokes at Wells' expense is to permit a luxury for young Rossum, son of the original scientist Old Rossum, who takes over his father's laboratory. The son enjoys the luxury of an effortless ease in his creation of a type of impossible human totally free of pain from the outset: what Moreau complains about finding impossible to accomplish, Young Rossum achieves with expressionist speed. No vivisection of already-living creatures is necessary to start up the mass production of perfectly pain-less laboring machines. Whereas Moreau, a demented sadist, broods alone, the cheerful team at R.U.R. enjoy fame and huge wealth: "the whole world worshipped you" (1:30), as Helena reminds Domin. Domin excels, for a while, at playing God. Robots are a market sensation. For robots, here in their originary dangerous incarnation as in their later friendlier forms, quite unlike the solitary monster of Frankenstein and the misbegotten hybrids of Moreau—robots are guaranteed to be useful. The android automata of young Rossum, prior to being altered by Dr. Gall, the team's head physiologist and research director, are practical, cheap, efficient, safe, indifferent, obedient—so obedient that it is impossible for them even to conceive of disobedience. Old Rossum—who, unlike Young Rossum, declared his full-on desire to play God—was an academic, a maker of botched artificial humans who belonged in a university, unsuited to the world of commercial enterprise (Pro. 8). Young Rossum makes humanoids only to make life for real humans easy (and so to make money). Properly served by robots, people no longer suffer in jobs they hate. They do not need jobs at all. (What would happen to politics in our time if elected leaders found themselves no longer obliged to promise to put the economy and *jobs, jobs, jobs* first? Imagine.)

Meanwhile, continuities in the Frankenstein myth persist. Just as Capek remains faithful to Shelley's model in not giving the robots any sexual reproductive capacity, he follows Wells in that he centralizes the madness of the dream of a species freed from pain. What the curse of pain was to Wells' Moreau, the curse of labour is to Capek's vain scientist. The vanity belongs to Domin. Domin dreams as Moreau does of being responsible for the creation of a new species of superhumans: he calls them "Supermen" (2: 54).[3] Forget about biological evolution, let us look at the history of labour instead, hints Capek. No torturous exploration of animal physiology is necessary; let us imagine instead the commercial multiplication of a supreme human-shaped machine to transform the world economy forever. The thing that the robot does—Capek's brother came up with the neologism, "from the archaic Czech *robota,* meaning 'drudgery' with strong feudal connotations of the serf's compulsory work on the master's property" (Suvin, *Metamorphoses* 270)[4]—is work. The robot means freedom for humans from the necessity of physical labour. Rather than the enemy "Nature" being experienced as the revelation of an animal ancestry embedded in a crushing cosmological temporality (Wells) or "Nature" being experienced as a cosmological void by the excessively ambitious romantic experimentalist (Shelley), "Nature" in *R.U.R.* is believed by Domin to be that which condemns humans to demeaning labour (Pro. 21; 1:30; 1:46; 2:54). It is the resentful desire of Domin to create a paradise in which the history of necessity in labour has been abolished and thus to create indirectly, as a side-effect of this abolition, human beings who have nothing to do with their mortal time but "perfect themselves" (Pro. 21) or "perfect [their] being" (Pro. 21). Capek creates a free-market utopian who is as fanatically mistaken as the most deranged communist ideologue. For Dr. Moreau, the religious delusion one must eradicate (a scandal for the pious) is the denial of our animal ancestry and the thinness of the veneer of civilization. For Domin, one must eradicate the moribund morality that sacralises work: Domin hates the curse that condemned the human to earn its bread by the sweat of its brow and bring forth its offspring in the agony of labour pains (Genesis 3:16–19).

To Believe in an Economy of Mechanical Value

There is a preoccupation in *R.U.R.* with labour. In terms of generative anthropology's model of a minimal language event, the component of the human that preoccupies Capek in this play is *value,* the economic category of the exchange of things. It is no coincidence that Wells' detour into beast-people has spawned few imitations, while twentieth-century stories of robots are innumerable. As Gans among others has argued, the modernist esthetic is a response to the rise of consumer society and to the decisive triumph of the bourgeois marketplace, the liberal democratic dethronement of ritually protected centers of attention. *The Island of Dr. Moreau* returned us to the isolation of the Robinson Crusoe myth and mixed it with the ethos of *Frankenstein* to make a point about the revelations of Darwinian theory. But in *R.U.R.,* despite the re-iteration of the English tradition of an island setting for the mad scientist, Capek is concerned with models of production and consumption, with the struggle between the perennial desire for political freedom and the universal resentment of economic necessity. Unlike Wells and Shelley, Capek is not animated by resentment of God (Matuska 103). My thesis is that in this originary robots play, Karel Capek aims to administer an ethical shock to those secular political believers who have a scientific faith in political solutions to dilemmas of human suffering. He directs his satire at spectators who believe in an economy of mechanical value.

What do I mean by "an economy of mechanical value"? To sketch this idea along the lines suggested by the originary hypothesis, we need to return to the hypothetical event in which the human and God (the center of the scene conceived as a Being) emerge together; and we need to recall Gans' account of the originary experiences of the sacred and the esthetic. The communal exchange of words, of signs that intend the central object as something re-presented as desirable rather than merely appropriated as edible, generates a minimal experience of the sacred. On the scene of representation, peripheral humans all equally resent the central object for seeming to withhold itself. The human self is experienced in its resentment of the sacred center. The exchange of ostensive signs generates also the minimal esthetic experience. The minimal esthetic experience is the oscillation of our attention between now the sign itself (gesture or "name") as a re-presentation of the object, a representation that defers our appropriative violence; and then the object as something made-significant by the sign, transformed into a figure of the desirable—oscillation to and fro. Desire (as distinct from appetite) is a result of the scenic activity of symbolic exchange, peripheral humans exchanging signs that transfigure the central object from a material thing indexically-designated to a center of paradoxically unstable attention, symbolically named. In esthetic experience, the center (or the central object) is simultaneously resented-as-desirable (as something recognizably inviolable) and desired-as-represented (as something one can imaginarily possess).

In the sacrificial consumption that completes a minimal language event, we consume that which we signify and hold sacred, having divided it up in an exchange of things. The division of the central object returns us to the concrete world where we satisfy our bodies with food and drink, cover ourselves with garments and ornaments, make shelter from storms and weapons against enemies. Sacrifice requires tools and technology, not just language. During the sparagmos, each participant considers his or her individual portion of the previously-inaccessible object exchangeable with all the others. The acceptance of the portions of the divided-up object *as equivalent* is the originary experience of *economic* evaluation. Parts are compared to the (remembered) whole and compared to each other. The sparagmos, recall, is thus the opening to scientific quantification. But for our purposes as readers of Capek's version of the Frankenstein myth, the essential point to recall is this: *this exchange of things takes place only in the scenic space that contains the exchange of signs, and thus the memories of sacred and esthetic experience.* Before there can be an economic exchange and consumption of "things of value," there needs to have been a human exchange of signs that represented the object as sacred and esthetic.

As long as we consider economic objects "meaningful" in the sense that the exchange of words and things is chosen and intended, then our economic activity will be carried out with some consciousness of the model of the exchange of signs. By contrast, a belief in an economy of purely mechanical value—hospitable to scientistic doctrine—assumes that such choices and intentions may be explained according to a non-human level of operation. Religious and moral, esthetic and ethical considerations are irrelevant. It matters only that physiological (biological) needs be satisfied. At the level of seeking knowledge, meanwhile, that we are not consciously aware of the reality of this or that genetic or neuronal or materially calculable "operation" in a given instance of physical exchange does not matter. In making measurements in an economy of mechanical value, scientism does not care what we think we know thanks to introspection, common sense, tradition, intuition, or "folk psychology." Likewise, according to a belief in an economy of mechanical value, the values in an exchange can be calculated in terms (for example) of the quantifiable pleasures and pains which the utilitarian philosopher may, with sufficient information and skill, be able to figure out. The values may be calculated in terms (for the Althusserian Marxist) of the surplus labour extracted by an exploiting capitalist from an exploited worker-victim.[5]

Why must such mechanical reductions be dehumanizing? Because the human is co-extensive with the transcendent, with the presence of sacred centers.

> However obvious it may appear to us that value is constituted in the exchange process itself by the interaction of supply and demand, in all premodern societies, value is established not at the periphery, in the process of exchange, but at the center—at first purely ritual, then, in hierarchical societies, at the same time ritual and political ("Culture vs. Exchange").

I do not underline Gans' assertion that premodern societies establish value at the ritual center in order to suggest we should return to premodern forms of social organization or to reject modern science as the enemy of the harmless primitive or the dethroner of the noble savage. On the contrary, generative anthropology proposes that we affirm the greater omnicentricity and freedom of exchange at the periphery characteristic of modern economies. *Value is constituted in the exchange process itself* assumes, however, the relevance of the question in the first place of what is being supplied and what demanded. For real humans, sacrality and significance, not just nutritive or materially quantifiable value, are being supplied and demanded in any economic exchange. Capek knew that to play God the creator by trying to make the world—including the humans in it—into things mechanical is to abolish the human altogether. Where nothing (no thing that might be exchanged) is sacred, significant, esthetic, or erotic, there are no humans. If we abolish labour, we have abolished all sacrality and significance and desirability.

Experiences of the object as sacred and experiences of the object as beautiful are never useless, for such experiences defer the violence that would consume the appetitive object too soon. They add value to the appetitive object by making it desirable in social terms rather than simply necessary-to-consume in material terms. Humans value food for its esthetically mediated taste and the pleasures of its presentation at the dinner table, not only because it contains nutritive value. Human societies engage in competition, but to explain the history of such competition as determined by material considerations such as (for example) the quest for digestible protein, is to explain by means of an economy of mechanical value (see Marvin Harris, *Cannibals;* Gans on Harris, *End* 73–97). Humans value erotic partners for their beauty and character and "secret" charms, not merely for the invisible force of the gene that contains the seed of reproductive success, that value so beloved of sociobiologists, which the purportedly erotic partner communicates by means of indexical signals (the broad shoulders or big hips, the strut, the smile, the come-hither look). Esthetic and erotic responses do not translate into a rational model of human praxis; the effects of artistic or erotic performance are unpredictable. Originary thinking defends the principle of the priority of desire over cognition. Therefore, the object valuable-to-humans will always carry some residue, however minimal, of the "irrational" sacred and the "impractical" esthetic, because the object valuable to humans enters the scene of economic exchange already imbued with some quality of transcendence.

Karel Capek has the various creators of the robots in *R.U.R.* represent in delightfully vivid, often comical fashion, various forms of the deluded worship of an economy of mechanical value. For Harry Domin, for Dr. Gall the head of the physiological and research division, for Dr. Hallemeier, for Fabry the engineer, and for Busman the lawyer, the clean severing of the economic from the sacred, esthetic, and erotic is done with little thought. They each in different ways dismiss the institutions of the sacred as wasteful misallocations of precious resources better managed by rational economic decision makers.[6] (Alquist is an exception.) Likewise, they mostly demote art as useless or impractical, an impediment to the real business of satisfying material needs. An economy of mechanical value leads us to be oblivious to the full range of human values; it degenerates into a foolish flourishing of the inhuman. Just as it is impossible for a human to imitate the inimitable originary Being to whom we have historically given the name-of-God, it is impossible for the freedom and equality we exercise in the exchange of signs to be concretely incarnated in the exchange of things. Such a concretization could mean only the abolition of the difference between words and things. The robots remove the necessity of labour and deliver that freedom and equality. The utopia they deliver is a disaster. The price of the abolition of the difference between words and things is the annihilation of difference between the human and the Divine, center and periphery. In *R.U.R.,* Karel Capek represents in fantastic dramatic form an annihilation of such fundamental differences.[7] For the android automata, nothing is desirable, let alone sacred. Humans, relocated into the position of their enemies, are certainly not sacred.

I will elaborate on these theses in an analysis of the text of R.U.R. in three segments. As most commentators on the play have correctly noticed, the text baffles us with sudden changes in the behavior of the robots. Indeed, the changes are so remarkable as to make the robots in the Prologue require classification as a fundamentally different type of creature than the lover robots (for example) in the final scenes of Act Three. I will argue that the pure mechanicality of the robots in the Prologue baffles Helena Glory the human rights activist in such a way that Capek foregrounds the essential link between human being and erotic exchange. Those original robots, impervious to pain and indifferent to death, I name *android automata*. A second section will examine the new type, which I name *altered androids:* the intermediate machines who have been secretly altered by Dr. Gall—given a capacity for "irritability"—at the request of Helena. The capacity to resent a single difference (labour / idleness) generates their war against humankind. A third section analyses the big differences made by Primus and Helena, the final pair of "robots," whom I will classify as *primary humanoids.*

Android Automata and the Comedy of Baffled Activism

The Comic Prologue of the play may be divided into five segments of action: a brief moment during which Harry Domin dictates letters to his secretary before the announced entrance of Helena Glory; the history lesson, during which Domin informs Helena about the origin of robot production; the demonstration, during which Domin orders the android automata Sulla and Marius to answer questions so that they might prove to Helena that they are not human; the group interview, during which

Domin gets help from the other directors in convincing Helena that robots are not humans; and the marriage proposal, with Helena and Domin alone on stage before the other directors return with lunch prepared. Capek names this part of the script a "Comic Prologue" both because Act One begins ten years later and because by then (tragically) the marriage has turned out to be childless and the robots worldwide have rebelled. My argument in this section on the android automata is that Helena is mistaken in her description of the automata but the Directors are correct: the Prologue's comedy follows from Helena's almost involuntary inability to accept robot being as non-human. There is a legitimate objection to the robot industry in Act One, but Alquist makes it, not Helena: the claim that the value of the work performed by real humans is being lost as people indulge in leisured idleness and robots do everything for them. The absurd law defeated when Helena seems to accept Domin's proposal of marriage and the audience applauds, as at the end of a comedy, is the principle of victimary politics that founds Helena's error: an absurd belief that would dissolve the difference between android automata and humans in a vain desire to humanize the poor exploited automata. It is impossible to exploit that which cannot be victimized.

The opening interview between Domin and Helena begins in a comically mechanical fashion. Because all of Europe has been talking about the robots, because everybody is buying and using them, Helena Glory, like all visitors to the factory, wants to see close-up for herself the "secret" way the robots are made. When Domin repeatedly finishes her sentences, he is performing a "mechanical" act that treats her as if her questions are identical to those of all other visitors to his office who arrive bursting with curiosity about robots—because, in fact, all such visitors are very similar and her activism does make her predictable. Domin is correct about the ordinariness of the anxiety that Helena the activist wrongly takes to be profoundly exceptional. Her desire to be given the privilege of seeing "the factory production of people," which implies her desire to be the liberator, is ordinary. It is the rule; romantic revolt is the norm. Capek opens his comedy of the baffled activist with an exposure of the oxymoronic desire to be the universal exception. The mimetic mirror image of a self-righteousness that wants to save the world by ruling it is a self-righteousness that wants to save the world from being ruled.

Nevertheless, the routine that Domin has learned to follow with other activist visitors, of which there have been many (Pro. 17), is broken by erotic interruption. In the course of their exchange, Helena removes her veil and shows her beautiful face: when Domin is struck by love at first sight, Helena does after all turn out to be an exception, inasmuch as her beauty affects him in a way that no other woman's has.

> HELENA: This is d-r-readful! Marry some female Robot.
> DOMIN: A female Robot is not a woman.
> HELENA: Oh, that's all you want! I think you—you'd marry any woman who came along.
> DOMIN: Others have been here, Helena.
> HELENA: Young ones?
> DOMIN: Young ones.
> HELENA: Why didn't you marry any of them?
> DOMIN: Because I've never lost my head. Until today. The moment you took off your veil.
> HELENA: I know. (Pro. 23)

The neoclassical trope of being elected to suffer love at first sight has more power to upset than does the romantic trope of the liberating rebel. The trope of love at first sight has consequences for the script: Domin provides a historical account of the robots only because he wishes to delay the obligation to share Helena with others. In somewhat the same way that Dr. Moreau's explanations of his vivisections must not be detached from the violent context of Prendick sitting across the table armed with two pistols, Domin's account of the creation of the robots by Old and Young Rossum must not be detached from the erotic context of his wooing of Helena. In short, he narrates because he wishes to linger alone with her, not take her on a tour of the factory—thus his exasperated aside, "Damned factory!" (Pro.5). His factory has become his rival. Unlike Victor and Moreau, Domin shows himself interested in erotic exchange and the private sacred. His self-demonstration counts as preparation for the play's later elegiac preoccupation with the end of intergenerational exchange.

Capek deploys ironic allusions to Shelley and Wells in Domin's history lesson to Helena. Domin closely identifies with Old Rossum, the scientist who played God, even in mocking him. It is to Old Rossum that the company owns the "secret of production." He is the founder, the discoverer of the gold, the striker of oil. When, in Act Two, Helena burns the crucial manuscript that could perhaps have been traded with the army of robots in exchange for human survival, she is burning the work of Old Rossum, not the work of his son (2:64). What is the content of the "secret of production"? One of the most enduring ironies of the fame Karel Capek achieved with *R.U.R.* is the fact that a play he intended as an attack on the idolatry of machines became a source for the primary symbol—the robot—of the hopes of those technocratic enthusiasts who dream of really creating a perfect mechanical humanoid. That is, Capek wanted robots to represent an attack upon reductions of the human to the mechanical, but the robot evolved into a symbol of the desirability of the mechanization of the human.

In a 1935 text, after the proliferation of metal-mechanical robots in film and literature, Capek felt compelled to compose a piece titled "The Author of the Robots Defends Himself." In it, he protested that he could not "be blamed for what might be called the worldwide humbug over the robots." Speaking of himself in the third person, he disowns any parental connection to the "plate-metal dummies stuffed with cogwheels, photovoltaic cells, and other mechanical gizmos" that one sees everywhere. Remembering the writing and staging of *R.U.R.* fourteen years before, Capek argues: "For his robots were not mechanisms. They were not made of sheet-metal and cogwheels. They were not a celebration of mechanical engineering. If the author was thinking of any of the marvels of the human spirit during their creation, it was not of technology, but of science." Capek goes so far as to express "outright horror" at the "thought that machines could take the place of people, or that anything like life, love, or rebellion could ever awaken in their cogwheels. He would regard this sombre vision as an unforgivable overvaluation of mechanics or as a severe insult to life." He insists that the "mystery of life" was not one with which he wished to deal "so frivolously" as to give to mere scientists "the ability to produce, by artificial means, a living cell in the test tube"; he claims that he "would regard it as an act of scientific bad taste if he had . . . created life in the test tube."

Now, there are two distinct claims in "The Author of the Robots Defends Himself." There is a verifiable and justified assertion (first) that his originary robots were not composed of metal, cogs, wires, and batteries. There is a much more problematic claim that in his literary invention of the robot, he did not provide a model for mechanical humans. I will support the first claim, but reject the second as a disingenuous evasion.

Let me defend the first, that his robots were quasi-organic. In the 1935 piece Capek offers a careful and lengthy description of the curious "living" status he had in mind for his robots. The description should not have been necessary inasmuch as Harry Domin's account of Old Rossum's discovery (as narrated to Helena) is as clear as can be. It is a part of the text that later appropriators of the robot symbol in the years leading up to 1935 would choose to ignore or happen to overlook.

> DOMIN: [Old Rossum] moved away to this remote island to study marine life, period. At the same time he was attempting to reproduce, by means of chemical synthesis, living matter known as protoplasm, when suddenly he discovered *a substance that behaved exactly like living matter although it was a different chemical composition* [emphasis added]. (Pro. 6)

The point my added emphasis tries to bring out is that the android automata, the altered robots and the primary humanoids (all the robot types in the text) do *not* have the composition of living matter that other organic life on earth has. Old Rossum does not discover "living matter" as modern biology about life on earth knows it; rather, he synthesizes "a substance that behave[s] exactly like living matter" without being living matter. I will use the term *quasi-organic* to differentiate it from organic matter as we know it on planet earth. For Helena's benefit, Domin quotes from Old Rossum's scientific diary, a portion he has memorized and can rattle off in a "mechanical" fashion: "'Nature has found only one process by which to organize living matter. There is, however, another process, simpler, more mouldable and faster, that nature has not hit upon at all. It is this other process, by means of which the development of life could proceed, that I have discovered this very day'" (6). Old Rossum here reports having discovered something that Nature herself had not afforded us: *Nature has found only one process.* The superiority of the mad scientist over Nature herself as a discoverer of things implies the assimilation of "Nature" to "the history of the evolution of organic life." His science has discovered something Nature herself did not; no wonder then he believes he can play God, having outdone God's Creation. We need to grasp that Old Rossum has discovered an alternative possible world of life, a partial parallel evolutionary universe, and that he has imported something of that evolutionary universe into ours. I cannot help but wonder if Capek here is not making another joke at Wells' expense. Whereas Moreau had to sweat and grunt within the framework of Darwinian evolution, stuck with materials provided by the real evolution of life on our planet (his mammal victims), Capek's Old Rossum steps outside terrestrial evolution altogether to find a "process, by means of which the development of life could proceed," one quite unlike the cosmic process thus far revealed by Darwinian Nature. Old Rossum therefore plays God in a far more spectacular manner than Victor or Moreau. Victor discovered only the secret of the generation of *life as we know it;* Old Rossum has discovered the secret of the generation of a never-before-known-in-the–universe form of life, an alternative quasi-organic matter.[8] So the android automata in the Comic Prologue, even after being re-designed radically by the Young Rossum, remain examples of "Man made from a different matter than we are" (6).

This "different matter" poses, however, a logical problem for the thesis of "The Author of the Robots Defends Himself." Capek's second claim—that he did not provide an originary model of the mechanized human—is not defensible. The assertion that the matter from which his artificial humans are composed is "organic" will prove unable to do the rescue work he wishes it to do. The form of "organic" chemicals from which Old Rossum draws creates a "life" that lacks any will to live. Can life lack a will to live and yet be life? Curiously, it is a form of "organic" matter that has no location in the ecological food chain: the robots can eat, digest and find nutrition in almost anything, as Hallemeier informs Helena: "That's very nice, Miss Glory, but nothing makes Robots happy. Dear God, what would they buy for themselves? You can feed them pineapples or straw or whatever—it's all the same to them. They have no interest in anything, Miss Glory. By God, no one's ever seen a Robot smile" (Pro. 18). Furthermore, this quasi-organic material may be fashioned into humanoid body forms that need not possess pain-sensitive nervous systems. As Dr. Gall explains to Helena in the Prologue, that is a problem and they "must introduce suffering": "For industrial reasons, Miss Glory. The Robots sometimes damage themselves because nothing hurts them. They stick their hands into machines, break their fingers, smash their heads, it's all the same to them. We must give them pain; it's a built-in safeguard against damage" (Pro.19). Finally, organic matter that does not have the ability to reproduce itself sexually or asexually is rather bizarre. It is life therefore incapable of participating in any evolutionary history of its own, inasmuch as evolution requires at a minimum that creatures that reproduce can also vary. The robots are no more capable of "natural" reproduction and morphological variation than statues, dolls, clocks or automobiles. Therefore, Capek may in his 1935 "Apology" name the imaginary substance discovered with Old Rossum's "organic matter," but such a label does not remove the mechanical quality of their being from his robots: it is wholly appropriate that his literary imitators made their successor robots out of copper, bronze, rubber, vinyl, and electrical components, because the organic matter of the robots in *R.U.R.* never really bore the features of life. Capek's originary android automata might as well have been made of metal or plastic. That they have humanoid body shapes and move and speak in a human-like way does not make them terrestrial organisms as we know then any more than the "plate metal dummies" being built today by robotics experts are living.

Regardless, neither of the impossible humans whom we see on stage in the Comic Prologue is a creature following the pattern set by the Old Rossum who unequivocally tried to play God and make humans. Domin calls Old Rossum a raving lunatic who "actually wanted to make people" (7); Helena counters, "But you do make people!" In response, Domin explains the guarantee that the robots they make now, after the design of Young Rossum, are not humans. Here the dream of the mechanical human, in all its gloriously potent absurdity, emerges. The excerpt is worth quoting at length.

> DOMIN: More or less, Miss Glory. But old Rossum meant that literally. You see, he wanted to somehow scientifically dethrone God. He was a frightful materialist and did everything on that account. For him the question was just to prove that God is unnecessary. So he resolved to create a human being just like us, down to the last hair. Do you know a little anatomy?
> HELENA: Only—very little.
> DOMIN: Same here. Imagine, he took it into his head to manufacture everything just as it is in the human body, right down to the last gland. The appendix, the tonsils, the belly button—all the superfluities. Finally even—hmmm—even the sexual organs.
> HELENA: But after all those—those after all . . .
> DOMIN: Are not superfluous, I know. But if people were going to be produced artificially, then it was not—hmmmm—in any way necessary–

HELENA: I understand.
DOMIN: I'll show you in the museum what all he managed to bungle in ten years. . . . (Pro. 7—8)

Capek may be poking fun here at both Victor and Moreau by bringing into the sunny daylight of Domin's tour-guide discourse their brooding, only-implied rivalry with the Biblical God. What Victor and Moreau will not express outright, Old Rossum declares in his diary as an explicitly intended policy: the hubristic literalism, the scientistic fantasy, the project of the technological homunculus as the expression of a rivalry with God. The science of revolutionary atheism comes into its own (Old Rossum), only to have the work of its unique "genius" transformed by engineering and absorbed by market exchange, thanks to a patricidal producer of consumer goods (Young Rossum). This passage indicates the way that Old Rossum's playing God leads us, more importantly, into a space where the question of the difference between real and impossible humans is inseparable from the "delicate" question of erotic exchange. The passage ends with Domin's assertion that for artificial humans, sexual organs are not *necessary,* which awkwardly implies the corollary that sexual organs *are* necessary for real humans. Must we be reminded that sexual organs are necessary if the community of real humans is to continue?

Yes: in a world where robots are being mass produced, the economy of human sexual reproduction has a new default value. Capek opens up an ironic textual current that operates at many levels. Playing God the enactor of human origins entails ceasing to value the human as erotic object. At the level of theatrical action, Domin and Helena are real humans at this instant on stage: we witness their sudden consciousness of their own (obvious) possession of sexual organs, but the sensation and significance of such possession are inaccessible to sexless android automata. At the level of the conflict between Helena as an advocate of rights and Domin as a seller of machines, the decision in favor of the unsexed quality of robots connects to the economic principle of the legitimacy of profit-seeking. The maker of metahumans, if he wishes to make a profit, must control the production of them, which includes their reproduction. He certainly cannot give the means of reproduction away to the impossible humans themselves; if he did, they would cease to behave mechanically and begin "naturally" to have sex and reproduce, freely, outside the factory, post-production. It follows that robots must *not* be manufactured in such a way that they may perform sexually (or reproduce asexually), because the R.U.R. factory would no longer make money. The rights of property must be asserted; the patent must be protected, copyright preserved. At one level of the play's philosophical reflection, the joke is the radically simple notion that the human can never be reduced to the mechanical as long as one of the ingredients of the essence of the human is the erotic-procreative: the android automata of the Prologue are not human because they are incapable of heterosexual passion. Domin, even as he feels manly and attracted to the beautiful Helena, self-contradictorily displays pride in his promotion of a sexless creature. For her part, Helena is knocked off-kilter by the scandal in the very proposition Domin takes to be self-evidently justified: that sexual organs could only be superfluous for robots. The sexlessness that Domin sees as a marvellous technological virtue is to the woman he woos an anti-erotic scandal. He is blundering.

Domin says he looks forward to showing Helena "what all he [Old Rossum] managed to bungle in ten years. The thing that was supposed to be a man lived for three whole days. Old Rossum had no taste. What he did was dreadful" (Pro. 8). Old Rossum is a failure in the way of Frankenstein and Moreau, creating abortions and monsters. By contrast, Young Rossum, "an engineer, the son of the old man" (Pro. 8), simplifies his goals and thereby achieves success: "'This is nonsense! Ten years to produce a human being?! If you can't do it faster than nature then what's the point?'" (Pro.8). Young Rossum's model is that of a perfect worker, a quasi-organic machine in human shape designed to work, but only work. As the history lesson continues, Domin blithely explains the economy of mechanical value that Young Rossum championed, exposing the indifference to the sacred and the esthetic presupposed by such an economy.

> DOMIN: So then young Rossum said to himself: A human being. That's something that feels joy, plays the violin, wants to go for a walk, in general requires a lot of things that—that are, in effect, superfluous.
> HELENA: Oh!
> DOMIN: Wait. That are superfluous when he needs to weave, say, or add. A gasoline engine doesn't need tassles and ornaments, Miss Glory. And manufacturing artificial workers is exactly like manufacturing gasoline engines. Production should be as simple

> as possible, and the product the best for its function. What do you think? From a practical standpoint, what is the best kind of worker?
> HELENA: The best? Probably one who—who—who is honest—and dedicated.
> DOMIN: No, it's the one that's the cheapest. The one with the fewest needs. Young Rossum successfully invented a worker with the smallest number of needs, but to do so he had to simplify him. He chucked everything not directly related to work, and in so doing he pretty much discarded the human and created the Robot. My dear Miss Glory, Robots are not people. They are mechanically more perfect than we are, they have an astounding intellectual capacity, but they have no soul. (Pro. 9)

The quality "superfluous," previously attached by Domin to sexual organs and erotic experience—such organs are superfluous for artificial humans—is here attached to esthetic experience (walking for pleasure, playing music). Capek defines the mechanical value of robots in stark contrast to erotic and esthetic value. Helena's attempt to inject the effects of the moral model implied by the exchange of signs into the discussion, as shown by her appeal to qualities such as honesty and dedication, is flattened by Domin's prioritization of cheapness, a quality which, by appealing to the quantifiable *price,* expels the impossible-to-coerce esthetic effect from consideration. Helena piously ventures the suggestion, "It is said that man is the creation of God." Domin replies, "So much the worse. God had no grasp of modern technology" (9). At that point, Capek would have us notice, the category of the sacred has been expelled from the account of the realm in which young Rossum operated and in which the company named after him still thrives.[9]

One of the most laughter-inducing moments in the Comic Prologue occurs when Helena refuses to accept Domin's staged "demonstrations" of the robotic nature of the secretary Sulla and the doorman Marius. We do well to keep Capek's stage directions in mind, which would influence any performance of the Sulla role: the android automata of the Prologue are "dressed like people. Their movements and speech are laconic. Their faces are expressionless and their eyes fixed" (2). Domin has been amused by Helena's politeness toward the feminine android typist, who responds to "Oh, you were born here?" with "I was made here, yes." Robots are not hired, they are bought; they do not die, they wear out; they are not born, but "made." Domin orders Sulla to stand, to turn around like a fashion model, to "chat with our guest . . . a distinguished visitor" (Pro. 10). Helena is appalled at his request that she study the manufactured complexion of Sulla and touch Sulla's expressionless face. After Sulla has recited some nearly random nautical statistics and proven her ability to say "Dear Sir" in four languages (German, French, Czech, English), Helena's activism bursts forth, her patience used up.

> HELENA [jumping up]: This is preposterous! You are a charlatan! Sulla's not a Robot, Sulla is a young woman just like me! Sulla, this is disgraceful—why do you play along with this farce?
> SULLA: I am a robot!
> HELENA: No, you are lying! Oh Sulla, forgive me, I understand—they've forced you to act as a living advertisement for them! Sulla, you are a young woman like me, aren't you? Tell me you are! (Pro. 11)

It is a wonderful touch that Helena, desperate to preserve the liberationist purpose of her visit to the island, concedes the artificiality of Sulla's monotonic speech and fixed gaze in the attribution to her of the actions of a "living advertisement." That is, Helena is imagining a human who is perfectly imitating a robot in an advertisement, but the reality onstage is a robot perfectly imitating a human.

Helena persists in her accusations that Domin is lying; Domin is offended, so he calls Marius, the robot doorman, to demonstrate his masculine robot non-humanity.

> DOMIN: Would you put Sulla in the dissecting room?
> MARIUS: Yes.
> DOMIN: Would you be sorry for her?
> MARIUS: I do not know "sorry."
> DOMIN: What would happen to her?
> MARIUS: She would stop moving. She would be sent to the stamping-mill.
> DOMIN: That is death, Marius. Do you fear death?
> MARIUS: No.
> DOMIN: So you see, Miss Glory. Robots do not cling to life. They can't. They don't have the means—no soul, no pleasures. Grass has more will to live than they do.

HELENA: Oh, stop! At least send them out of the room!
DOMIN: Marius, Sulla, you may go.
[SULLA and MARIUS leave.] (Pro. 12)

Not to know what "sorry for [another]" means is not to be capable of empathy. But even the absence of empathy is beside the point, given the lack of any "will to live" in these android automata, of a piece with the lack that rules out any risk that they might reproduce sexually. The mutual empathy of self-sacrificial desires displayed by the humanoids Primus and Helena in Act Three, elicited by Alquist's orders that they go to the "dissecting room," will recall this scene with striking contrasts. At this point, most noteworthy for my argument is the way that the object for which Helena is purportedly expressing generous concern is the very object whose presence she cannot abide. To acknowledge the reality of the robots as they are in fact would be to confess that her dreams of liberation are—superfluous. Helena is more attached to her own resentment of the persecutors of an imaginary victimized robot than free to acknowledge the manifest reality of what we might call the non-victimizable robot. Why is it so horrible that this thing of human appearance is *beyond victimization?*

Helena, Capek shows, inclines toward a view of exchange which requires that there be victims before there can be values, which assumes that production of goods must necessarily entail production of human victims, must necessarily entail the violent sacrifice of exploited or coerced labour. In the Comic Prologue and specifically in the errors of the activist Helena, I suggest, Capek is having much fun at the expense of the dogmatic politics of rescuing exploited labour. Capek is satirizing, in amazingly prescient fashion, the victimary politics of labour, the rhetoric of the champions of pure culture who suspect, despise and denounce impure commercial exchange.

This satire of victimary politics emerges more clearly in the group interview, during which, by comic reversal, Helena now refuses to believe that real humans—Domin's co-workers—are not robots. Some readers of the play have suggested that Helena's mistaking of the Directors for robots exposes the robotic behaviour of the directors themselves: they have already been dehumanized (Harkins, *Karel* 87; King 103). I would argue, on the contrary, that Helena is the mechanical one in her inability to let go of her fixed idea of liberating the automata. Even the most cursory comparison of the directors' speech patterns beyond their first half-dozen lines shows that, for example, they speak *voluntarily,* unlike Marius and Sulla, who offer nothing without being externally prompted. Nor would it be an easy performance task for actors playing Gall, Fabry, Hallemeier, Alquist and Busman to deliver *all* the lines preceding Domin's explicit announcement "These gentlemen are people, just like you" (Pro. 15) in the laconic fashion specified in Capek's stage directions as appropriate for robots. The stage direction "They all look at each other, puzzled" (Pro. 15)—they are puzzled by Helena's question about whether the way they are treated by humans is not "hurtful sometimes"—gives away their nonrobotic status. No, the laughter rises at Helena's expense: we laugh at the bafflement of her incitement to robot rebellion: "Brothers, I have not come as the President's daughter. I have come on behalf of the League of Humanity. Brothers, the League of Humanity has more than two hundred thousand members. Two hundred thousand people stand behind you and offer you their support" (Pro. 16–17). Helena's blunder is to address this impassioned speech to real humans, whom she is mistaking for impossible humans. She will be corrected; she will be embarrassed.

The one concession I make to the critical commonplace is that at the very opening, when the men are all struck by Helena's womanly beauty just as Domin was, when they draw up easy chairs in unison and surround her, there is a uniformity in their response (Pro. 14). But the uniformity is brief; moreover, it shows the directors being "mechanical" by exposing the tendency in libidinal excitement to something "mechanical." The sexual desire of the five men, mimetically reinforced, is of an entirely different order than the mechanical sameness of the asexual indifference of Marius and Sulla. The uniformity of the erotic impulse in the men highlights the comical liveliness of human sexuality as opposed to the inert sexlessness of android automata. Meanwhile, Helena's anxious discomfort at the androids' non-erotic nature establishes her special contribution to the R.U.R. community. Her esthetic misgivings will continue to have a certain moral dignity, even after Gall's response to them contributes to the disasters brought on by the next type of robots being, the altered androids. She marries Harry Domin, but she persists in pursuing her subversive intuitions. Capek's generous spirit will not reduce Helena to nothing but a laughing stock.[10]

Altered Androids and Mechanical Resentment Mobilized

Having abandoned the untroubled comedy of the Prologue, Capek in the bulk of the play (Acts One to Three) develops instead a series of confrontations with the extinction of humankind, the end of history. Capek scholarship has uncovered two quite distinct authorial accounts of the moment of inspiration for this main action of the play. In one, Capek pointed toward the fear of the android automata of Act One as his source of esthetic intuition; in the other, he claimed that the vision of a group of people facing imminent extinction was his object of meditation. Regarding the first account, we might note that Capek's memory of the moment of inspiration strangely resembles the vision of Edward Prendick on his return to London: both are situated in the public scene of anonymous mass transit, and in both, faceless workers in the labouring masses figure as precursors of degenerate humanity—for Prendick, a humanity returning to bestiality; for Capek, a humanity that has lost its ability to think. This curious conjunction fits with the suggestion that the modernist esthetic finds in the "mass" behavior of industrial production and consumer society not the possibility of omnicentric multiplicity but only condemnation to a mindless uniformity. [11] In the other account, Capek claims he had been seeking a form through which to examine what it would be like to be faced with the extinction of the human species.[12] Scholars have remarked on Capek's fondness for scenes of the "siege," scenes of a small group of sensitive people surrounded by dark barbaric forces that threaten the destruction of all civilized life (e.g., Naughton 73; Klima, "Introduction" 16). Acts One and Two of *R.U.R.* present such a siege scenario. In both of Capek's moments of inspiration, the mood evoked fits with the sombre, meditative tone dominant after the Comic Prologue. The "collective" in this collective drama is represented by a group of people whom we are to take, because of their opposition to the robots, as emblematic of humankind itself.

In the Prologue, one of the jokes about the android automata is that their inability to feel pain renders them defectively unable to behave in a way compatible with self-preservation. The "suffering" that Dr. Gall at that point intended to introduce into robot physiology was to provide only a bio-mechanical feedback system intended to decrease the self-endangerment and increase the self-protection of robot bodies taken as atomistic units. It was not to be "suffering" embedded in a sacrificial exchange system where the destruction of one body is intended somehow to benefit transcendentally the religious solidarity of a group of other bodies. By contrast, the pain that will be felt by the altered androids who produce the worldwide war of rebellion will have a pseudo-social meaning. In its psychological manifestations, it mimics human resentment on the comically reduced scale of an awareness of one difference. The one difference that the altered androids *know* is the one between robot and human: robots work, humans do not.

Clues accumulate in Act One until we have learned that Dr. Gall, at the request of Helena, has performed secret experiments on the robots. The goal was to make the robots more human and to give them "souls." Nana, the elderly "nurse" to Helena, has been cursing the robot Radius in Helena's presence as a monster inferior even to the household dogs that hate it, partly because (unlike dogs) it cannot bear young (Pro. 26—27). Radius, the only robot featured in Act One (he will return in Act Two as the commander of the killer androids [2:66]), returns to the stage; he has been smashing statues rather than reading books in the household library. To begin to have a soul is to begin to be resentful and to break things. The irony here may be that a robot who smashes statues is destroying idols of his ancestors: the most ancient humanoid automata were talking or moving statues (John Cohen 18–22). Helena calls for Radius and tries her best, in an ineffective exchange of words.

> RADIUS: You are not like Robots. You are not as capable as Robots are. Robots do everything. You only give orders—utter empty words.
> HELENA: That's nonsense, Radius. Tell me, has someone offended you? I want so much for you to understand me.
> RADIUS: Empty words.
> HELENA: You're talking that way on purpose. Doctor Gall gave you more brains than he gave the others, more than we have. He gave you the greatest brain on earth. You're not like the other Robots, Radius. You quite understand me.
> RADIUS: I do not want a master. I know everything.
> HELENA: That's why I put you in the library—so you could read everything. Oh, Radius, I wanted you to prove to the world that Robots are our equals.
> RADIUS: I do not want a master.

> HELENA: No one would give you orders. You'd be just like us.
> RADIUS: I want to be the master of others.
> HELENA: Then they would certainly approve you as an official in charge of many Robots, Radius. You could teach the other Robots.
> RADIUS: I want to be the master of people.
> HELENA: You're out of your mind!
> RADIUS: You can send me to the stamping-mill. (1:37)

In this hilarious exercise of futile therapeutic outreach and sulking talkback resentment, Helena projects and Radius rejects. He repeats references to the single difference that has taken total possession of his programmed brain: the difference between the peripheral robot-who-works (mastered) and the central human-who-does-nothing (master). Helena imagines him having taken offense, imagines him understanding her, tries to flatter him with reminders of the special brain Dr. Gall gave him and the library privileges she has given him; she imagines she can flatter him with promotion to the role of leader. Although altered to resent only one difference, Radius shows that one is sufficient to create an insoluble problem.

It would be a mistake to say that Radius has been humanized, because he is quite incapable of grasping the significance of Helena's suggestions; indeed, I should speak of indifference rather than rejection. Radius here does not occupy an intentional scene, in which voluntary symbolic exchange takes place. He cannot imitate Helena's intentions, because he cannot understand them. He has been altered in such a way that he reduces even her ability to make suggestions as evidence that she does no work; to him, all her suggestions amount to "empty words": "You only give orders—utter empty words." Paul Selver's translation of this crucial line is "You only give orders. You talk more than is necessary" (44). Camilla Kinyon points out that there is a strangeness in the expression "utter empty words" (Novack's translation), a strangeness that obtains even in the Czech original: Kinyon's own translation reads "You make unnecessary words." She shows in her exposition of the nuances in the Czech original that Radius has contempt for the labour of making "words" and that he values only manual labour, the work of robot hands (397n6). I would place Kinyon's observation in the context of the altered android's mechanical severing of the exchange of things from the exchange of words. That mechanical severing starts up the horrors of an economy of mechanical value. Radius in predictable resentment speaks as if there are only two possibilities: humans must die (and robots become masters) or robots must die (and Radius go to the stamping mill). There is no in-between. By signing a note to Dr. Gall so as to prevent Radius from self-destruction via the pursuit of punishment for misbehaviour, Helena unintentionally destroys any possibility of the pseudo-freedom that would permit him to commit robot "suicide," getting himself sent to the stamping mill.[13] We ought to avoid the victimary temptation to humanize the altered androids prematurely and thereby miss Capek's insight into the originary force of resentment. Capek brilliantly simplifies resentment to a mechanical periphery-center structure, a minimal opposition that exists only in the minds of the altered androids, an opposition between the robot worker and the idle human being.[14]

Dr. Gall is the scientist who usurps covertly the place of Old Rossum, Young Rossum, and Domin. But Dr. Gall himself, Capek instructed his actors, is not to be played as an example of the gloomily single-minded mad scientist: he is to be played as "trifling, vivacious, suntanned, with a black mustache" (2). Gall certainly plays more flirtatiously with Helena than do any of the other men (Pro. 17; 1:29; 1:32; 2:62). Enchanted with Helena, he wishes to please her; in following that wish, he hastens the end of world civilization. There are serious hints of an affair between him and Helena.[15] He does not take the violence Radius has displayed by smashing statues as seriously as Helena does (1:37). When Helena asks whether Radius now has a "soul" thanks to the alterations they have performed, he replies: "I don't know. He's got something nasty" (1:38). The flippancy, as things turn out, will prove to have been fatally misplaced, given the eschatological twist that Dr. Gall has made the robots into something more human than either Young Rossum or Domin would have desired: "I did it secretly . . . of my own accord. *I transformed them into people*" (257) [emphasis added]. He will confess: "I changed the way they were made. Just certain physical details, you see? Mainly . . . mainly their . . . temperament" (2:56). Paul Selver's different translation is again sharply revealing: "Chiefly—chiefly their—irritability" (3.70). In that word "irritability," I submit, we find Capek's comical allusion to the primordiality of resentment.

The alteration of the robots so that they move and act *as if* they feel the one human-robot difference as something to be "resented," opens the possibility of their conceiving of total war against humankind. Capek sees to it that Gall's altered androids have been given the blunt thought that the human itself is the "unnecessary," in keeping with the thematization of the superfluous, useless, and beautiful as essential components of the human. Capek is suggesting the paradox that freedom is necessary for humans: human being of necessity requires the freedom to choose (at least once in a while) to use the sign that defers violence. The robots have no human consciousness of the violence they do. Gall's robots have not "chosen": they are mechanically "free" only to destroy their human makers. These robots are no more human that heat-seeking missiles or bomb-dropping drones. The robotic discovery of center-periphery difference does not defer but completes violence, in the annihilation of the human-at-the-center that they now resent, find imitable, and hate in the mode of mob violence.[16]

Capek persistently drives home both the identity of "robot" and "worker," and its counterpart—the identity of "human," "useless," and "superfluous." Domin in the Comic Prologue prepared Helena for the mindlessly unceasing quality of robot labour: "The Robots don't know when to stop working" (Pro. 13). Likewise, Fabry then answered Helena's question about why the company would bother to make such creatures in the first place as follows: "For work, Miss. One Robot can do the work of two and a half laborers" (Pro. 17). Dr. Gall calls them, in a nutty oxymoron, "living work machines" (2:53). Domin dreams of the day when robots will have freed global society from the need to work: "But within the next ten years Rossum's Universal Robots will produce so much wheat, so much cloth, so much everything that things will no longer have any value. There'll be no more poverty. Yes, people will be out of work, but by then there'll be no work left to be done. Everything will be done by living machines" (21). In Domin's failure even to begin to catch the implications of the prophecy *things will no longer have any value,* Capek hints at the self-destructively oblivious creed of those who believe in the desirability of an economy of mechanical value. The mechanical is no more truly economic than the material is truly spiritual. The category of economic value is nothing that the human will ever be able to transcend by way of the creation of a land of material plenty, whether communist or capitalist.[17] Robots doing all the work is robots doing too much work: total plenty guarantees the apocalypse of economic annihilation by emptying human scenes of their center-periphery structure and collapsing the differential tension between the exchange of signs and the exchange of things.

Much of my analysis thus far has insisted on the lighter side of Capek's theatrical art.[18] But the theatrical moment in which the humanexecuting soldier robots take over the stage at the end of Act Two casts a chill. Even given the limits on audience sympathy for the play's characters that have been set by Capek's expressionist anti-realism—it is true that each character here is to some extent just a mouthpiece for attitudes (Matuska 343–50, 353; Klima, "Introduction" xxi)—we feel no small attachment to the characters being summarily murdered by the robots. Any audience witnessing a successful performance would be acutely aware of the unsparing finality of the murder of Hallemeier, and the murders of Domin and Helena, hiding in the room through the door on the left. We witness the end of the human species. There is nothing for us to envy in the victory of the robots. The "life" they will have for themselves after we are gone is nothing that, upon reflection, we could desire.

Capek's fantastic premise in the unceasing labour of the robots (they continue to work long after humans need anything from them) is the same one that he develops in different, more elaborate ways in his novels *The Absolute at Large* (1922) and *War with the Newts* (1936) (see Matuska 166, 271). It is the fantasy of a capacity for excess production that renders consumption itself insignificant and destroys the possibility of a price system. It is a nightmare of production without purpose, without the constraints of finitude imposed by the scene of human representation (which finitude is not to be confused with an extra-human ecological limit). The nonhumanity of this identification of robot being with "work" reaches its *reductio ad absurdum* in the final act, when the Third Robot (in Novack's translation) informs Alquist that the robots are continuing to produce consumer goods despite the fact that the humans who would have consumed them are long gone: "We have increased productivity. There is nowhere left to put all we have produced" (3:73). As Robert Philmus observes, the equivalent passage in Paul Selver's Oxford translation is lengthier and richer in images, "giv[ing] concreteness to an abstraction" ("Matters" 22); in Selver's version, the absurdity of an economy of purely mechanical value is transparently vivid.

> RADIUS: Sir, have pity. Terror is coming upon us. We have intensified our labour. We have obtained a million million tons of coal from the earth. Nine million spindles are running by day and night. There is no more room to store what we have made. Houses are being built throughout the world. Eight million Robots have died within the year. Within twenty years none will be left. Sir, the

> world is dying out. Human beings know the secret of life. Tell us their secret—if you do not tell us, we shall perish. (Selver IV.93)

"The secret of life" ironically turns out to have been human erotic desire, not any exceptional scientific formula. The meaningless of the quantitative measures that Selver's robot recites (a million million tons, nine million spindles), is one of Capek's jokes, as is the absence of any logical link between houses being built and robots dying out. Instead of access to reproductive sexual coupling, the robots will demand from Alquist the "secret of production." They will seek the formula contained in the manuscript of Old Rossum that Helena burned in the fireplace early in Act One. The robots are never free of their total dependence on the mad scientific discovery of Old Rossum, just as the Monster is never free of Victor and the replicants are never free of Tyrell. A small hope of robot self-perpetuation will emerge in the Primus and Robot Helena scene, but I will argue that its size has been exaggerated by more than one interpreter. Moreover, the hope that will appear with Primus and Helena will owe nothing decisive either to the formula itself or to the intended designs of any of the men, from Old Rossum to Dr. Gall, who have played God in creating the robots.

When negotiating with Alquist in Act Three, the committee robots wish to perpetuate their kind; but their wish does not mean they have become human. Damon's imaginary scenario of guaranteed species perpetuation by maternal machine is designed to elicit laughter, not victimary sympathy for a dignified proto-humanity. Damon, the leader of the Central Committee of the Worldwide Union of Robots, prophesies to Alquist in a manner somewhat reminiscent of Domin (the proximity of the names itself suggests such a parody): "We will give birth by machine. We will build a thousand steam-powered mothers. From them will pour forth a river of life. Nothing but life! Nothing but Robots!" (3:75). This mock-prophecy should take us back to the fact of the *quasi-organic* basis of the altered androids' being: even their form of "life" is not comparable to the normal organic life on planet earth. Nothing could be further from a human mother than the "steam-powered mothers" of Damon's sexless imagination. The committee robots' urgent demand for a means to species self-perpetuation should not lead us to amalgamate a mechanical movement toward self-preservation with a human desire for communal immortality, with ideas of a shared, intended history. The joke is that such a species perpetuation would not be something that humans could value. Today we witness Hans Moravec and his collaborators begin to dream of the post-human spiritual machines of their robot utopias (Winner; Schanze). Eighty years ago, Capek provided a demolition of such fantasies with an anti-fantasy of his own. We are invited to laugh at the outrageous contrast between the humans in Acts One and Two who face extinction with some measure of mutual care, and the vacuity of the robots' desire to re-produce mechanically as they beg, boss, and threaten the bricklayer Alquist into playing the vain scientist and making them into something like the very beings that they have exterminated.

The price for a small increase in robot self-awareness has been the obliteration of the human. Having successfully destroyed the real human, the impossible human catches itself, stumbles, realizes that it was a human scientist who made the impossible human in the first place, and pauses. It becomes aware of its finitude only when human vulnerability to extinction has been absolutely, irreversibly demonstrated. It is therefore appropriate that Alquist despairingly cries, at the very end of Act Two, looking through a door at the offstage corpses of the murdered Domin and Helena, "What have you done? You'll perish without people!" (2:70). Alquist is correct. But Radius, in the last line of Act Two, as a creature quite incapable of sacred or esthetic experience, monotonically declares: "There are no people. Robots, to work! March!" (2:70). He makes explicit the implication of Alquist's too-late warning in his affirmation that people no longer exist. Obviously, however, Radius operates only inside an economy of mechanical value; he does not know what he has done. Unlike the humans in the originary event who preserve a memory of the sacred center, Radius is simply . . . oblivious. His is not a human consciousness. Capek's inspired fear of human-like beings incapable of *thinking* thus finds a vivid embodiment. It is a hopeless scene, Hallemeier's dead body on stage and Alquist distressed beyond any hope of recovery. One is almost tempted to wish Capek had ended the play there—an unequivocal revelation of the horrors of an economy of purely mechanical value, a theatrical dramatization of the horrors of scientism.[19]

Committee Robots and Primary Humanoids

In this closing section of our exploration of *R.U.R.*, my focus will be on the awkward Act Three of the play. Directors, actors, audiences and critical readers may feel that Act Three presents a problem because it cannot help but seem to take back some of the esthetic effect of the preceding drama. I have already pondered the risks Capek has taken in constructing the drama in such a form that a Comic Prologue of erotic courtship that ends with a marriage proposal is followed by two acts of tragic movement that end in mass murder and human extinction. We may feel discomfort in Act Three with yet another thematic re-orientation and strain on our imaginative reach. Many critics acknowledge this discomfort (Harkins, *Karel* 90; Matuska 240, 275; Naughton 86; Kinyon 381, 386; Klima, "Introduction" xix). Having let loose into Western culture the symbol of the robot, Capek seems already to repent of his success (in somewhat the way he would later repent in "The Author of the Robots Defends Himself"). He tries to diminish our fascination with the violent mechanical humanoid by substituting a model of lovable artificial humanity, represented in this act by Primus and Robot Helena.

Primus and Helena, it has been suggested, represent the robot species somehow becoming somewhat human (Wellek 51; Harkins, *Karel* 88; Matuska 228; Suvin, *Metamorphoses* 272). Certainly, they differ sufficiently from the other robot types in the previous scenes of the play to merit being named apart. I refer to them as *primary humanoids* so as to distinguish them from Marius and Sulla and from Radius, Damon, and the unnamed army and committee robots. While conceding some of the assertions that might tempt one to describe Primus and Helena as "humanized," however, I will persist in my suggestion that the originary robot even at this third stage is best grasped as an impossible human, if for no other reason than given its quasi-organic origins it has no relation of biological continuity with the human community. The mutual willingness of Primus and Helena to suffer death sacrificially in the name of the individual erotic other does prove effective, certainly in giving Alquist hope for robot re-generation and perhaps as an index of changes in their quasi-organic physiology. Capek is suggesting that the human-as-mechanical or the human-as-robotic should horrify us not primarily because the reduction of humanity to the robotic would equal the proof of scientistic materialism or proof of the expulsion of "spiritual" or "supernatural" ghosts in the machine. Rather, the human-as-mechanical should horrify us because it kills the human erotic.[20]

Alquist's part in Act Three contributes new vividness to the audience's impression of the ugliness of the mechanical. Since the robots have killed off the human species, the humour in the novelty of robotic behavior has evaporated. The crudity of the technological apparatus behind the manufacture of the robots, hitherto hidden off stage in the unvisited factories, is displayed onstage: test tubes and chemical flasks on a long bench, a microscope under a hanging naked light bulb, a tool cabinet, a door leading into a dissecting room and another door leading into other laboratories (3:71). This is where the end of the human world began, we in the audience taking in the stage set think; the end began in the cold laboratories of scientific overkill. At one level, much of Act Three can be grimly enjoyed as Capek's satire on scientism itself: Alquist's impotence in the face of the task of re-finding Old Rossum's original robot-making formula parodies the vanity of all such human attempts to play God the creator of the human.

The ugliness of the mechanical is further elaborated by means of the irony that Alquist, humble and suspicious of scientific hubris in Acts One and Two, is now forced by the altered androids to occupy the very position he least desires to take: that of the human playing God. This is involuntary centrality with a sadistic vengeance. Along with those of Nana, Helena's nurse, Alquist's opinions about modern market society were always reactionary, close in content to the "peasant" or artisanal values that some Capek scholars identify as some of those most cherished by the playwright himself (Matuska 253; Suvin, *Metamorphoses* 271; Klima, "Introduction" xv, xx). In somewhat the same way that Prendick was forced to become one with the Beast People whom he alone found pitiable, in somewhat the same way that Victor was forced to construct a female companion for the monster only under blackmail and with great gnashing of teeth, Alquist carries out his tests with histrionic protestations about their futility. The challenges for an actor playing Alquist would be immense: the role calls for a more daunting range of vocal, gestural, and bodily nuances than that needed for any other part; it also requires a sudden transformation from the posture of a dignified elderly artisan to the energy of a speaker of demented soliloquies. It is no wonder Capek scheduled an intermission before Act Three, even though the audience would return to their seats for the briefest of the four acts. The Alquist of Act Three seems to be going mad from emotional exhaustion, approaching death. His condition should limit any inclination we feel to concede the validity of his belief that Primus and Helena represent a promise of the perpetuation of the human.

The two primary humanoids possess a whole new set of capacities: they appreciate beauty, have dreams, and feel sexual appetite. They play, they laugh, they weep, they flirt. It is quite impossible for us to picture the stern, imperious members of the Robots' Central Committee doing any such things. Capek emphasises that the primary humanoid Helena is a robot made in the specific image of Helena Glory, the wife of Domin and the adored feminine center of the R.U.R. community of directors (1:38–39; 3:78; 3:83). In possession of an *individual* model, Robot Helena already differs significantly from the thousands of look-identical anonymous robots. When she first appears on stage, she shows a distaste for cruelty in not wanting to see Damon cut. She turns pale and feels faint at his screaming under the scalpel of Alquist. When she explores the lab while Alquist is sleeping, she displays an ability to banter with Primus and expresses a preference for the natural beauty of the rising sun as opposed to the mathematical formulae on the open pages of Alquist's books inspected by Primus. She is interested in the singing of birds. She tells Primus that she has wandered into an abandoned house once occupied by humans and taken delight in some puppies there (Capek loved and wrote about his dogs). Robot Helena feels bodily sensations that Capek surely expects us to take as signs of a sexual appetite finding its voice: "I feel so peculiar, I don't know what it is. I'm so silly, I've lost my head—my body hurts, my heart, I hurt all over—and do you know what's happened to me? . . . No, I can't tell you! Primus, I think I'm dying!" (3:80). This surprising-to-the-audience Helena is certainly not an altered android designed on the plan of Radius or Damon.

Primus, for his part, has experienced dream-filled sleep during which he spoke "beautiful" things to Helena: "I didn't understand it myself, and yet I know I've never said anything more beautiful . . . When I saw that my words touched you I could have died. Even the place was different from any place I've ever seen" (3:80). He woos Helena directly, making an esthetic-erotic judgment: "You're beautiful" (3: 80). She studies her face in Alquist's mirror, and then his face behind hers; she compares their body shapes and playfully combs his hair. The sound of her laughter wakes up the violence-traumatized, sleeping Alquist: "What—what on earth is this? Laughter? People? Who's there?" (3:81). Alquist induces in the humanoids a feeling of *shame* ("Primus, what could have come over us?"). Their becoming proto-human is likened to the embarrassing of Adam and Eve after the disobedience in the garden: robots are not to laugh and play, but only to work. (Alquist will boss them around shortly.)

Erotic playfulness followed by erotic self-consciousness heralds the proof of "love" between primary humanoids, a fall from robot innocence into the beginnings of something comparable to human sexual desire. They have performed the self-withdrawal essential to the erotic: "Oh, Primus why do you avoid me? Why must I run after you all day long? And still you say that I'm beautiful!" Primus' answer is a gently ironic reciprocal accusation: "You run away from me, Helena" (3:81).

However, the most powerful differentiation between the robots of the Central Committee and the primary humanoids is that signified by the contrast between the judgments Alquist makes of two desires. One is Damon's desire to serve the purpose of the perpetuation of the robot species in a will to self-sacrifice, a desire which Alquist renders *abortive.* The other is the primary humanoids' desire to serve the perpetuation of the robot species in self-sacrifice, a desire which Alquist renders *superfluous.* The interpretive key to Act Three lies in the different judgments of these desires made by Alquist.

Let me spell out this contrast between the differing judgments. Twice, Alquist is cast in the role of one who must take robots to the dissecting room and there, cut them up, take the scalpel to their robot flesh, in a futile pseudo-scientific search for the "secret of life." The first time, having carried out the threat and agreed to put Damon on the table, Alquist judges that he cannot complete the dissection: the performance of such cruel surgery falls outside the range of his moral compass. Alquist, the involuntary technician playing God to the robots, therefore renders Damon's sexless plan to save the robot species *abortive.* The second time, neither Primus nor Helena actually goes to the table; Alquist is only playing a game with the youngsters, testing them, not really intending to dissect them at all. The desire for self-sacrifice, thanks to their passing his test, is rendered superfluous. The value attributed by the primary humanoids to something like the private sacred renders superfluous the public imperative of robot-species perpetuation. Erotic value guarantees species perpetuation and imitates the real human, whereas the biological "law" that dictates reproductive necessity divorced from erotic experience finds expression only in the ridiculously mechanical movements of the committee robots. A certain investment in the value of the private sacred (one such as that symbolized by Capek's interest in the playful hopes of Primus and Robot Helena) paradoxically rescues the values on which the public economy depends.[21] Because Primus and Helena hold each other sacred as private erotic partners, Alquist sees in them the hope of robot species perpetuation. He hopes and believes they will bear offspring and life will thus continue.

In the first mock-sacrifice scene, we witness Capek's devastating satire on the brooding pretentions of Victor Frankenstein and Doctor Moreau bent over their respective victims at their respective dissecting tables. We witness a satire on their materialism and their reduction of the human to mechanism. Here at the conclusion of *R.U.R.,* Karel Capek places the erstwhile gentle, devout, humble Alquist in the position of last man on earth trying to find yet again the "secret of life," pushed into submitting against his will to the Robot Committee's demands that he "cut" Damon in the search for Old Rossum's formula (for the compassion, see 3:73, 75, 76). He has informed them in vain: "I am the last human being, Robots, and I don't know what the others knew." Mechanically, they fail to hear; mechanically, they insist he try, their indifference to his confession of scientific ignorance the sign of their ignorance. He is a very unwilling mad scientist: his hands shake and he weeps. His suggestion that a supernatural intervention may give the robots some way to keep living (3:76), even his bullying attempt to put some fear of torture into the robot who wishes to be cut open (3:76)—both fail. Nothing works. Damon mechanically commands: "Perform experiments on live Robots. Find out how they are made!" (3:75); "Take live bodies!"; "Live bodies!"; "Begin!" (3:76). The audience knows, as Alquist knows, that the experiment will prove to be a bloody farce, and yet it must go ahead. Such is the method of vain science, Capek implies: its materialism trusts in torturing physical matter to produce substitute metahuman objects cut off from the sacred and the esthetic.

We must imagine the stage emptied of bodies during the following exchange, one that deserves to be quoted at length.

> [*He goes off right, leaving the door ajar.*]
>
> [*Pause*]
>
> ALQUIST'S VOICE: Hold him—firmly!
>
> DAMON'S VOICE: Cut!
>
> [*Pause*]
>
> ALQUIST'S VOICE: You see this knife? Do you still want me to cut? You don't, do you?
>
> DAMON'S VOICE: Begin!
>
> [*Pause*]

DAMON [*screaming*]: Aaaa!
ALQUIST'S VOICE: Hold him! Hold him!
DAMON [*screaming*]: Aaaa!
ALQUIST'S VOICE: I can't go on!
DAMON [*screaming*]: Cut! Cut quickly!
[*Robots* PRIMUS *and* HELENA *run in through the center door.*]
(3:76–77)

As a variation of the archetypal animation scene in the Frankenstein myth, this exchange deflates the models of Shelley and Wells even as it shocks us with a new form of impossible human, cast in the mould of a masochistic perversity. In contrast to the corpse-body on which Victor works, the body of the robot Damon is terrifically alive and open to sensation: thus the screaming. Victor's Monster would have the option of blackmailing his creator to make him a mate from body parts lying around in graveyards and slaughterhouses; here, poor Damon must look to his own flesh. Not only is Damon infinitely more alive than the unanimated corpse-parts in *Frankenstein,* but also the "secret of life" here sought is far more mysterious than that signified by the electric jolts administered by Victor, powered by a Benjamin Franklin thunderstorm. The robots' bodies *look* human; the monster's looks of absolute ugliness guaranteed his metahumanity, the impossibility of his exchanging things with people. With respect to the desire for species perpetuation, ironically, Frankenstein's monster when performing his extortion did not at all have the succession of generations on his mind, but only erotic companionship. By contrast, Damon feels no desire for erotic companionship, but is moved by what seems to be a qualia-defying determination to keep the species going. Certain irresistible comparisons between this dissection scene and the laboratory of Dr. Moreau are equally illuminating. In *Moreau,* the screaming gave voice to utterly involuntary, undeserved suffering; the screaming of Damon is voluntary and ironically deserved. It is ironically deserved inasmuch as he was the one who took the robots into war. They have no other option because their military victory destroyed all other options. So Damon's self-inflicted pain provides the figure of a nemesis of sorts: the robotic warrior has become the victim of the consequences of his own victimization of the human.

His military-macho bravado, his will to self-sacrifice, forces him after the fact to feel something less mechanical and so to say things sounding a little more like those a real human might say. It is well worth noting, as

Kinyon (391) and others have done, Capek's ironic attribution to Damon of a minimal egocentricity just before he submits to being the one robot who must be cut on the table: "Me—why should it be me?" (3:76). Even though this "why me?" is only fleeting, by singling out Damon from the other members of the Robot Central Committee, Alquist has forced upon him a layer of self-consciousness, some intimation of his *individual* mortality. More comical is his re-entrance in the laboratory after Alquist has given up on the cutting and taken off his bloody lab coat in guilty despair. Damon's return is reminiscent of the ghostly return of the murdered Banquo in *MacBeth,* not an inappropriate suggestion given that Alquist has been talking to his bloody hands as Lady MacBeth did (3.77–78) and awaits robot Helena's bringing him water so that he can wash them.

> [DAMON *staggers in from the right, swathed in a bloodstained sheet.*]
> ALQUIST [*shrinking back*]: What are you doing here? What do you want?
> DAMON: I am al-alive! It—it—it is better to live!
> [SECOND *and* THIRD ROBOTS *run in after him.*]
> ALQUIST: Take him away! Take him! Quickly!
> DAMON [*helped off to the right*]: Life—I want—to live! It is—better—
> [HELENA *enters, carring a pitcher of water.*] (3:78)

Damon has learned to *value life,* but it makes a difference that here he seems no longer to be talking about the biological-mechanical perpetuation of robotkind *en masse* as a species, robot life. Now he is talking about his one individual, personal, embodied life. Capek's satirical point against mad science again cuts neatly: doing violence to the individual life in a crazy search for the fantasised "secret of life" of the whole species misses the point that those who can let go of the perfect-species fantasy already possess the "secret of life." What life should we humans be valuing in the first place, other than the erotic life of the personal body, our attention necessarily oscillating between it alone and the larger community of other embodied human persons? In other words, madness lies in that form of worship of the species body that abolishes respect for the personal, individual body. The species body is artificial when opposed to the communal-erotic body, and the *real* human is (to the healthy, self-aware being) always to be preferred to the impossible human. The first mock-sacrifice scene is soaked in the ethos of warrior self-destructiveness: Capek shows us the exhaustion of battlefield exchange, its unisexual sterility. Making love makes those who make war, but making war does not make those who make love. Military robots in the mould of Damon can make war, but nothing else. We know such robots are impossible humans because only real humans are capable of making love, capable of the scenic erotic.

In the contrasting mock-sacrifice scene that features the primary humanoids, the exchangeability of erotic gestures is foregrounded. After the laughter of Helena has awakened him, Alquist inspects the primary humanoids, not without a bullying tone. As the Central Committee pushed him the human around, so he will push these two robot lovers around a little: "A robot? Turn around! What, are you shy? [*He takes her by the shoulder.*] Let me look at you, lady Robot!" (3:81). Primus reciprocates with a hint of his willingness to protect Helena—"Heavens, sir, leave her alone!" (3:82). His protectiveness surprises Alquist and prompts him to order Helena from the room, presumably so that he can interrogate Primus privately. The audience quickly realizes that Alquist is testing Primus, having sensed his difference from the committee robots. We know the last thing Alquist desires for himself is a return to the role of surgeon dissecting flesh, yet he announces to Primus: "Well, then, dear Primus, I—I must perform some experiments on Gall's robots. Everything from here on out depends on that, understand?" (3:82). The goal denoted by "everything" refers again to species perpetuation. To the horror of Primus, Alquist selects Helena as the victim. Now it is important to notice that Primus does not at first question the necessity that experiments be performed in order to find the formula to perpetuate the robot species. Primus questions only the choice of Helena as sacrificial victim.

> PRIMUS: Helena?
> ALQUIST: Of course. Go get everything ready. Well, what are you waiting for? Do I have to call someone else to take her in?
> PRIMUS [*grabs a heavy mallet*]: If you move I'll smash your head in!
> ALQUIST: Smash away! Smash! What will the Robots do then?
> PRIMUS [*falls on his knees*]: Sir, take me instead! I was made exactly like her, from the same batch, on the same day! Take my life, sir! [*He opens his jacket.*] Cut here, here! (3:82)

There are continuities between Damon and Primus: they both leap to threats of violence against Alquist, which Alquist rebuffs by reminding them the human maker is their only hope: killing him kills the possibility of finding the secret of robot manufacture. (The audience knows the secret is long lost; Alquist is playing God.) They both betray a masochistic over-eagerness, the delusion that cutting open robot flesh will prove effective as a means to acquiring knowledge. The crucial difference is that Damon offers himself in sacrificial substitution *for the whole of robot-kind.* Damon's would be a public exchange; the value of his individual bodily pain would be weighed against the value of robot species perpetuation. Primus, by contrast, wishes to enact a sacrificial substitution that will save the individual Robot Helena. He values the body of Helena more than he values his own, and the value of robot species perpetuation is for him subordinate to that of Helena's welfare.

The Robot Helena is next put to the test. We have seen her laugh, but the android automata did not laugh; now we see her weep for the doomed Primus (but the android automata did not weep). She offers herself to be dissected in the place of Primus; Primus "blocks her way" (3:83) to the dissecting room door; then we see him "holding her back" (3:83) when she promises to jump out the window if he goes to death instead of her. Like Primus, Helena does not question the value of perpetuating the robot species. Like Primus, she obeys Alquist as authoritative human. Like Primus, she values the erotic body of the other more than her own. But the lovers' mutual exchange of professions of willingness to sacrifice the self for the erotic other could go on endlessly. ("I will die for you"; "no, *I* will die for *you.*") As long as Alquist continues to demand that one or the other be hypothetically sacrificed, the crisis continues. Capek, staging this mock-ritual, has not yet required that the private (erotic) sacred itself be opposed to the exigencies of the public (economic) sacred; so far, he has required only that erotic value be permitted to circulate in these exchanges as an accompaniment to the public sacred. The defender of the private sacred has not yet learned to *resent* the dictates of the political sacred.

Therefore, it makes all the difference when Primus defies Alquist in a wholly new way: he demands that Alquist recognize their value *as an erotic couple.* That is, Primus proposes that he and Helena as a scenic unit stand in opposition to the demand that one or the other of them die on the public scene. When Alquist hears "You won't kill either of us," he hears affirmed the value of the erotic partners together-as-one, the occupants of a private sacred scene, in opposition to the value of the military scene on which the perpetuation of the robot species is the sole bio-mechanical imperative to be obeyed.

> PRIMUS [*holding her back*]: I won't allow it. [*To* ALQUIST.] You won't kill either of us, old man.
> ALQUIST: Why?
> PRIMUS: We—we—belong to each other.
> ALQUIST: Say no more. [*He opens the center door.*] Quiet. Go.
> PRIMUS: Where?
> ALQUIST [*in a whisper*]: Whereever you wish. Helena, take him. [*He pushes them out the door.*] Go, Adam. Go, Eve—be a wife to Primus. Be a husband to Helena, Primus.
> [*He closes the door behind them.*] (3:83—84)

The primary humanoids wish to belong to each other rather than wishing to belong to the mock-community of the military altered androids. But we will fail to grasp the clever maturity of Capek's vision if we see in this concluding moment only a sentimental overvaluation of private affection as distinct from public duty. The deferral of crisis here entails no such adolescent consolation or facile worship of the romantic couple. It implies instead a tension between private and public value. Let me recall the assertion I promised to pursue above: the interpretive key to Act Three lies in *the different judgments of models of desire* made by Alquist. Alquist may well be deluded in his interpretation of Primus' declaration of desire, but he judges the desire to hold the private sacred above the demand of species perpetuation as more praiseworthy than the warrior violence of Damon. The mere species is meaningless without the scene of representation that generates *community,* and community implies some inclusion of the private sacred, not simply the biological—reproductive.

A human species deprived of the possibility of significant erotic exchange is not the one to which you and I, reader, happen to belong. Erotic exchange and the intergenerational history that it serves is not reducible to sexual reproduction, asexual reproduction, cyborg or prosthetic or mechanical reproduction. Indeed, erotic exchange is not reducible to reproduction: perpetuating the species is not extending the being of the historical community. Only the members of a human community exchange valued, valuable things. If we can exchange erotic values, the possibility of a properly human community continues.

Alquist's histrionic final speech, made up of prayers to his Biblical God, apostrophes to his dead friends, apostrophes to "nature" and to "love," declares his faith that "life will not perish." At one level, it seems strange that "life" should be the object of his praise, when the grass has continued to grow and other animal species are still roaming the earth in the absence of humans. By "life," Alquist seems to mean human life; his hope is that the offspring of the erotic attachment between the primary humanoids will perpetuate it. However, despite the hopes of Alquist, we ought to notice the multiple clues by which Capek discourages our identifying with him. The significance of the transition from committee robots (altered androids) to primary humanoids is limited. All the robots themselves remain rigorously respectful of the human/robot boundary; all defer to the authority of Alquist and place faith in the possible success of the experiments he might carry out, whereas we in the audience know that any such experiments Alquist performs will be a waste. The masculine robots in both scenes threaten Alquist with violence in a masochist-like willingness to be "cut" themselves.

As to the justifiability of attributing a fully human status to Primus and Helena, apart from the strong possibility they may be as physiologically sterile as any robot ever produced by the R.U.R. factory, we might concede that Alquist's mental condition subtracts from his reliability as an interpreter of the data that Primus and Helena present. Alquist's kneeling hope need not be ours; indeed, his manic-depressive oscillations throughout Act Three should lead prudent members of the audience to keep a distance from the predictions in his closing soliloquy. Our doubting his reliability is no insult to his humanity. The certainty we can enjoy is that Alquist will die more peacefully thanks to his hope. The basis of that hope—the affirmation of the erotic as the vanquisher of the mechanical, grounded in his discovery of a human-like capacity for erotic exchange in Primus and Helena—has dignity regardless of what will happen after his death.

Capek's Originary Script and the Popularity of Robots

In closing, let us ponder why Capek's robot turned out to be an infinitely more popular, powerful, exchangeable symbol of the impossible human than either the Monster of Shelley's Victor or the Beast People of Wells' Moreau. Why is our world of images so much more heavily populated with robots than with Frankenstein-Monster figures or talking ape-men? The argument I have pursued in trying to make sense of the play is twofold. First, I have claimed that the robots of Capek's founding drama represent the horror of an economy of mechanical value. *Mechanical value* means value vainly, self-contradictorily attributed to objects that are exchanged in ways severed from the anthropological roots of real value in experiences of scenic objects as sacred, esthetic, erotic, and economic.[22] Economic exchange implies the inclusion of the paradoxical *necessary freedom* (for humans) to choose to participate in exchange rather than to be programmed, hypnotized, brainwashed, coerced, or otherwise deprived of the freedom to say "no" to exchange itself. (In our model of the origin of the human, there must be a minimal sense in which it makes sense to say we choose to exchange the sign.)[23] Second, I have argued (more specifically) that in *R.U.R.* Capek privileges history and intergenerational community over biology and species perpetuation. In modern market society, the sphere of the oxymoronic private sacred opposes the desacralizing commercial economy that seems to threaten mechanization. The ability of robots to work in our places and take over the center from us makes them frightening: the robotic by definition threatens to usurp the centrality of the human, particularly that of human labour. But Capek invites us in the audience, walking home from the theater, to laugh at robots when they try to do the many human things they will never be able to do. A first answer to the mystery of the great popularity of robots may be the security we feel when confronted with their fantastic non-humanity. The manifest artificiality of robotic mock-humanness reassures us of one essential component of real humanness in particular. Robots will not be erotic.[24]

The private erotic sacred (like any form of the human deferral of violence through representation) takes work; it takes energy, it takes time. The robotic lacks the ability to do this slow work, to expend this meaningful energy, to participate in historical time. The erotic defeats the robotic and the mechanical, because the human erotic must remain sacred to us as long as we continue to find ourselves in love, continue to make homes and care for children (our own or others'), continue to create our private worlds in the midst of the wider public economy regulated by political exchange. The private sacred generates those exceptions to the general rule of free exchange without which culture would cease to operate. In proportion as the freedom to create such private worlds is extinguished in the name of utopian political schemes, in just such proportion do we risk the wholesale mechanization of the human. The human erotic is mobile, democratic, accessible: it may assert itself anywhere and everywhere. The erotic now generates the most portable, easily expressible, and readily exchangeable sacred values in modern market society. Modern market society opens up the pathways for the expression of the private sacred and liberates its power.[25]

A second reason for the great popularity of the robot may emerge if we attend to the pseudo-danger the robot presents in its being easily reproducible, but never assimilable to a human being. The origin of the robot is necessarily mechanical. Capek started things off in such a way that the robot as a distinct form of impossible human would always be threatening the real human with its immense powers of mechanical self-replication, the risk of its being made in too-great quantities, mass-produced on an industrial scale with a speed that bears no comparison to the slow rhythms of personal love and social life and human gestation and child-rearing. Although a metallic imperviousness to bodily injury is the more celebrated feature of the descendants of Capek's originary robots in movies, comic books, plastic playfigures and hundreds of novels, it is thanks to its perverse ability to outperform human self-reproduction, to reproduce itself mechanically rather than naturally, that the robot is most terrifying. We can always find ways to kill robots, but what do we do if they produce themselves faster than we can kill them? That is the nightmare Capek proposes in *R.U.R.*.

Capek suggests in Act Three with wise, witty humour that we can awake from any such nightmare by recognizing anew the modest, imperfect glory of the human erotic. Although our erotic lives and the private worlds they engender are tied to vulnerability and mortality, android automata cannot choose to have such lives. The production and re-production of robots presents us with a model of an inhuman economy of purely mechanical value: that is the esthetic revelation of Capek's instantiation of the Frankenstein myth. This chapter's essay has only given you a few segments of the whole play; it deserves to be read from beginning to end, by itself. The real-world science and technology of robots has grown exponentially since 1921. But I am not sure there is a single idea of human significance about humanoid robots that Capek in his prescient genius did not foresee—no idea that you will not find contained already in this variegated, dazzling, disturbing originary robots play.

FIVE

Blade Runner: Minimizing the Difference of the Impossible Human

Blade Runner as Postmodern *Frankenstein*: Contesting the Nondifference Thesis

The two most influential modernist versions of the Frankenstein myth represent the impossible human as a figure of animality or as a figure of the mechanical. As it has persisted into our postmodern era, scientism seems to have split in roughly similar directions, which attests to the esthetic prescience of Wells and Capek. The discourses of sociobiology, evolutionary psychology and genetic determinism sometimes endorse Frankensteinian attitudes (the reductionism, the materialism).[1] Wells himself was not immune to fantasies of how we might improve the human via state-programmed eugenics that removed the power to self-determine from individuals; in our time, a purportedly harmless eugenics has been entrusted to the free market (Evans; Paul 154). When the control of the genetic bases of behaviour becomes a holy grail justifying the allocation of research dollars, the secular faith of those dreaming the impossible human today falls under the sign of *a designed animal.* In the other direction, the Frankensteinian reduces the human to a *mechanical humanoid.* The idol worshipped here is the computer; the fantasy seems to be something about the digitization of human consciousness. It is curious that while most scholars in the humanities vociferously resent the spectre of genetic determinism, perhaps because of the shortness of the leash that ties it to the stake of the eugenicist horrors of World War II, no such resentment is directed toward the equally dehumanizing faith in the mechanization of mind.

I choose as an exemplary instance of the Frankenstein myth in the postmodern era Ridley Scott's film *Blade Runner* (1982; Final Cut 2007). In claiming that *Blade Runner* is an artwork that exemplifies postmodern trends and values, I follow the crowd (e.g. Bukatman 60; Booker 178; Marder 88).[2] More particularly, I have chosen *Blade Runner* because its figure of the impossible human brilliantly merges the two streams of scientism just outlined. Scott intentionally created a new type of metahuman distinct from robots and androids, but captured by the biological revelations inherent in the Crick-Watson discovery of DNA. The new impossible human is the replicant.[3] The replicant is a victim of Darwinian evolutionary scientism and of robotic-mechanical scientism, of both at once. The replicant mediates between animality and mechanics: as a product of *genetic* engineering, a replicant is nothing-but-an-animal, made from DNA; as a product of genetic *engineering,* a replicant is nothing-but-a-machine, programmed to be controllable and predictable in body and brain, without a mind of its own.

Many interpreters have observed that *Blade Runner* operates in the philosophical shadows cast by the Frankenstein myth.[4] My thesis in this chapter is that *Blade Runner* exemplifies the post-modern Frankenstein myth because it minimizes the difference between real and impossible humans: the only thing that makes the replicants finally other-than-human is the historical fact that Tyrell has created them and never lets them go. In other words, *Blade Runner* reduces the differences between real and impossible human to nothing but those produced by the peculiar relationship between the vain scientist and his creature: Tyrell as creator may be the only human left standing, but his historical influence over the impossible human uniquely determines its humiliation. As a result, *Blade Runner* maximizes an investment in its decision to dramatize the origin of the real human as historical and ethical, not biological or "natural." Tyrell as genius bioengineer can make human-like creatures, but he cannot substitute for the One other of humankind. I will support this thesis by exploring three corollary suggestions about what *Blade Runner* in exemplary fashion does. First, it converts the scientific usurper into a figure of corporate malfeasance and thus represents the postmodern trend of collapsing the effects of all scientific work into the oppressive effects of scientism; it participates in the general contempt for realist scientific epistemology prevalent in much postmodern cultural theorizing. Second, it minimizes the difference between real and artificial human by showing a human who falls in love with a replicant. Third, the difference between us and impossible humans is minimized further in that, perhaps surprisingly, Eldon Tyrell is the least scandalous of the usurpers we have studied: the leading replicant's murder of his maker is more luminously deliberate than the vengeance of Shelley's Wretch against Victor, the Beast People's mobbing of Moreau, or the extermination of humankind regretted by Capek's robots. Likewise, Roy Batty's dying moment is peculiarly resonant.

The moral confusions in the film result from the reduction of the key event in the Frankenstein myth to the staging of an arbitrary historical difference: as the human scientist must manufacture the synthetic impossible human, so must real humans be recognized as coming first—chronologically and ontologically. Our "coming first" seems an unjust priority. The sense of injustice implies that perhaps only an originary model of human signification can definitively refute the delusions of scientism. If and when we do create clones, robots, or designer babies who prove themselves capable of genuine human resentment, then the question of their entrance into society will be the question of the replicants: how can such creatures be integrated into the human community as *equal to us* on the scene of exchange, in spite of their being mediated at the very core of their existence by scientific choice rather than natural chance? Before I proceed to the elaboration of the three corollary suggestions just sketched, I must spell out two assumptions and discuss one problem, a problem I will dub the nondifference thesis.

First assumption: although it floats in the atmosphere of speculations about humanness created by Philip K. Dick's explorations of the android/human boundary, *Blade Runner* is an artwork separate from Dick's novel *Do Androids Dream of Electric Sheep?* The lines of influence extending from *Do Androids Dream of Electric Sheep?* to the movie have been studied elsewhere; it would extend beyond the scope of this book to detail every overlap and disconnect between the two works.[5] I accept Robin Wood's claim that the film stands alone: "But *Blade Runner* is not really an adaptation: rather, the film is built upon certain ideas and motifs selected from the novel. Its aim, argument, and tone are so different that it is best to regard it as an autonomous work" (161; see also Kolb 132; Kerman, "Private" 75). With a restless persistence, Dick explored the problem of the synthetic human, as *Robots, Androids, and Other Mechanical Oddities*—a collection of some of his most brilliant short fiction—makes clear. Dick's attentiveness toward to the contingent ethical fragility of the human matters most to me, for it helps us situate *Blade Runner* in the context of the postmodern ethic as Eric Gans theorizes it. For generative anthropology, the postmodern ethic is structured by the consciousness of victimary politics subsequent to the revelations of Auschwitz and Hiroshima. The inspiration for *Do Androids?* connects it to Auschwitz.

In a 1981 interview, Dick noted that that it was one of the three of his novels he liked best (the others, *Martian Timeslip* and *The Man in the High Castle*). He offered a concise definition of *android:* "'*Sheep* stemmed from my basic interest in the problem of differentiating the authentic human being from the reflexive machine, which I call an android. In my mind android is a metaphor for people who are physiologically human but behaving in a nonhuman way'" (quoted in Sammon 16). Furthermore, we can trace Dick's idea of the "android"—in other words, a form of agency captured by the phrase *physiologically human but behaving in a nonhuman way*—to an experience he had when doing research for *The Man in the High Castle.* Reading some journals kept by Gestapo officers working in Hitler's extermination camps, Dick was brought to a full stop.

> "That sentence read, 'We are kept awake at night by the cries of starving children,' Dick explained. There was obviously something wrong with the man who wrote that. I later realized that, with the Nazis, what we were essentially dealing with was a defective group mind, a mind so emotionally defective that the word 'human' could not be applied to them." (quoted in Sammon 16)

For Dick, therefore, the term *android* denoted an artificial human whose externally visible lack of empathy indicated the menacing difference of its concealed, internal inhumanity or amorality.[6] However, for Ridley Scott and the viewer of *Blade Runner,* the behavior of the replicants certainly cannot be assimilated to that of Gestapo officers such as the one whose journal Dick had read. Nor are the replicants in *Blade Runner* anything like the emotionless androids in *Do Androids Dream of Electric Sheep?* On the contrary, *Blade Runner* makes us feel for (if not fall for) the type of beautifully tragic impossible human we come to know as the replicant.

Second assumption: *Blade Runner* is best approached as a creation of Ridley Scott rather than as a result of a fortuitous collaboration of writers, actors, designers, and producers. Hampton Fancher wrote the original drafts of the screenplay that would later evolve into *Blade Runner,* starting in 1979, at first collaborating with producer Michael Deeley in the latter's Cape Cod home and later working long shifts in London with Scott. Tentative titles varied, from *Android* (the second draft) to *Mechanismo,* to *Dangerous Days* (Sammon 38). Fancher continued to write and revise when shooting began in 1981, working closely with Scott. But tensions between them forced his replacement by David Peoples (who would later go on to pen the script for *Soldier*, itself an interesting artificial human science fiction film starring Kurt Russell). Peoples did not ride roughshod over what Fancher had created.[7] Nor did Ridley Scott ever lose affection and respect for Fancher. The impression one gets is that Scott wanted more than Fancher did both to humanize the replicants and to intensify the film noir detective aspects of the story.[8] Peoples was charged with inserting *clues* into the narrative mix, foregrounding for the policeman Deckard threads to follow in an investigation. Scott Bukatman credits Peoples with "tightening the mystery aspects of the screenplay and deepening the humanity of the android adversaries, now known as *replicants*" (17).

So, the script preparation increasingly moved the screenplay away from the novel-as-blueprint (so that the shape of the film would belong to Scott, not Dick). Additionally, the more deeply one reads in the *Blade Runner* archives, the more frequently one hears the opinion voiced that both as commercial product and esthetic text, *Blade Runner* is the result of Scott's visionary powers and his tenacious will to defend the integrity of his vision in an obstacle-strewn Hollywood setting. Fancher describes Scott as "the reason it [the movie] got made."[9] When asked if Scott had any influence over the script, David Peoples exclaimed in a 1995 interview: "Absolutely! Ridley *is* the author of the movie in every sense of the word. He is a complete author. It is his movie and he dominates every frame of it with his way of looking at things" (L'Officier 32). Actor William Sanderson (J. F. Sebastian) made this comment: "*Blade Runner* was really Ridley's show; if there is such a thing as a visionary, Ridley's the guy. He's got the eye and the passion for that" (Sammon 142). Scott answered a question from Sammon about the coolness of his relationship with Harrison Ford with this memory.

> I said [to Harrison Ford], "Listen, this is my movie, I have my performance as well as you have yours. And, you know, both will be brought together. That's all I can promise." Because if I hadn't [neglected Ford], a lot would have gone out the window. To put that kind of thing on screen requires enormous attention to detail. And *it can finally only be accomplished through one pair of eyes.* (387) [emphasis added]

Paul Sammon's amazingly thorough, careful account of every step, stage, scene, cut, and question in the long history of *Blade Runner* leaves no doubt that Scott was a fighter. For example, disoriented by the labour codes in Hollywood that reserved for cinematographers alone the right to look through the camera, he insisted on getting things right: "Scott's habit of filming the same take fifteen or twenty times before he was satisfied with the result quickly became a point of major contention with the Tandem financiers" (206). To add to the pressure from above, the unionized workers below expressed frustration. Crew members resented Scott's perfectionism to such an extent that, as Hampton Fancher reports, one of the insults circulated against Ridley Scott was—ironically in the context of a film about artificial humans—"'he's a fucking machine, he's not a human being'" (Sammon 52). The crew revolted at one point, printing and sporting teeshirts with an anti-Scott logo expressing resentment of his British managerial style: "'Yes Guv'nor / *My Ass!*" The long composite interview in Sammon's book ends with Scott's affirmation of the artisanal value of singleness of purpose.[10] Meanwhile, Scott avoids probing the philosophical questions the film raises and refrains from heeding any calls to the duty of interpreting its murkier enigmas for us.[11] What we do know is that, building on the success of the "pretty dark" science-fiction film *Alien,* Scott had (in his own words) "decided to make *Blade Runner* a further inversion of Hollywood values" (Sammon 383).[12] His own description of the movie identifies some signposts that, if taken to heart, help one's orientation through the many difficulties of analyzing it.

> *Blade Runner* works on a level which I haven't seen much—or ever—in a mainstream film. It works like a book. Like a very dark novel. Which I like. It's definitely a film that's designed *not* to have the usual *crush-wallop-bang!* impact. (Sammon 393)

We will have occasion to return to the potent suggestion that *it works like a book . . . a very dark novel* and to the fact that the film (despite appearances) renounces stock figures of Hollywood violence.

The two assumptions shared, I turn now to the postmodernity of the ethic of *Blade Runner.* It emerges most strikingly when we register the temptation the movie creates for viewers and critics to accept what I will call *the nondifference thesis:* the idea that there is ultimately no significant difference between replicants and humans. I propose that we both resist the temptation and reject the thesis. Many writers on *Blade Runner* find the replicants beautiful (e.g., Dempsey 36; Kerman, "Private" 72–73; Wood 163; Slade 14). Vivian Sobchack situates them amid a larger trend in science fiction film of the 1980s, a movement toward "altering the spatial and temporal boundaries that limit the meaning of both 'human' and 'existence'" (253). The 1980's trend assimilates aliens and humanoids, Sobchack claims, to the human itself: "Alien Others have become less Other—be they extraterrestrial teddy bears, starmen, brothers from another planet, robots, androids, or replicants. They have become our familiars, our simulacra, embodied as literally alienated images of our alienated selves. Thus, contemporary SF generally embraces alien Others as 'more human than human' or finds it can barely mark their 'otherness' as other than our own" (293). Notice Sobchack's curious, perhaps careless, decision to appropriate as a banner of inclusionary desire the slogan of the Tyrell corporation—*more human than human*—as if that slogan were something other than Tyrell's justification of biotechnology at its most degrading. She is not alone in her misappropriating the slogan *more human than human.* The desire somehow to rescue the replicants by reclassifying them "human" is one that many thoughtful viewers of the movie wrestle with.[13] However, I contend that our appreciating the human-like beauty of the replicants threatens our tipping over into a forgetfulness of the fact they have been *manufactured* to be beautiful, strong, attractive and the like. If we are moved by their beautiful human-likeness into positing their identity with us, then we have been duped by Tyrell into overlooking the divisive primacy of the invisible defects that enslave and terrorize them. No less than Shelley's Wretch, no less than the Beast People or Capek's originary modern robots, the metahumans in *Blade Runner* suffer the fate of all impossible humans in the Frankenstein myth.

The off-world replicants Roy, Leon, Pris and Zhora have not mutinied and voyaged and murdered their way to Los Angeles, planet earth, because they were made to be just like us. The truth is the opposite. They seek their vain scientist creator Tyrell because he made them to be unlike us—most obviously, to die after only four years of life. When Tyrell speaks the slogan *more human than human,* he speaks as a business executive who means only that these products have certain instrumental properties that make them wonderfully easy to display and to sell as slave labour. Humans are naturally adaptable to new environments, given their intelligence and bodily flexibility; we might translate *more human than human* into the name of a promotional feature Tyrell is implying: as products, replicants are *more physiologically adaptable than ordinary humans* to working in ecologically hostile off-world conditions. Humans are naturally a species that can be subjected to techniques of mind control; *more human than human* means something to the effect of *more pliable and manageable as employees than ordinary humans.* We ought to recognize that Tyrell aims his slogan at the buyers of replicants, the Los Angeles 2019 equivalent of the slave trader. To appropriate *more human than human* as if it were a correct description and a compliment is to evade the horror of replicant misery. They are tormented to the core in their knowledge that Tyrell has made them *less* human than human. The memories of replicants are implanted, the emotional foundations of replicant experience are pre-programmed, and the replicant as genetically engineered organism is condemned to premature expiration. If we forget these facts, then we are not sufficiently attending to the depth of their suffering or to the motives for their murderous contempt *of us.* For the replicants *hate* humans. They hate us because we are more like Tyrell than them. We do no justice to Roy Batty, Leon, Pris and Zhora by pretending that they are just like us when everything they do for themselves ought to be making us aware that they feel they differ from us, and differ so strongly that mutual co-operation is almost unthinkable.

A feeling that the replicants are empathetic and beautiful, then, may grow imperceptibly into a compulsion to label them "human." The idea legitimating the label is the nondifference thesis. Critic after critic, in reading after reading, will make a remark that moves from indisputable observations—such as that they are beautiful, or that their memories are fluid—to the contestable inference that they *are human.* I limit what could be a lengthy survey to two examples. In an article in *Literature/Film Quarterly* (1998), Peter Lev comments: "The film suggests that the replicants, despite differences of genesis and history, are emotionally and morally human. This point is made by Rachael (Sean Young), a replicant who does not know her origins and is therefore completely human in behaviour. It is reinforced when Roy Batty, who seems to be *Blade Runner's* arch-villain, ultimately saves Deckard's life in a Christ-like gesture of compassion" (33). Notice how the question of membership in the historical human community is shunted aside as inconsequential, with the setting-off of the phrase *despite differences of genesis and history.* I would suggest that the story of one's belonging to the human community is no trivial matter: religions have had just the task, historically, of universalizing the human community, offering myths of creation to narrate the origins of the communities that people cherish as their own. The replicants have no such communal origin, just as Shelley's monster was unable to tell DeLacey the story of a communal origin, the Beast People had only incomprehsible laws, and Capek's robots learned too late to read books (3:74). The replicants are victims of scientism (the religion of science) precisely to the extent that Tyrell has set them from the beginning altogether outside humankind as manufactured products, biomechanical slave labour, outlaws. If they are caught having smuggled themselves onto earth, home of the Tyrell corporation headquarters, they are to be *retired* (executed) on sight. Things are no different for the non-mutinous, here-on-earth replicant Rachael. When she learns that what she has always believed to be authentic childhood memories are "implants" taken from the brain matter of Tyrell's nieces, she is tormented by her lack of an authentic *genesis and history.* And Roy Batty, in his final confrontation with Tyrell, presses hard for access to a new genesis, a new history, an opening to a fair future—that which has been denied him since his beginning. It makes little sense to insinuate that *differences of genesis and history* are inconsequential to the human.

Nor is Lev's "emotionally . . . human" a helpful phrase. The sufficient cause of being human cannot be the possession of an emotional life. To the extent that replicants can be distinguished from the android automata of *R.U.R.* on the basis of their emotional lives, the fact that they seem to feel resentment and affection matters profoundly. But it does not provide much insight to argue as follows: *all humans have feelings; replicants have feelings; therefore, replicants are humans.* This fallacy makes its way into interpretations of *Blade Runner* rather frequently. [14] Perhaps those who make the argument are operating on the subtler presupposition that the enigma of the replicants dissolves when "morally human" is set up against "amorally inhuman." But such a move only returns us to Dick's empathy criterion. Yes, *empathy* does name a trait necessary to human flourishing. But it is not a sufficient criterion or cause, ontologically or historically, of the human. Judith Barad wisely observes that it is a problem for the replicants that they are not given enough time to grow up into adult beings capable of adult empathy. Anyway, for the defender of the nondifference thesis who invokes empathy as that which makes the replicants human, there are many embarrassing facts on file. In the film's opening scene, when the replicant Leon, disguised as a janitor working inside Tyrell's pyramidal headquarters, is interrogated by the blade runner Holden, Leon murders him without empathy. Without empathy, Roy Batty and Leon tease, torture and murder the genetic designer of eyes, Chew, in his refrigerated laboratory, trying to find a way to get close to Tyrell. Batty and his female partner Pris terrorize Tyrell's employee J. F. Sebastian at his home, without empathy; without empathy, Batty murders Tyrell and Sebastian; without empathy, Batty breaks the fingers of Deckard one-by-one and tauntingly hunts him nearly to his death in the rooftop rain, only at the last minute letting him live. Other relevant facts are these: by contrast, Rick Deckard the blade runner shows reluctance, bafflement, horror, fear, guilt, and remorse in performing violently; he also shows unequivocal empathy when he risks his status for, and falls in love with, the despised replicant Rachael. Although the last-minute rescue of Deckard and dying speech of Batty are often cited as evidence of his "human" superiority to Deckard, nothing compels us (as I will later argue) to set aside the very *differences of genesis and history* that motivate and make comprehensible to us his tragic fate. I will argue that it is best to believe Batty dies as he wishes to die, not reclassified as a human but as a defiant replicant.[15]

My second example of the nondifference thesis is a variation presented by certain psychoanalytic interpreters. Kaja Silverman argues that our human memories are just as implanted as those of the replicants: "although their borrowed memories represent the feature which might seem to distinguish them most decisively from the human characters in *Blade Runner,* those memories in fact make them almost hyperbolically human" (124). Marleen Barr in somewhat the same way would wish away the concrete cause of Rachael's suffering: "Her man-made memories are no less real than the ones humans manufacture for themselves. For example, sometimes people choose to recall only the positive aspects of ambivalent past experiences. Like replicants, we too have manufactured pasts and finite futures" (28). The analogy between human and replicant memories here seems weak. *Sometimes people choose to recall only the positive aspects* is very well; but the replicants do not so choose to recall—the point of their being victims of Tyrell is that, as in all the predestinated features making them less than human, the memories they recall cannot be trusted as authentically *theirs.* Rachael's memories of her mother are implants, fake, false, unreal: to give that cruel truth a new name, and then to lump it with facts of normal human psychology changes nothing. To imply that when Deckard shocks Rachael with the revelation her memories are only implants, she has no reason to feel humiliated because ordinary human memories too are "constructed," is a little like reasoning there is no difference between having been involuntarily subjected to a lobotomy and growing forgetful in old age, there is no real difference between brainwashing and education, nor between involuntary subjection to techniques of mind control and the unconstrained expression of selfhood.

Likewise, Vivian Sobchack seems insufficiently interested in the difference between memories implanted by a vain scientist to make his products usable and memories made contingent by the vicissitudes of ordinary human psychic development: "And nowhere before in the SF film (if in Mary Shelley) has such a fully self-conscious longing for life and eloquently ferocious challenge to humanity been articulated as in *Blade Runner.* Its 'replicants' not only have human 'memories'—*given to them (as to ourselves) in 'imaginary' constructions documented and conserved as the referential 'reality' of photographic images.* Supremely self-conscious and reflexive, 'more human than human,' they are also capable of irony and poetry" (239–40) [emphasis added]. The scare quotes Sobchack puts around *memories* and *imaginary* and *reality* hint perhaps at a sense that such blurring of borders amounts to the stirring of moral quicksand. Nor is her pluralizing correct: only Roy the replicant leader is "capable of irony and poetry." The others are not. Leon twitches and stammers, then taunts and bullies; Zhora laughs evasively, cautious and fearful; Pris oscillates between the cutely self-effacing and the awkwardly duplicitous. Roy Batty alone speaks poetry. His deployments of irony express his will to transcend the horror of his entrapment. That he ironizes his nonhuman condition does not make its metahuman status dissipate into unreality. On the contrary, his speeches express resentment at the very ineradicable limits that Tyrell has built arbitrarily into his replicant being.

Some parameters of the nondifference thesis and the ways it depends on a curious evasion of some of the more painful facts of the tragedy should be visible by now; I could add many other voices (Senior 10; McNamara 423; Tiitsman 39–40). By insisting they are not human, I do not aim to victimize further the already-victimized replicants by insisting they are not human. On the contrary, I believe the prevalence of the nondifference thesis is comprehensible. I am arguing that postmodern versions of the Frankenstein myth in general minimize the difference between real and impossible human. But my goal is to study Scott's vision in *Blade Runner* in the company of the visions of Shelley, Wells, and Capek—which means I am less inclined to overlook the scenic centrality of the mad scientist Eldon Tyrell. Tyrell is a human who has played God; we might direct more suspicion toward him rather than slipping into a victimary defense of the replicants and despising the police whom Tyrell seems to have in his pocket.

Corporate Science and Postmodern Paranoia: Tyrell as Scapegoat

The first corollary of this chapter's thesis is the suggestion that *Blade Runner* participates in the general contempt for realist scientific epistemology prevalent in postmodern theorizing, insofar as the movie converts the vain scientist into an isolated, excessively powerful figure of corporate scientism. To describe the ethos of *Blade Runner* as paranoid would not be inappropriate. Experts credit Dick with shifting the direction of science fiction away from the technology-celebrating optimism of Isaac Asimov toward a vision of technology as dangerous and dehumanizing. Patrick Parrinder made the observation several decades ago: "Post-1960s SF responds to *the era of multinational corporations* with its vision of a neo-feudal world in which the individual is condemned to servitude. Such visions reflect the apprehensions and fears of an ever-broadening social group in today's society" (*Science* 34) [emphasis added]. Later in the book, Parrinder names Dick as a leader in promoting what he names the paranoid vision: "Like Vonnegut's comedy of the absurd, the fiction of the paranoid vision today spans both the 'mainstream' and the science-fiction categories. In science fiction, its representative (and highly prolific) exponent is Philip K. Dick" (119). Warrick and Greenberg, editors of the *Robots, Androids, and Mechanical Oddities* anthology, imply that Dick in his blanket mistrust of centrality did not discriminate between forms of social order: "He was opposed in principle to all authority figures, just as he was opposed to war and violence" (84). Notice in such a formulation the victimary assumption that "authority figures" will cause and create "war and violence," but can never have anything to do with containing or limiting violence. Questioned about the famous opening shot of the movie in which a single human eye fills the entire screen, Scott himself suggested to Sammon that *Blade Runner* is suffused with paranoia: "You hit it. *Blade*

Runner, in a sense, actually is about paranoia. And that eye underscores Deckard's dilemma, because by the end of the film, he believes he may be a replicant himself." This exchange occurs just before.

> [Scott:] That the world is more of a controlled place now. It's really the eye of Big Brother.
> [Sammon:] *Or Eldon Tyrell?*
> [Scott]: Or Tyrell. Tyrell, in fact, had he lived, would certainly have been Big Brother. (382)

Scott's comments mix metaphors here. George Orwell's Big Brother in the futuristic dystopian novel *Nineteen Eighty-Four* (1949) was a figure of totalitarian dictatorship and a statist economy; by contrast, Tyrell is a figure of over-powerful independent capitalist accumulation, served up to crystallize resentment of a system of market exchange. Scott makes the two into one. As Parrinder observes, postmodern science fiction generally collapses distinctions between statist central authority (at the extreme, the disorder of a disastrously overplanned economy) and decentralized capitalist markets (at the extreme, the disorder of a disastrously unregulated economy). In the postmodern ethos, the sturdy faith that democratic institutions might provide ways to negotiate individual and collective resentments in the social order has more often than not been abandoned. The substitute for that faith (from my perspective, a victimary substitute) seems to be a generalized distrust of authority, more specifically, a conviction that "corporations" rule the world as if detached from every other brute and institutional fact on the planet; and this conviction is combined, not without awkwardness, to assumptions about the viability of an undefined utopianism. In *Blade Runner,* Tyrell is elevated and condemned at once to the status of mythical scapegoat; the elevation and condemnation go to extremes farther than those suffered by Victor or Moreau or the team at Capek's robot factory.

This victimary fusion of resentment of science with resentment of corporate power characterizes almost every postmodern version of the Frankenstein myth (Atwood; Bay; Cook; Darnton; Hyman; Koontz; Mathews; Nova; F. Paul Wilson). The making of the vain scientist into hireling or C.E.O. of an evil corporation may well be the most significant addition-alteration in the basic pattern of the Frankenstein story during the postmodern era. Why? To the extent that the fusion collapses all science into scientism and all power into oppression, the postmodern version of the Frankenstein myth inspires attitudes in its readers and viewers less politically mature than those of its predecessors, by creating an *us and them* division between the reader-spectator and the Victor figure. For nothing in the real world warrants the fictive reduction of scientific and economic institutions into one figure of central control. There is a certain evasiveness of the real in such paranoia, to which Hans Bertens alludes in the following critique of postmodern attitudes.

> The battles have mainly raged . . . over blueprints for impossible societies and over the political potential of artistic strategies that were mainly known to the readership of *October, Art in America, Parachute,* and other highbrow journals. There is virtually no sense of real, concrete politics in these discussions. One finds no theories of the (postmodern) state, no theories of power, no discussion of, say, the legitimate use of violence, no theories of the macropolitics that highly complex late twentieth century societies obviously cannot do without. The postmodern Left has great trouble to let go of the paralyzing idea that power is repulsive. (112)

In Bertens' phrase *the paralyzing idea that power is repulsive,* we find described a counterpart to Dick's mistrust in principle of authority and a correlate to the scapegoating of Tyrell. To organize a chorus of cultural voices around a scapegoat, as Rene Girard has taught us, means to cease thinking for oneself, as a human who might defer his or her violence by differing from others in the mob (by exiting the mob). Bertens' claim that postmodern thinking is victimary to the extent that it condemns all violence rather than asking difficult questions about *the legitimate use of violence* also shows that he realizes the postmodern is the victimary. To ask questions about the legitimate use of violence in *Blade Runner* is to resist the false dilemma of siding either with the replicants or with their executioners.

To fix on Eldon Tyrell is to feel our postmodern resentment crystallizing. Many interpreters of *Blade Runner* invoke the Marxist heavyweight Frederic Jameson, relying on his pronouncements about the weight of corporate oppression in postmodern times. Joe Abbott's 1993 study is one example, with its claim that the corporation is a "determining essence" in the movie's futuristic fictional world: "But the determining essence in *Blade Runner's* postmodern world is the impersonal and monolithic Corporation with its sullied capitalist ethic, an institution that Frederic Jameson has located firmly within the whole world system of present-day multinational capitalism" (343; see also Slade 12; Wood 162; Byers 331). Tyrell himself alone is corporate capitalism; Tyrell merges scientism and executive political power into one fantastical figure. The romantic Victor worked alone in his laboratory; the modernist figures Moreau and Rossum exploited the distant solitude of south Pacific islands; but the postmodern mad scientist occupies the top floor of a gigantic pyramidal skyscraper, wholly insulated from the vulgar vagaries of market exchange, as far removed from sweaty social entanglements as any Babylonian demigod ruler who existed millenia before Christ.

Many contributors to the *Blade Runner* literature concur that Eldon Tyrell seems almost omnipotent. What has not been sufficiently emphasized is that Tyrell's mythic figuration incurs the cost of a discursive oversimplification. Giving way to a certain thoughtlessness, we end up satisfying ourselves with as a scapegoat. The merging of socioeconomic power and technological expertise into one figure is symptomatic of postmodernism's loss of faith in historical progress and its mistrust of authority as oppressive and dehumanizing. Making the scientist into an unapproachable demigod cashes in on the postmodern tendency to seek persecutor-victim models of social interaction: Tyrell persecutes, the replicants are his victims. His pyramid tower invokes the unilateral imperialism and slave-labour of the archaic empires—Mayan, Sumerian, Egyptian. An image representative of a rigidly hierarchical ancient social formation is superimposed over images representative of the consumer capitalism that seems to have demoralized the inhabitants of Los Angeles 2019.

Precisely inasmuch as Tyrell is a *mythic* figure, not represented as one situated in a network of nuanced economic and social relationships, *Blade Runner* is less sociologically mature than Shelley's novel, Wells' *Moreau,* or Capek's *R.U.R.* A few observations may bring into view the comparative maturity of the romantic and modernist texts instantiating the myth. His economic uselessness (caused by his ugliness) condemns the Wretch to infinite isolation; Mary Shelley invests in the paradox by which Victor's self-destruction mirrors the unexchangeability of the creature itself, in that Victor refuses to *share* the burden of his technological know-how: there is mutual influence between scientistic experimenter and victim, not unilateral oppression. Wells similarly gives Moreau, the scientist most dedicated to "pure" research, the motivation of returning to England to prove wrong those rivals who hounded him from the country as a charlatan: Moreau's revenge motive ties him to a larger network of exchange. Capek explicitly details the mutual complicity of consumer and producer in creating the success of the worldwide robots, especially in the argument of Busman: "Demand controls production. The whole world wanted its Robots" (2:59). However, in *Blade Runner,* there is nothing resembling an equivalent dialogic exposition or an unequivocal allusion to the exchange system in which the replicants function as slave labour. We accept that Tyrell is a narcissistic commercial agent who makes and sells these creatures, but we do not see people buying them or employing them (we do not go off-world). Tyrell sits isolated and aloof, mounted weirdly atop the overcrowded environment of jumbled styles and strangely beautiful wastelands below, as if he alone could have created an economy in which nobody who consumes is connected to anyone who produces. By making all those except the mad scientist, humans and replicants alike, into victims, such a mythic structure contributes to the minimization of the human-replicant difference.

Falling in Love with the Impossible Human

I turn now to the second of my three supporting suggestions: another way *Blade Runner* minimizes the differences between the real and impossible human is by having a human fall in love with a replicant. Here we encounter one interpretation of *Blade Runner* that keeps the nondifference thesis internal to the structures of a conventional Hollywood romance. This line of analysis foregrounds what we might call the *redemption narrative:* Deckard's decision to protect Rachael redeems him, and the human Deckard (as opposed to the replicant Batty) is the hero the movie wants us to celebrate. Even so, a human's falling in love with a replicant props up the nondifference thesis. At the beginning, one observes, Deckard is the cynical and disaffected detective of conventional film-noir. However, he witnesses the trauma Rachael suffers at the revelation she is a replicant; he feels gratitude that she has saved him from Leon; he appreciates her strange self-possession so much that he falls in love with her—and she reciprocates. From one who submits to the order that all replicants on earth must be "retired," Deckard changes into one who defiantly admits the "humanity" of at least one replicant—Rachael.

One point in favour of this view is that it highlights certain atrocities of Tyrell in a way that privileging the Batty—Tyrell conflict does not: it highlights the way that Tyrell abuses and betrays Rachael in her first scene. When we first meet Rachael, Deckard and she herself both believe she is human. The dialogue between them quickly comes to the topic of the boundary between natural and synthetic creatures. Rachael asks Deckard whether he likes their artificial owl; she insinuates he is a throwback.

> RACHAEL: It seems you feel our work is not of a benefit to the public.
> DECKARD: Replicants are like any other machine: they're either a benefit or a hazard. If they're a benefit, it's not my problem.
> RACHAEL: May I ask you a personal question?
> DECKARD: Sure.
> RACHAEL: Have you ever retired a human by mistake?
> DECKARD: No.
> RACHAEL: But in your position, that is a risk.

Her questions extract from Deckard a crude antithesis that he will later heroically subvert: Rachael will become for him both benefit and hazard: benefit as rescuer and lover, hazard as the clandestine object of his protection. Rachael probes further: she asks "Have you ever retired a human by mistake?" and insists in response to Deckard's negative that such regrettable mistakes must, despite his clean record so far, be a possibility: "But in your position, that is a risk?" There are multiple ironies in her teasing him about the inevitability of the risk. At a first level, she elevates herself as one close to the science of Tyrell and demotes Deckard as a hired gunman. Speaking from her assumption that she is human, Rachael is already arguing for the minimality of the difference between replicant and human: it is impossible to say whether the minimizing means she is aloofly amused at the difficulty the corporation has created for employee executioners, or it means she would take a "progressive" stand (reminiscent of that of Hellena Glory defending R.U.R.'s android automata) that defends Tyrell's science for its creation of "real" humans (she may be insinuating that to kill a replicant is as wasteful as to kill a human: they are ontologically equivalent). At another level, that of narrative irony—given the later revelation she is a replicant—Rachael in her hauteur errs by believing that the risk of being killed as a human counts most, for only as a replicant will she feel that fear.

The slow administering of the Voight-Kampff test on Rachael increases our sense of both her self-possession and her vulnerability. Tyrell tells a lie even before the test is administered. He demeans both Rachael and Deckard.

> DECKARD: Where's the subject?
> TYRELL: I wanna see it work on a person. I wanna see a negative before I provide you with a positive.
> DECKARD: What's that gonna prove?
> TYRELL: Indulge me.
> DECKARD: On you?
> TYRELL: Try her.

Tyrell's lie is the insinuation that since Rachael is human—"a person"—the test will prove harmless; there will be nothing cruel in its outcome. His lie is the promise that he will provide a "negative" result, that the test is guaranteed to fail. But the Voight-Kampf machine succeeds: it exposes Rachael. His indifference to the humiliation that Rachael is bound to undergo is captured in the command, "Indulge me," an appeal to the value of his personal pleasure, in keeping with his much later command to Roy Batty, "Look at you." Rachael smiles, unafraid, and she agrees to take the test. Tyrell also humiliates Deckard. As in his dealings with his boss Captain Bryant, Deckard is forced to be cruel here, made party to a deception without his consent.

The editing tells us many moments pass before the questions stop and Deckard shuts down the machine. Sean Young as Rachael betrays the slightest uneasiness in a sideways glance before rising from her chair and walking out of the room.

> DECKARD: She's a Replicant, isn't she?
> TYRELL: I'm impressed. How many questions does it usually take to spot one?
> DECKARD: I don't get it, Tyrell.
> TYRELL: How many questions?
> DECKARD: Twenty, thirty, cross-referenced.
> TYRELL: It took more than a hundred for Rachael, didn't it?

> DECKARD: She doesn't know?
> TYRELL: She's beginning to suspect, I think.
> DECKARD: Suspect? How can it not know what it is?

Jason Vest is correct to praise Harrison Ford, playing Deckard, in this scene: "Ford's performance in this scene is masterful, moving from the bored indifference of his initial conversation with Rachael through neutral efficiency while administering the Voight-Kampff test to quietly contained shock at Tyrell's methods" (13). The dialogue quickly conveys a conflict between the blade runner whose conscience is troubled and the vain scientist whose conscience is not. For his part, Deckard is scandalized by this new kind of replicant unaware that it is one. Dick explored that type of synthetic human in some of his most brilliant short stories: "Imposter" and "To Serve the Master" take the dazzled reader inside the consciousness of a very humanlike robot programmed by his creators both to be constitutionally oblivious and in denial of his own robot nature, and yet fated to discover the shocking delusion through his own concerted efforts. I would go so far as to say these horrifying stories outdo that of Sophocles' Oedipus in the inscrutable cruelty of the ironic fate assigned their benighted robot protagonists. The newness of the Rachael-type replicant that *does not know what it is* therefore adds a dimension of psychological horror to the administration of the test, for now the test destroys a creature's whole sense of personhood. Deckard's switch from *she* to *it* is best read not as a sign of heartlessness, but as a linguistic self-shielding from the pain he has without consent inflicted on another sentient, self-conscious being. By using the pronoun *it,* Deckard betrays feeling stunned by the unwitting character of his deception and distances himself from the victim he wishes he had not helped to hurt. Meanwhile, Tyrell wants to know only the numbers—*How many questions?*—a quantification of the qualitative difference Deckard appreciates.

In response to Deckard's question *How can it not know what it is?* Tyrell presents an account of his scientific and business practices.

> TYRELL: Commerce is our goal here at Tyrell. "More human than human" is our motto. Rachael is an experiment, nothing more. We began to recognize in them a strange obsession. After all, they are emotionally inexperienced, with only a few years in which to store up the experiences which you and I take for granted. If we gift them

with a past, we create a cushion or pillow for their emotions and consequently we can control them better.
DECKARD: Memories. You're talking about memories!

I analyzed *more human than human* above. Everything implied by its context here, by Tyrell's reductionist declaration *Rachael is an experiment, nothing more,* supports my contention that he intends no tender-hearted compliment with the slogan. For Tyrell, to be more human than human is to be in fact less human than human, to be an *experiment,* a product built for controllable efficiency and planned obsolescence. For Victor, the Monster is an experiment, nothing more; for Moreau, the tortured puma is an experiment, nothing more; for Dr. Gall, the robot injected with resentment is an experiment, nothing more. The speech suggests that for Tyrell, their having *only a few years in which to store up [their] experiences,* that is, their subjection to the corporation's built-in four-year lifespan, is likewise nothing that might require a moral justification. Apparently, there are no well-funded bioethics commissions in the world of Tyrell. The accidentally "natural" self-humanizing of the replicants (somewhat like the natural evolution of Capek's primary humanoids) creates an obstacle for his business plan, but he manages it as a technical problem. Less likely to be dissatisfied when real memories are cushioned inside false ones, the existential hunger of emotionally developing replicants gets programmed out. In other words, the Tyrell Corporation can "control them better" by implanting the delusion of a past. The "gift" is a con. In reality, replicants will always only have existed for four years; in their controlled minds, they will feel they have lived longer than that.

When he invites Rachael by videophone to join him at Taffey Lewis' Snake Pit Bar, Deckard begins to show a romantic interest in her. On-screen, Rachael turns down his offer, but then appears on the street after he has executed Zhora and his boss Bryant has praised him as a "one-man slaughterhouse." She shoots Leon in the head with Deckard's kicked-away blaster. The action returns to Deckard's apartment, for the longest quiet scene in the film: "sixteen minutes of screen time take place with Deckard and Rachael in Deckard's apartment where there is sparse dialogue, much silence, and a subtle and skillful integration of music and visuals" (Morrison 8). Deckard promises Rachael that he would not hunt her if she fled to the North, but warns her that another blade runner would. He falls asleep after washing and making himself a drink: Rachael gets no answer to her question of whether he has ever taken the Voight-Kampff test himself. Rachael studies his vast array of old photographs and, imitating one feminine portrait, slowly unties her hair and spreads it out. She places her fingers over the piano keys, hesitates, but then plays a soft melody. The Vangelis soundtrack is perfectly attuned to the yearning mood of the moment: a moment of introspective self-gathering after a hectic collision with brutal, random violence. Deckard wakes up and joins her, seating himself beside her on the piano bench and studying her face.

> DECKARD: I dreamt music.
> RACHAEL: I didn't know if I could play. I remember lessons. I don't know if it's me or Tyrell's niece.
> DECKARD: You play beautifully.

These three lines prepare for Deckard's kiss of her cheek; in my opinion, it is the moment in the Rachael-Deckard scenes where Harrison Ford most clearly gets his character to convey in a convincing manner some tenderness for Rachael. Although here she remains a creature shocked and fragile, no longer the self-possessed *femme fatale* of the scene in Tyrell's shimmering office, Rachael has begun to reconcile herself to the fact her memories were implanted. Without melting into tearful distress, she acknowledges the impossibility of recognizing the authenticity of any particular memory; she may now live with the impossibility of distinguishing real and artificial. Likewise, Deckard can live with the impossibility of telling the difference between real and artificial human. I have already emphasized how essential replicant beauty is to the minimizing the difference of the postmodern impossible human from the real human. Deckard's esthetic judgment *You play beautifully* implies the irrelevance of the historical origin of the playing: its beauty suffices, rather as the beauty of the replicants suffices to make them all but indistinguishable from humans. Rachael here and now is really, beautifully human in every respect that counts for Deckard (and for us in the audience).

The narrative arc comprising the love-interest is marginally reanimated by the film's closing scenes. Having come in his spinner to verify the death of Roy Batty on the rain-soaked rooftop, Captain Bryant's assistant, Gaff, congratulates Deckard: "You've done a man's job, sir. I guess you're through, huh?" The phrase *a man's job* is ironic, given that Gaff's tinfoil unicorn will later imply that he knew Deckard had done *a replicant's job:* he was a replicant killing replicants. Gaff pauses before exiting, turning to speak this enigmatic line: "It's too bad she won't live. But then again, who does?" He thus informs Deckard that he knows of the love-interest, of Deckard's outlaw insubordination. Edward James Olmos, who played Gaff, informed Paul Sammon that he believed Gaff had changed: "[he has] come to a point where he empathizes with [Deckard]. That's why he lets Rachael live" (198). More importantly, the line minimizes the difference between replicant and human: Gaff approves of Deckard's human love for a replicant partner with whom he will enjoy only limited time. A short, intense life can be a complete life; the replicants' being condemned to a built-in immaturity is one of the ways *Blade Runner* invests in postmodern youth culture. (I cannot help but hear Debbie Harry of the 1980s New Wave band Blondie singing their hit "Die Young, Stay Pretty.")

Any complete account of *Blade Runner* ought to acknowledge the interpretive option of emphasizing the love affair of Deckard and Rachael. One thing in favour of the option: without this romance, the film seems almost unbearably bleak. Deckard would have nowhere to go in the end. The killings of Holden, Chew, Sebastian, Leon, Zhora, Pris and Roy would pile up to create an effect as sordidly disheartening as that of the most blood-puddled final act of a third-rate Elizabethan revenge tragedy. However grimly uncertain we must suppose the fictional future of Rachael and Deckard to be, in the final image of the elevator door closing on them to take them down to earth from the 97th floor, we can find a sign of tense hope in the private sacred that *R.U.R.* deployed (in the Primus-Helena exchanges) to offset the horror of universal mechanization. But the waste and horror we in the audience endure before seeing the new couple standing side by side on the brink of their descent ultimately outweigh any joy we might feel. Nobody dances at the end of *Blade Runner*. To try to fit *Blade Runner* into the model of a story where Deckard's "transcendence" or personal spiritual growth is key accounts for some data, but ignores much more. If the movie's mythos is that of tragedy, we must turn to its tragic hero to find the dramatic center. We must turn to the leader of the replicants who did not have the delusional "cushion" that mellowed Rachael into a lovably harmless being. We must turn to Roy Batty.

Victimary Thinking and the Human/Replicant Boundary

No quality of harmless, easy lovability is attributable to the replicants whose journey to earth is delivered along a plotline more emotionally charged than Rachael's drama. Roy Batty and his brave mutineers fail in their quest. The victimary tendency in the critical literature has built a near-consensus that seems to authorize pointed expressions of outrage at the violence of Deckard while passing over with forgiving glances the violence of the replicants. Many analysts tend to moon over Batty's dying lines while neglecting to wrestle with his unsettling earlier confrontation with Tyrell. The victimary attitude leads one to avoid analysing the murder of Tyrell because the more one tries to understand it, the more one faces the uncomfortable questions raised by Roy's violence. It is easy to map the ever profitable persecutor-victim binary onto the Tyrell-Batty conflict. The conclusions about *Blade Runner* the most suspicious of the "humanity" of the replicants appear in the studies that ask the hardest questions about the justifiability of replicant violence. From the perspective of generative anthropology, the clarifying effect of a focus on violence should not surprise us: the originary hypothesis posits that the primary function of human culture is the deferral of violence through representation. My thesis in this section and the next continues to be the claim that the *postmodern* Frankenstein myth, in minimizing the difference between real and impossible human, returns us to the scenic source of culture and frees us from the delusion that the human is an objective "natural" category. It is not true that by technological means, just by rearranging DNA, humans may get "made" without reference to the scenic exchange with a sacred center that is the source of language and our contact with a spiritual Other. More specifically, I will in the next section claim that the failure of Tyrell, far more than that of Victor or Moreau or the R.U.R. executives, is a moral rather than a technological or epistemological failure. Tyrell is held to account for playing God as a moral judge rather than a physiological fabricator.

Considering the replicants Scott has imagined for us, we feel pressed to judge them as morally responsible agents to an extent that we do not even consider judging the beast people or Capek's android automata. Nor do they cut the special figure that Mary Shelley's monster cuts, unique in one-of-a-kind loneliness. Unlike Batty, Victor's Wretch does not have a Pris to love. The replicants approach us as closely in their human-like beauty as Shelley's monster hangs distant from us in the exile of his ugliness. The curse suffered by Shelley's monster was that of not being able to participate in the human economy, not even as an invisible servant; the curse suffered by the replicants is one of being forced despite their wills into the center of a hideous techno-slave economy. In Los Angeles 2019, the most advanced technology has created a mode of production reminiscent of that of the most "primitive" archaic empires because it is, perversely, one built on the primitive and the cutting-edge at the same time: cutting edge because of the "genius" of Tyrell in creating androids, but primitive because the incarnations of that technology are reduced to slaves. And these android slaves, unlike the unfeeling android automata of *R.U.R.,* face us as really human-like. The crawltext at the beginning of the film chills: that impossible humans returning to earth are to be shot on sight suffices to reveal the totalitarian violence of the world we enter. As the movie unfolds, we find nothing in the organic design of replicants that marks a biological difference adequate to justify such execution without trial. I wish to review just how much "like us" the replicants are, so as to clarify that my rejection of the nondifference thesis does not entail contempt for them or obliviousness to their human-like features.

Many details accumulate to create the impression of an oscillation in the replicants between adult behaviour indicative of an intelligence dangerously superior to that of the average human, and (in contradiction) behaviour indicative of a vulnerable immaturity and innocence. When Roy joins the only surviving replicant Pris, he appears in Sebastian's apartment behind the seated Sebastian; with the trembling relish of a wowed boy, he says "Ah. Gosh. You really got some nice toys here." But he suddenly reverts to the self-possession of his menacing warrior posture: "Sebastian. I like a man that stays put. You live here all by yourself, do you?" Batty and Pris next kiss with a staccato hesitation, as if they are adolescents experimenting with the idea of being girlfriend and boyfriend—kids trapped in adult bodies. Another example of the vulnerable immaturity: only seconds later, when Roy informs Pris that Zhora and Leon have been killed—"There's only two of us now"—he rolls his head, glances upward, and seems on the verge of crying like a child. One more example: Leon, oversized and goofy, reveals to Batty that he has been unable to retrieve his photographs because Deckard has searched his apartment and confiscated them as clues; his hurt is betrayed in a pouting expression of silenced chin-down sadness. But this indrawn sulking melancholy follows his gunblasting murder of Holden, and it vanishes as quickly in the sadism of his tauntingly slow attempted murder of Deckard in the street. Defenders of the nondifference thesis might claim that such behavioural oddities alone do not legitimate our withholding human status from the replicants; I would agree. Tyrell makes the replicants dangerous in the same way children with loaded guns are dangerous, but a loaded gun alone does not make the child holding one inhuman. The dangerousness of a humanoid creature with an adolescent self-concept and a great capacity for violence resembles the dangerousness of many fully human inmates in our prisons and psychiatric wards, not to mention that of countless young soldiers trained to kill and take their state-paid wages all over the planet.

I analysed the problem of replicant "memory" above. Replicants suffer the peculiarly scandalous condition of being endowed with seemingly authentic but actually implanted memories—memories subjectively true but objectively false, true for the internal victim but false to the external judge. Would I argue that the replicants are nonhuman because they have been deprived of "real" childhoods and designed to intrude into human history as parasitical entrants, their blank brains injected into adult bodies and their bodies thrown into the off-world workforce? I would not. In suffering these defects, replicants can no more be disqualified as human than people suffering accidental amnesia, or people who have lost their memories and identities due to Alzheimer's disease. To classify them as expendably non-human because Tyrell has endowed them only with inauthentic memory implants would be unfair. With such a line of reasoning in their defense, I agree again.

The mutual "empathy" the replicants are said to demonstrate, used as a counterweight to the violence they display in their quest for longer life, raises more invidious questions. Here, I believe the arguments for nondifference become weaker, because in the critical literature some victimary proposals aim not only to elevate the replicants by "inclusion" in humankind, but also to demote the humans in the movie. For example, the cliché that Rick Deckard behaves like an automaton or machine runs throughout the *Blade Runner* literature. The truth is that Deckard feels horror at the violence that he must perform, which feeling opposes him to the replicants, who appear to delight in exercising violence and who, when threatened by it, display fear but never esthetic responses of horror. Harrison Ford is on record claiming that he was "anxious to make sure that this character [Deckard] represented an abhorrence of violence" (Kolb

152).[16] I believe the representation was a success. Equally invidious is the claim that the replicants are emotionally closer to one another than the humans in the movie. Details that would compel qualification of this claim are routinely ignored: for example, Batty *mocks* Leon's attachment to his photos by calling them "precious"; Leon is not shown mourning (not even for an instant) the murdered Zhora, but shown moving without hesitation to attack Deckard; at the news that Leon and Zhora have been killed, Pris shows less sadness than childlike fear at the prospect of her own death. Conversely, it would not be difficult to perform a counter-analysis making a case for recognition of the submission to an ethic of concession and co-operation shared by the boss Bryant, Gaff, Deckard and Rachael, an ethic opposed to the incautious individualism that follows the leadership of Roy Batty. Deckard works *with* Gaff despite their rivalry, obeying Bryant's demand for a visit, searching Leon's apartment. Gaff communicates with Deckard by means of origami; Gaff lets Rachael live, defying the orders of Bryant; Deckard takes risks protecting Rachael. Those left alive in the end owe their survival in part to a self-preserving deferral of resentment and violence. By contrast, the replicants' built-in possession of a capacity for superior physiological violence does not protect them.

I outline such details less to reject the pro-replicant sentiment than to suggest that taking sides for replicants or against humans is something to avoid. I do contend, however, that accounting for the replicants' violence (their disposition to it and the outlandish persecution that provokes it) is the crucial problem in drawing the replicant-human boundary. To defend the replicants as killers requires the observation that the violent action they take seems all too human. Most humans similarly enslaved, denied civil rights and legal protections, and objectified as exploitable labourers, would certainly be tempted to react and rebel as Roy and his collaborators do. The replicants' suffering might not justify the killings they carry out, but we can understand their rage against the human species whose abuse or indifferent complicity has caused them such pain. We can understand the contempt that motivates (for example) Leon's taunting of Deckard or Roy's intimidation of Chew (see, for example, Gwaltney 33; Abbott 345). Some critics even describe the replicants' revolt as a "revolution": the "blame" belongs to the human oppressors, who came first; the replicant oppressed (who come second) are no way to blame for their violence. [17] A different argument claims that we may hold Roy and his rebels to a moral standard other than that applicable to humans because they have been "programmed" to do violence and therefore, in an important sense, they are not free to choose (e.g., Morrison 6; Barad 23; Vest 23). But such reasoning ironically "dehumanizes" them: if we cannot hold replicants morally responsible for the violent acts they perform, by the same token we cannot respect them for a free will they do not have.

Let us return to this chapter's thesis: the only thing that makes the replicants other-than-human is the historical fact that Tyrell has created them. I would suggest our accepting a certain ratio. In proportion as one justifies or excuses the violence of the replicants by appealing to their status as creatures manufactured by the Tyrell corporation to suffer in unique ways, one concedes their *meta*-humanity. Either the mediation of Tyrell as maker grants to the replicants special considerations at the price of conceding their ontological difference, or their ontology is one with ours at the price of removing from them any circumstantial concessions following from the way Tyrell has thrown them into history. In short, those critics who seek to justify replicant violence should not simultaneously posit the irrelevance of the replicants' ontological difference from us: replicant violence is justified, the defense argues, because replicants are just like us. Ironically, however, I would point out that replicant violence may be more justified if we argue that they are unlike us. If we take into account the mediation of Tyrell as the vain tyrannical scientist who created them, then we can apply concessions that lessen our readiness to blame them. Regardless, I insist only that one's argument be consistent with respect to its decision regarding the presence of Tyrell as vain scientist. If one wishes to make moral exemptions derivative from their relationship to Tyrell, then one must concede the ontological fact of the derivative nature of the "humanity" of the replicants.

On the Vanity of Eldon Tyrell

Almost everything in the crucial dialogue between Roy Batty and Eldon Tyrell emphasizes the ways in which Roy is *not* "human" but—is it not obvious?—a special case, a unique being, different from us as viewers precisely because he alone faces Tyrell, while we cannot fully imagine what it would be like to have been Batty. The point is to grasp that in some way Batty, like all impossible humans, has suffered a condition that we find unimaginable, that comes to us from a "beyond" precisely because he does not share our genesis and history. When he speaks, we hear a voice from the other side of the curtain that separates us from a totally godless world where scientism has overwhelmed religious wisdom and anthropological sense.

Let us return to the narrative action of the movie. Pris has charmed herself into the apartment of Sebastian, a lonely genetic designer who works for Tyrell. Roy has joined her there. Without making any explicit threats, Roy has forced Sebastian to deceive Tyrell and thus help him gain access to Tyrell's private rooms in his gigantic corporate home. Sebastian's ruse of pretending a late-night wish to continue a longstanding game of chess with Tyrell has worked. Batty whispers to Sebastian the moves to make; Tyrell has left his bed and crossed to his chessboard, inviting Sebastian in. He does not know he is letting in Roy Batty as well.

> SEBASTIAN: Mr Tyrell? I . . . I brought a friend.
> TYRELL: I'm surprised you didn't come here sooner.
> ROY: It's not an easy thing to meet your maker.

It is painful to witness the harmless Sebastian, who loves his "toys" and loves to "make friends" (a pun he repeats when proudly displaying his humanoid artifacts to Pris), being forced to betray Tyrell, especially given that Tyrell seems to be the only real human friend Sebastian has. Scott in directorial advice to William Sanderson described Sebastian as "totally innocent" (Sammon 142). "I brought a friend" is a deadly irony: Tyrell and Sebastian both know Roy is an enemy. Tyrell (played by Joe Turkel) pulls back with the slightest upward tilt from the hips, just a little, tightening the belt on his long white robe. He feels fear. As viewers, we share that fear. The contrast between an elderly, wrinkled, bespectacled scientist and a tall, vital, Aryan replicant in leather coat with a vampire-like collar does not promise that if things come to blows, it will be a fair fight.

Roy's opening line *It's not an easy thing to meet your Maker* is one of the most brilliant puns in the script (Schwartz 46). It is powerfully ostensive. His assertion that the meeting has been difficult to achieve ties the mock-God-impossible-human encounter to its previous moments in Roy's quest: Roy taunted the freezing Chew in his lab with the line "Not an easy man to see . . ."; he menaced Sebastian with the understatement, "From what I understand, he's sort of a hard man to get to." With the phrase *not an easy thing* Roy hints to Tyrell that the difficulty has required intimidation, extortion, and murder preliminary to this here-and-now unscheduled appointment: the culmination is this meeting, here, now. Tyrell has had to suffer and risk nothing for the meeting; his creature has had to risk and suffer everything; a reproach is implied. Meanwhile, the phrase *to meet your maker* plays with at least three levels of implication. *To meet your maker,* meaning *to die.* Roy has been fighting death, fighting his four-year lifespan, from the outset of his mutiny. He does not find it easy to go gentle into that good night. He would shame Tyrell, implying something like *the prematurity of my death is painful, and you are the cause of it.* Second, to meet your maker means *to meet God your eternal judge after you die.* But Eldon Tyrell is not God; Eldon Tyrell is pinned up by the ironic label *maker* as a paltry substitute for God. He is a vulnerable old man, all the state apparatuses that he might have expected to insulate him in his safe-house having failed. In other words, the pun suggests this: *you may be my maker but you are certainly not God, the omnipotent "maker" humans name in this expression.*

A third level of implication plays with *maker* as God, the one who, now reaching into eternity or the afterlife, can redeem beings from death. The cliché *prepare to meet your maker* refers to the God whom humans meet after they die; Tyrell's not being God is plain in that his metahuman victim confronts him now—here is a God one may meet (illogically) *before* dying. The immeasurable difference in responsibility, in capacity, in appropriateness, between the hubristic maker of impossible humans that Tyrell really is and the absent divine Creator whom he has mocked, is weirdly self-evident. An apocalypse is imminent, one in which meeting one's maker will amount to a sound and fury signifying nothing. *It's not an easy thing to meet your maker.* Finally, meeting this maker does not mean redemption. Just as Victor tears to pieces the mate that his monster has requested he provide, just as Moreau cannot cut from the puma a new superhuman being, just as Alquist fails to find the secret of robot production when he cuts Damon, so does Tyrell fail to provide the biomechanical secret to expand Roy's lifespan. The vain scientist succeeds as initiator but fails as sustainer of his creations. He is a mock God who cannot redeem his lost creature. Like the meetings between Shelley's Victor and the monster, between Moreau and the puma, between Alquist and Damon, this one stages the violence of a non-reciprocal waste. The mad scientist forced to face the empty end of his mechanistic materialism is no substitute for the scenic center (or central Being) that mediates the origin of real human community and human exchange.

Over the next few lines, Tyrell in nervous self-preservation shows uneasiness, as Sebastian is relegated to the background and Roy closes the physical distance between them.

> TYRELL: And what can he do for you?
> ROY: Can the maker repair what he makes?
> TYRELL: Would you like to be modified?
> ROY: [to Sebastian] Stay here. [to Tyrell.] I had in mind something a little more radical.
> TYRELL: What . . . what seems to be the problem?
> ROY: Death.
> TYRELL: Death. Well, I'm afraid that's a little out of my jurisdiction. You—
> ROY: I want more life . . . father!

Tyrell's second question (*What seems to be the problem?*) implies that at least some minor changes might be possible, while Batty's wish for "something more radical" implies likewise a distinction between minor and major, cosmetic and radical changes. But Roy's naming the problem with the word *death* obscures things; it muddies the specific injustice inherent in replicant-human difference. For as surely as the impossible humans for whom Tyrell is responsible must die, so must "real" humans for whose mortal destinies Tyrell is not responsible similarly die. Tyrell senses an out: immortality is beyond his ken, and he shrugs off the request as a thing beyond the ability of any scientist, even the fabulously rich genius at the top of the Los Angeles world that he is. For an instant, we side with Tyrell inasmuch as we feel the unreasonableness of an imperious desire for immortality.

Ultimately, however, the clear focus of Roy's desire returns with the line he speaks when getting close up to Tyrell, in a shot where his glowing eyes and shining face fill the screen. *I want more life . . . father* insults Tyrell; it expresses a righteous indignation. But in the context of the alliance between youth culture and *Blade Runner*, it casts the replicants as creatures whose essence is that of a painful yearning: although they have tasted enough life to want more of it, their maker has always already programmed them to die young. Shelley's monster suffered the curse of ugliness. Wells' beast people suffered the tortures of Moreau's scalpel; Capek's committee robots suffered the fate of extinction. Replicants suffer the curse of always, everywhere, dying young. The special curse endured by the replicants is their knowing that *Tyrell* has designed them to die young; he could have made them otherwise, but he made them this way. The feeling is that Tyrell either should have made them to enjoy a normal human lifespan or not have made them in the first place. But Roy is less voicing the lament of Job, wishing that he had never been born, than he is wishing for more of what he already has: he wants more life.

The dialogic sparring that follows stands alone in the movie as a sudden, one-and-only flurry of pseudo-scientific jargon. The exchange is meanwhile suspenseful, as the faint hope of a new future for the replicants dies out with each negative answer that Tyrell gives. Roy's uppity courage quietly coils into resentful resignation. Tyrell has already attempted to step back from the tension by speaking of himself in the third person (*What can he do for you?*); he distances himself further by switching into lecture mode, as if reciting from a textbook.

> TYRELL: The facts of life. To make an alteration in the evolvement of an organic life-system is fatal. A coding sequence cannot be revised once it's been established.
> ROY: Why not?

Notice that the dialogue classifies Roy as primarily a biological being; the question is *life,* not being human. Roy asks not for more human credentials, but more life. Tyrell's paternal opening *the facts of life* ironically mocks the normal information about the birds and bees a father might give a son; here, nothing about reproduction or sex could enter the discourse. Tyrell's claim that "alteration" in a "life-system" is fatal and his claim that "revision" of a "coding sequence" is impossible together clash with all we might have hoped from the one whom Chew and Sebastian called a "genius." Tyrell is speaking of what cannot be done, when he is most famous for violating the technological imperative, doing anything that can be done. Roy's *Why not?* reminds us of imbalance in this knowledge-as-power relation, the vulnerability of the hopeful youth asking the aged expert for a solution.

> TYRELL: Because, by the second day of incubation, any cells that have undergone reversion mutations give rise to revertant colonies like rats leaving a sinking ship. Then the ship sinks.

> ROY: What about EMS recombination?
> TYRELL: We've already tried it. Ethyl Methane Sulfonate is an alkylating agent and a potent mutagen. It created a virus so lethal the subject was dead before he left the table.
> ROY: Then a repressor protein that blocks the operating cells.
> TYRELL: Wouldn't obstruct replication, but it does give rise to an error in replication so that the newly formed DNA strand carries a mutation and you've got a virus again. But this, all of this, is academic. You were made as well as we could make you.
> ROY: But not to last.

The allusion to an experimental humanoid "subject" who gets no life at all (*dead before he left the table)* reminds us we are looking into a world where no ethical prohibitions govern research on "human subjects." The biotechnologists in the Tyrell laboratories work trial and error with human DNA and bodies as they please; subjects dead on the table are routine. The self-excusing *you were made as well as we could make you* is false. Captain Bryant informed Deckard that the four-year lifespan was built into the replicants as a "fail-safe" measure against the uncontrollability that would have accompanied their inevitable emotional development. Tyrell chose *not* to make them as well as he could; he is lying. He made replicants to die young because, like all mad scientists, he wanted control; he wanted to reduce the human to nothing but a cosmological object. At this point in the action, Batty's sitting down signals his deflation and sense of defeat. Inasmuch as the exchange has constituted a debate between expert and amateur, original researcher and experimental subject, Batty has lost: there is nothing to be done. His qualifying line *But not to last* re-injects some irony into the dialogue and breaks from the uncharacteristic earnestness that has possessed him, for a few intense seconds.

I endorse whole-heartedly a motif in the critical literature that registers Tyrell's next move as the one that somehow tips him from vulnerable neutrality into the overconfidence of a foolish narcissist, pleased with himself, oblivious to dimensions of the relationship he has with his creature (Senior 7; Richard Schwartz 48; Sammon 174). One would think that the very success of Roy's questing, his presence in front of his maker, would suffice as evidence that the lead mutineer is too dangerously intelligent to be mollified by a compliment. But Tyrell seems incapable of thinking so.

> TYRELL: The light that burns twice as bright burns half as long. And you have burned so very, very brightly, Roy. Look at you. You're the prodigal son. You're quite a prize!
> ROY: I've done questionable things.
> TYRELL: Also extraordinary things. Revel in your time.
> ROY: Nothing the God of biomechanics wouldn't let you into heaven for.
> [*Roy kisses Tyrell on the mouth, then crushes his skull between his hands.*]

Tyrell lies again: replicant early death is not the result of replicant special abilities. There is no ratio between their superhuman qualities and their expiration dates. It is false to say that Roy's having burned half as long is the biomechanical consequence of his having burned twice as bright; there is no such interdependency. The four-year limit was (on the contrary) designed to contain and manage developing emotions—to stop them from "burning brightly." Tyrell's verb tense—*You have burned*—indicates he knows Roy is soon to die. He asks Roy to see in himself what he sees in Roy, but any such reciprocity of perspective is beyond thinking. The careless uncaring of the imperative *look at you* makes the violation of reciprocity explicit: Tyrell wants Roy to see himself as Tyrell sees Roy.

With the prodigal son metaphor, Tyrell extends the twisted misappropriations. Far too often, interpreters have proposed that Roy is a Christ-figure, a Biblically and theologically tone-deaf notion floated on non-sequiturs inspired merely by the presence of nail and dove imagery (Booker 176; Begley 187; Doll and Faller 95–96; Heldreth 47, 50; Instrell 169; Kellner *et al.;* Morrison 4, 9; Pyle 229; Richard Schwartz 47; Shapiro, 79; Wilson 36; Vest 24).[18] Is it not possible that Roy is not a Christ-figure? Postmodern pastiche is a desacralizing ironical mode that dissolves and breaks connections with the historical past, rather than revalidating it with analogic binding. Likewise, we need to grasp that Tyrell's *You're the prodigal son* is an appropriative pastiche that parodies the Biblical parable, evoking dissimilarities rather than directing an allegorical convergence. In the Biblical parable, the father's rejoicing at the voluntary return of his wayward son is authentic and outreaching; here in the pastiche, Tyrell's rejoicing at the involuntary "return" of his unnatural "son" is shallow and narcissistic (not to mention Tyrell hardly deserves the title *father*). In the parable, it is for the repentant son to ask forgiveness of the father. Here, the unrepentant scientist playing God should ask forgiveness of the creature (but does not).

The only aspect of the pastiche that might elevate a figure from the movie by association with the cultural gravity of the gospel source is this: as the prodigal son sought forgiveness for wrongdoing, Roy will now speak of his sins: *I've done questionable things.* However, Tyrell's response to Roy's attempt at moral dignity only dehumanizes Roy further. The confession *I've done questionable things* opposes the compliments of bright-burner, prodigal son, prize. Roy is alluding to the violence he has done in getting to meet Tyrell, and perhaps to crimes he has committed while following orders as a warrior slave in the off-world. If we assume there is a logical connection between what Tyrell says and the timing of the instant at which Roy decides to kill him, then we sense something serious happening in Tyrell's cancellation of the "questionable" things Roy has done in the light of the "extraordinary" things: *Also extraordinary things.* One cannot overemphasise, in my opinion, the way that Tyrell is advising Roy to adopt the amoral indifference to violence, to adopt the non-reciprocity, which has been his hallmark as scientistic authority. Tyrell is implying, in effect, "I have the spiritual-moral authority to absolve you of questionable deeds because I am the god-like source of your extraordinary deeds." It is yet another bid for playing-God control over Roy, an attempt to determine and to restrict Roy's self-concept through manipulative dialogue. All that Roy has suffered, and all that his victims have suffered—the victims of the *questionable things* Roy feels ashamed of having done—mean, to Tyrell, who just wants to keep living, nothing. Tyrell is foolishly hopeful at this instant. He is less crudely arrogant than absurdly self-insulated, detached from moral reasoning.

Therefore, when Roy meets Tyrell's gaze as he, smiling ironically, delivers the line *Nothing the God of biomechanics wouldn't let you in heaven for,* his sarcasm targets less the materialist hubris and more the amoral indifference of Tyrell. To get the depth of the irony, we need to understand that Roy is offended at Tyrell's usurpation of the position of God not as creator of matter or life, but *as judge of good and evil.* The judge who permits entrance to heaven is a judge of moral performance. For the vain scientist has suggested that the impossible human should smother his conscience, should let go of qualms about having done *questionable things,* and should substitute for such qualms a self-regard that would shine glory on Tyrell as a creator, the maker of the *prize* replicant, the maker of the best of the warrior slaves. Tyrell is hinting it would shine glory on Roy himself, too, should he accept the description. In deciding to kill Tyrell, Roy is not just venting rage at the failure of his quest for life. More profoundly, he is rejecting Tyrell's insinuation that he and Tyrell might share a metahuman moral sphere unanswerable to the human sphere evoked by Roy's feeling he has done *questionable things.* Tyrell as the *God of biomechanics* is not the God who has the power to forgive sin. Tyrell as the *God of biomechanics* is not the Divine allegorical counterpart of the father in the parable who welcomes home and forgives the prodigal. Tyrell as the *God of biomechanics* has tried to appropriate for himself the power to absolve Roy of his sins, his morally questionable deeds. Killing the God of biomechanics, Roy is (despite himself) alluding to the absence of the real God whose firstness does include the power to defer resentment, whose firstness does open a way to the reciprocal arbitration of moral considerations and the forgiveness of sins.

I believe it is a mistake to take the murder of Tyrell as a triumph, or as a liberation, or as an opening for Batty into existential forlornness (Picart 126; Wilson 36–37; Barad 30). That the killing of Eldon Tyrell is itself no more dramatically satisfying than the deaths of Victor, Moreau, or the humans at the R.U.R. plant, hints at the emptiness of the postmodern collapsing of science into *nothing but* scientism, institutional authority into *nothing but* an unlocatable source of omnipresent oppression. When we watch Batty crush Tyrell's skull, we are awakened by the shocking reality of the vulnerability—the anti-climactic *human* ordinariness—of even this most politically powerful of all the Victor figures. For what does Roy accomplish by this killing, and the killing of Sebastian that follows? Next to nothing. Witnessing it, we witness the violence of a meta-human who is almost totally like us but not quite like us: the horror we feel, palpably different from the puzzled, breath-holding respect we are compelled to feel when Roy pauses before dying, is the source of that "irrational, nightmare slant" that Jack Boozer intuits as more-or-less beyond representation: "his [Tyrell's] death opens up a dark abyss in the narrative" (221). The killing of Tyrell blocks our full attachment to Roy in roughly the way that the murder of young Ernest and framing of Justine creates a barrier of mistrust between us and Shelley's Wretch. When we see Roy's face contort and hear Tyrell whimpering as his skull is crushed, the mad scientist becomes now just another human body being destroyed by an outraged enemy. Roy loses in open comprehensibility that which he gains in radical otherness. I agree with Richard Schwartz's position that Scott is too humane a visionary and too astute an ethicist, to serve up Tyrell as a cheap scapegoat for a mythicized techno-capitalism.[19] Tyrell resembles Shelley's Victor in his narcissistic obliviousness more than he does Moreau the defender of efficacious pain-infliction or Old Rossum the seeker of a conclusive proof of the non-existence of God. Tyrell is the Victor figure most definitively, intimately destroyed of all those we have studied. But there should be little to satisfy us in his being destroyed.

Batty's Enigmatic Gesture of Rescue

In the closing sequence of the movie, Scott continues to subvert Hollywood conventions. The audience is not given a triumphant victory of obvious good over self-evident evil: Deckard does not kill Batty; Deckard is no "hero" and Batty too attractive and pitiable to function as melodramatic villain. It is impossible to map onto standard Hollywood narrative the bizarre game that Batty plays with Deckard in the closing moments of the film. His creator dead, no way out of the expiration curse discovered, the creature must die, regardless of what he might do now that he has returned to Sebastian's apartment to reconnect with Pris. In the interval, Deckard has saved his own life but blasted away that of Pris. Many viewers may find something oddly anticlimactic in the closing sequence where Batty plays with Deckard and pursues him in the rain onto the roof of Sebastian's building.

The domination of replicant over blade runner in this last act is striking: "A reversal of roles . . . another characteristic of *Frankenstein,* shows their opposition and their unity: Deckard, who has been the hunter throughout the film, now becomes the pursued, and Batty, the former quarry, becomes the pursuer" (Heldreth 50). Deckard has been humiliated in combat: thrown, slapped, and battered by Leon; punched and strangled almost to death by Zhora; drop-kicked and squeezed between the legs of a shrieking Pris, his nose ripped upward and his skull drummed between her hammering palms. Batty humiliates him more. Much of this closing scene was worked out between Ridley Scott and Rutger Hauer.[20] Many of the lines were made up by Rutger Hauer (who played Batty), appearing neither in the Fancher nor the Deeley-and-Fancher screenplay. Ford as Deckard had almost no lines. Whereas the blade runner is all but speechless with fear, the replicant is exuberantly, whimsically talkative: "I can see you!"; "You better get it up, or I'm gonna have to kill you! Unless you're alive, you can't play, and if you don't play . . ."; "Six, seven—go to hell or go to heaven!" Batty enjoys himself. One need only recall the clipped, emotionless reports of Capek's murderous robots at the end of Act Two of *R.U.R.* to register the emotional quotient in Roy's game-playing, which has been celebrated by several interpreters, most extensively by Eric Wilson. Notice the irony that Batty is following the advice of Tyrell: whether conscious of doing so or not, he certainly seems to "revel" in his time (Barad 30–31). The irony counts as further support for the claim that the killing of Tyrell is no simple victory of replicant good over human evil. And yet it would not do to exaggerate this revelry so much as to miss two interrupting moments that remind us of the tragic mood and context: Batty is dying. In one, Roy briefly mourns in shock, and, bending over Pris' lifeless body, kisses her goodbye, pressing her tongue back into her mouth. In the other, cramps in his hand indicate that he feels his time running out, and he pulls a nail from the floor and drives it through his own palm in an attempt to prolong sensation.

The viewer catches on quickly that Rick Deckard, the "one-man slaughterhouse," has decided to give up on the fight, to escape if possible, just to survive. As we watch him grope slowly up the rain-soaked side of the building, aware that his adversary could easily catch up to him at any second, we may well ask why he is risking a fall to his death. When Batty sticks his head out a window, looks up at Deckard's precarious perch and asks "Where are you going?" in a quizzically amused tone, his question is ours. Only when Deckard tries to jump from one rooftop to the other can we infer he was hoping to go down through the other building, flee back down to the street below and get away. Deckard's jump fails—he does not make it to the other side, but ends up hanging precariously from a rain-soaked girder, his grip slipping and a horrible long fall to a shattering death guaranteed if he lets go. Batty makes the jump to the other roof easily and approaches Deckard. The camera gives us the view of Batty's approach from Deckard's perspective, Batty looking down now, indifferently.

> BATTY: Quite an experience to live in fear, isn't it? That's what it is to be a slave.

The sardonic line echoes that which Leon spoke to Deckard: "Painful to live in fear, isn't it?" as Leon prepared, slowly, to kill Deckard by poking out his eyes. But we need to pause over the object and content of the "fear." Hanging on for life, Deckard does not know if he will survive. Roy therefore means it is quite an experience to live in the ceaseless suspense of not knowing *whether* one will live. The replicants' living in fear is not about eventual death in a future governed by normal organic patterns of growth and decay (childhood, adolescence, young adulthood, middle age, senior years). The replicants' fear is one of a death that could come *any day* because they do not know the moment at which their mad scientist maker has programmed them to die. In this way, Tyrell the measurer of their lifespan is an oppressively omnipresent pressure on their self-consciousness. It would be better even if the replicants knew they had a full four years to live, but they know less than that; they know nothing at all about how long they have to live. To imagine replicant experience, we must imagine living all day, every day, as if we were machines on which the batteries at any moment might without warning run out. Roy wants to know from Chew the maker of eyes how long they are to live: *Morphology—Longevity—Incept dates.* But Chew can only reply he does not know *such stuff.* Leon's first question to Deckard is *How long do I live?* When Pris and Roy press Sebastian for the chess-game ruse with Tyrell, Pris zeroes in on the indefiniteness of their knowledge by labelling *accelerated decripitude* the problem they share; Sebastian, like Chew, responds with unhelpful ignorance: *I don't know much about biomechanics, Roy. I wish I did.* To be a replicant slave, therefore, is not merely to have one's labour power exploited, to be turned into an object. To be a replicant slave is to be the machine-puppet of a vain scientist who has designed you to live in moment-to-moment uncertainty about the moment of your death. You may cease moving this morning, this evening, tomorrow, in an hour.

Why insist on such a reading of that one line? Because Batty's rescue of Deckard, his catching Deckard when he falls, gains mystery and intensity to the extent that we understand the rescue as one performed at the instant he sees in Deckard the restrained panic, the running on the blade of fearful uncertainty coextensive with replicant self-consciousness. About to fall, Deckard has become completely "other" to Batty, in the sense of potential victim now at Batty's mercy. At the same time, Deckard now resembles Batty in experiencing the fear Batty has experienced every day. Batty can now "play God" to the extent that he controls Deckard's chance at life. Deckard slips, spitting defiantly. Batty reaches and pulls Deckard upward to safety, tossing him like rag doll against a pillar on the rooftop. That is the act of rescue over which such a surprising quantity of analytical ink has been split.

The claim that Batty's movement constitutes an act of rescue is the first premise of the victimary argument that Batty has become human, or has allowed his latent humanity to manifest itself, or has broken with his previous inhumanity (meanness, coldness) so spectacularly as to erase any reasons to keep him on the far side of the replicant-human boundary. A first interpretation of the rescue gesture sees in it *compassion* and *empathy* and therefore humanity. Claiming in one of the most important early articles on the film that "Roy's transformation stands as one of the narrative's most bizarre and interesting aspects," Douglas Kellner and his colleagues elaborate: "Roy, who had been presented as a Nietzschean poet-warrior, renounces his program as a ruthless killer and instead chooses pity and compassion; the Nietzschean *ubermensch* (superman) thus becomes a *mensch,* a human being" (see also Francavilla 11; Vest 24; Spinrad quoted in Bukatman 70). Ridley Scott, in a 1982 interview in *Starlog,* claimed that Batty's saving of Deckard's life was "an endorsement, in a way, that the character is almost more human than human, in that he can demonstrate a very human quality at a time when the roles are reversed and Deckard may have been delighted to blow his head off. But Roy Batty takes the humane route" (quoted in Sammon 193).

A variation on the compassion thesis adds the suggestion that the uncharacteristic gesture—uncharacteristic because it breaks the pattern set by his murders of the mutiny victims, Chew, Sebastian, and Tyrell, and the pattern of his treatment of Deckard until that instant—signals an exercise of free will. The idea is that Roy now defies the combat-model programming of the Tyrell corporation. Joseph Slade reasons that Batty's nonviolence ironically extends his violence against Tyrell by extending his defiance of the creator beyond his killing of the creator: "But Batty refuses to kill Deckard because it is the only act of negation left open to him. He will *not* kill precisely because he has been programmed to. A true romantic, Batty cherishes his hate, but he will not waste this last life, because that [wasting of life] is what humans do" (17). In her Sartrean analysis of *Blade Runner,* Judith Barad more extravagantly proposes that the switch indicates Batty's sudden access to choosing his essence: "Batty's emotional maturity, his choice of empathy and compassion, is what makes a human truly human. In the end, it's not Tyrell or any genetic engineer who can makes Batty human—he must create this in himself. Being human isn't a particular DNA configuration but a state of mind, a feeling" (32). This approach to the gesture pervades the literature (e.g. see also Boozer 225; Abbott 341, 345; Schwartz 46–47).

Why so many critics, coming from so many different theoretical perspectives, have been so willing to see in the sparing of Deckard the definite evidence that Roy suddenly "becomes" human may ultimately matter more than the question of what the gesture itself means. Certainly, it testifies to our postmodern eagerness to see empathy in the replicants. The philosophical anthropologist Mary Midgley quite correctly observes that a person's lacking the normal human affections is among the most frequent reasons for the attribution of *inhuman* (*Beast and Man* 206). What I would point out, however, is the species self-flattery involved in making the quintessential traits of the human *empathy and compassion,* as opposed to the far less flattering emotional states of (say) resentment, envy, curiosity, or simply the state of being subject to insatiable desire—desire for knowledge, belonging, being—in general. Anybody who looks at our historical past will find quantities of violence to weigh against those of compassionate behaviour, so the proposition *to be human is to be compassionate* amounts to little more than ahistorical moralizing that would seek to overcome the problem of human violence simply by denouncing it. Human culture *defers* violence but it can never hope to abolish it. Loftily erecting as a universal imperative *be compassionate* is unhelpful; being commanded to *be compassionate* is not unlike being ordered to *relax:* the more we become aware of it as command, the less able are we to obey it freely.

I venture a second objection that builds on the familiar notion in ethics which separates evidence of motivation from evidence of consequences. The conferral of the status of compassionate human on Roy may be less an intended result of the character's motive (a representation "intended" by the actors and filmmakers), than an unintended consequence of our reading backward from the benefit to Deckard of his being rescued a motive in Batty of "empathy" for the human blade runner. The line preceding the rescue—*that's what it is to be a slave*—has no connotation in tone or content of compassion; it despises, it condescends. Despite a near-consensus in the critical literature, we might open our minds to the differentiating possibility that Roy saves Deckard for motives other than the altruistic. Once we have our minds so opened, we might be prepared to absorb two alternatives to what I have been calling the compassion thesis.

First, there is the *desire for a listener* thesis. This interpretation posits that Roy has a non-altruistic purpose in saving Deckard: Roy knows his time is almost up, he does not want to die alone, and he wishes for an interlocutor. Ridley Scott himself briefly endorsed this thesis: " . . . Roy Batty takes the humane route. But also in a way, because he *wants a kind of death watch,* where he knows he is going, dying. So in a sense he is saving Deckard for something, to pass on the information that what the makers are doing is wrong—either the answer is not to make them at all, or deal with them as human beings" (quoted in Sammon 193; see also Slade 17). Notice that Scott claims Batty has displayed a human quality and taken a "humane" route, but without declaring Batty has *become* human. Now from the point of view of minimal anthropology, this *wish for an interlocutor* itself has moral content, empathy aside: in recognizing Deckard as a partner in dialogue, Batty presupposes between them that mediation by the center of the scene of representation that (according to generative anthropology) lies at the minimal origin of the human. From my perspective, this less taxonomic but more scenic approach to the meaning of the act of sparing Deckard is promising. Now that Tyrell the mock-god at the center of their previous relationship has been destroyed, Deckard and Batty can negotiate *their own* scene of exchange (a little like Alquist and Damon, Prendick and the Beast People, Walton and the Wretch). The deferral of violence through representation substitutes for the mindless violence that would make of Deckard another scapegoat, a substitute for Tyrell.

Another argument against the compassion idea is one we might call the *unthinking reflex* thesis. Certain memories Rutger Hauer has of the shooting of the scene, as reported to Paul Sammon, are worth quoting at length.

> "But Ridley insisted that one thing Batty had to have was absolutely no sense of hesitation. He doesn't reflect, he reacts. He's faster than anybody. A characteristic of the Nexus-6. So, if you follow that thought, you reach a point where you realize that if somebody falls, Batty grabs. It has nothing to do with how he feels about Deckard; it's just a reactive moment. That's what Roy's built for.
>
> "In fact, while we were shooting this moment of Roy grabbing Deckard, we had a problem with the rain machines and had to wait around for them to be fixed. And I actually asked Scott, 'Ridley,

> what do you think? Why does Roy save this fucker?' and Ridley looked at me and said, 'It's purely a reflex. Other than that, I don't know.'
>
> "This reply might bother some people, because so many folks have read a lot of meaning into Batty saving Deckard's life. But actions always come first. *Then* we think about them, later. Roy doesn't know why he saves Deckard or grabs a dove. He just does it." (Sammon 194)

Like the desire for a listener thesis, Rutger Hauer's unthinking reflex idea—*he just does it*—avoids the inconsistency in making of the Batty who spares Deckard, a creature who has suddenly shed his replicant being like a snakeskin and burst into humankind by means of one atypical act. Does such a skepticism "dehumanize" Batty? It does, in the sense that it preserves the disturbing otherness of his impossible-to-assimilate replicant being. The challenge for us is to make sense of what Deckard can only experience as a rescue when the creature who has rescued him may have had no such "human" motive, feeling, or thought. But the purported dehumanization would not prevent our seeing the deferral of violence that Batty and Deckard enjoy, briefly, as he speaks his dying lines. To the extent that Batty and Deckard remain on the scene of representation as they did under the desire-for-a-listener thesis, it is Batty's participation in the *deferral of violence* that "humanizes" him rather than the attribution to him of a "human" motive of compassion or the possession of a trait such as a capacity for empathy. As I have argued elsewhere in this book, respecting the uniqueness of human scenicity does not prohibit moral concern for other sentient and human-like beings, whether animals or metahumans.

On the Dying Lines of Roy Batty

The second reason that many wish to argue Batty somehow becomes human at the end of his time is the tragic impact of his dying speech, a matter of some five sentences. Batty is seated across from Deckard. Hauer's blood-streaked, rain-soaked face fills the screen; a wry, teasing smile catches the corner of his mouth just before he begins; we sense now that he will probably not harm Deckard. As with the act of rescue, something in these lines moves commentators on the film to locate in them a transition.

In one of the best 1982 reviews of the film, Michael Dempsey wrote: "As his hands atrophy and release a pigeon he has been holding into the sky (in a kind of stop motion), the distinction between human and machine no longer has any real meaning" (38). Jack Boozer insinuates that only now has Batty's esthetic sense come to fruition: "In his last groping words he transcends the limitations imposed by his maker and becomes a witness to the beauty of lived experience" (224; see also Sobchack 272; Abbott 348–49). Here for our pondering is the text of the brief dying speech of Roy Batty.

> BATTY: I've seen things you people wouldn't believe. Attack ships on fire off the shoulder of Orion. I watched c-beams glitter in the dark near the Tanhauser Gate. All those moments will be lost . . . in time . . . like tears . . . in rain. Time to die.

From the perspective of originary thinking, I venture a few observations. First, Batty employs an impersonal tone of address: he does not speak to Deckard individually. Deckard belongs to *you people.* Hauer's delivery is couched with pauses pregnant enough to imply that the phrase *you people* opposes an implied *we replicants.* Deckard belongs to *you people.* Batty does not so belong, never has, and does not wish now to belong to *you people.* The phrase names those replicant-like creatures who would not believe what he has seen as a replicant. We can connect the line to one in Roy's very first scene, when he and Leon are extorting Sebastian's name and address. The stammering, terrorized Chew has confessed in shivering fear, "I just do eyes, see? Just eyes. Just genetic design. Just eyes. You Nexus, huh? I design your eyes." As unimpressed by the design achievements of Chew as he will later be unmoved by the genius of Chew's paymaster Tyrell, Batty replies: "Chew . . . if only you could see what I've seen with your eyes." But the clause *if only you could* implies *in fact you cannot.* In other words, it is impossible for humans to see as replicants. Because "people" cannot share replicant experience, they would not believe what Roy has seen in his off-world warrior labours. Allusions to c-beams and the Tanhauser gate, neither of which have any referent in our dictionaries of cultural literacy (unlike the constellation of Orion), leave us with a beautiful if vague image of light beams glittering in the darkness of space somewhere near a gate—perhaps a gate meant to be breached in battle. But our not really having any idea what a *c-beam* is, along with our not being able to locate *the Tanhauser gate* on our maps of the night sky, confirms Roy's claim that we have not seen the things he has.

The key line is the simple but oddly moving *All those moments will be lost . . . in time . . . like tears . . . in rain.* Those unacquainted with the fact might be surprised to learn that Rutger Hauer rather than Hampton Fancher or David Peoples penned the line, in his trailer during a break in the shooting, speaking it for the camera after getting Ridley Scott's approval (Sammon 195–196). Why does it strike so many of us as spot-on perfect? The childlike simile insists on a likeness in the ways two things are lost—moments lost in time, tears lost in rain. To register loss in such tones presupposes the value of things: the moments and tears have value. As organic secretions indicating uncontainable emotion, tears recall by close relationship the very element of replicant being that became their curse and the cause of their quest at once: growing emotions. Their developing emotions (the defect Tyrell failed to foresee) in turn created a desire for more such emotional experiences, experiences worthy of memory. A desire for more moments worthy of memory is a desire for more life. Single, discrete moments and valued experiences fall into time, losing their discreteness as tears in rain. "Time" and "rain" represent atemporal, shapeless indifferentiation; "moments" and "tears" point to historically meaningful events, valuable and exchangeable memories. I spell out these layers of meaning, but they get the power because they are quickly apprehended, ostensively immediate, "obvious" just below the threshold of consciousness. The rain self-evidently falls on the head of the dying Roy Batty even as he names it; the head self-evidently holds those memories about to be lost in time; the words express the memories that would not be without those words.

All symbolic exchanges perform the deferral of violence through representation. The deferral creates the space in which we can name the objects of shared human consciousness; the deferral makes the scene of representation. Roy's dying speech in like manner defers the violence between replicants and humans that has defined the horrifying, intractable conflict in *Blade Runner.* Deckard sits listening, witness to Roy's action. It is the pacific action of merely representing in words some "moments in time" rather than continuing the agon of physical violence, bodies colliding, matter in motion, kind against kind. Roy has not instantly become human because of a flash of empathy, compassion, existential self-choosing, or any other such borrowing from human ontology. Roy dies a replicant. His lines mean most if we let him remain an impossible human, unlike ourselves. We, unlike him, are not victims of mad scientists like Eldon Tyrell. We cannot be where he is: we mourn him best by respecting him as an impossible human.

The poignancy of his dying speech derives ultimately from character, from the ethical being of the strange other-than-us figure who speaks them, the impossible human Roy Batty. Their ostensive self-referential force takes precedence: his speaking of the line makes and is Roy's moment in time, the instant before his *time to die.* I would suggest that honour demands we recognize Roy's dying moment as the very moment Tyrell scheduled to extinguish all the desiring, resentful life-energy that has been Roy Batty up until now: the mad scientist's subjection of Roy imprints itself even on this instant. And yet the serenity of Roy's self-representation cancels the tyranny of Tyrell with a usurpation of centrality as minimally resentful as it is maximally memorable. He should not have been made at all (given the mad science that tormented him from beginning to end) and yet the very impossibility of his "full" humanity (impossible because Tyrell is always in the way) makes him glad he was made.

By recognizing our difference from the impossible human, we learn what the human might be. We do not, thanks to the One who quietly mediates our exchanges simply by being the Being in the center of the scene of human representation, suffer as impossible humans do. We are not the Wretch of Victor Frankenstein. We are not the Beast People of Moreau's realized fantasy. We are not the altered androids from the Rossum factory. We are not the condemned replicants of *Blade Runner.* Nor are they us. We are humans. We owe it to ourselves—and we owe it to the creative genius of Mary Shelley, H. G. Wells, Karel Capek, and Ridley Scott—to respect the space of difference these imaginary impossible humans endure, and by enduring, open up for us.

SIX

Afterword: Sharing the Human Scene

I close with a few considerations to situate our findings in the context of what I take to be scientistic tendencies in knowledge-seeking today.

Perhaps now, having come so far, we can affirm that the four classic representatives of the Frankenstein myth dramatize a truth that philosophers attentive to our paradoxical situation as language users have long defended. It is the proposition that humans are not reducible to cosmological objects; we cannot come to know each other as we should, as such objects. It is not a good thing to play God. The separateness of the God-like agency in the *playing God* metaphor links to an epistemological inequality between human scientist and human-as-object. That epistemological inequality in turn reveals something essential about human mind. If we can spell out the inequality and present it for inspection in a way that might be clearer given the work we have done in thinking through *Frankenstein, Moreau, R.U.R.* and *Blade Runner,* generative anthropology's investment in the notion of scenic exchange shows its value. Materialism in its scientistic version has been my polemical target; however, by now it may be clearer than at the beginning that I take *spiritual* to mean nothing but (reduction intended) a requirement and result of minimal linguistic exchange. The *spiritual* is an irreducible reality contingent upon the anthropological fact of the scenic structure of the human exchange of signs. I have no quarrel with substantive materialism. I have none with (for example) the biological naturalism of John Searle or the atheistic humanism of Raymond Tallis. I find it fascinating that some of the conclusions reached in their works, works of a very different style from those of Eric Gans, often resemble the conclusions reached by Gans in his brilliant originary thinking. Searle claims that the principal problem in contemporary philosophy is that of figuring out how humans with their social ontology fit into a universe that seems to be made up entirely of physical forces: "How is it possible that a world consisting entirely of material particles in fields of force can contain systems that are conscious?" (*Mind, Language* 45).[1] I would stand with Tallis when he queries Searle, suggesting it may be a mistake to take "the correlation between neural activity and reported experience to mean that there is an intimate causal relation between them" (*Aping* 89). However, any differences between Tallis and Searle and Gans (for example) are small when measured against the big differences between the scientism they reject and the anthropological stance they share. The anthropological stance respects the irreducible sociality of human linguistic consciousness.

The dangerous tendency of scientism, and of mad science as the Frankenstein myth spectacularly displays it, follows from a unidirectional inequality between the scientist thinker and his object of attention, particularly whenever that is a human-as-object. All the power and knowledge flow *from* the scientist *onto* his object. If the object is a human person or body, the "first-person" status of the human (or, in the myth, the impossible human), is denied from the outset. Its first-person otherness is reducible to the material. The mad scientist has contempt even for the object-as-economic, for he believes the object need not necessarily be transformed into anything usable, exchangeable, or valuable. To confront the full horror of the myth, recall the uselessness of the impossible humans. Shelley's originary usurper Victor is the most powerful representative of such contempt, in his obliviousness to the question *why* he is sewing pieces of dead bodies together to create this thing, what *use* it might have for other humans, how it might enter society and the marketplace. Victor's monster is too ugly; the Beast People do nobody any good; the robots are excessively useful for a while, but they annihilate value when they annihilate humankind; the replicants are useful only as enslaved, only as long as their owners consent to a project of unmitigated cruelty. Given what such victims of scientism suffer in futile and terminal existences (for none bears children or creates intergenerational history), the vain scientist who plays God by trying to re-enact the origin of the human does the involuntarily directed enactors no good. He wishes just to prove an "objective" cosmological point, goodness be damned as required by his experimental or commercial goal and the demands of his staunchly preserved ego.

Inequality and Mad Science: Imagining a Mind-Materializer

To clarify a little further the strangeness of the ambitions of materialist reductionism, let me indulge in a brief science fiction scenario inspired by the Frankenstein myth. Imagine that the work of cognitive neuroscientists working with human subjects has advanced to a stage where, by means of some machine—let us name it the *mind-materializer*—the complete inner workings of a human brain can be known. Imagine that by some application of fantastically advanced photographic nanotechnology and computer mapping, you can be fitted with electrodes along spots on your skull or injected with harmless vapours that suffuse your brain, or fitted with a super-microscopic particle reader in various orifices, so that a technologist using the equipment can predict each movement of every subatomic particle bouncing in your brain. Assume that the ancient atomist's dream of *proving* that the mind is nothing but matter in motion has therefore been accomplished, in that the mind-materializer has been demonstrated to have perfect *predictability:* you experience no mental event that the neuroscientific technician cannot predict. All we have are atoms and the void; but lo and behold, there is no swerve. In this universe, LaPlace's demon rules: everything is determined, as if in a finite universe of causes and effects. For all your sense of your interior life, you might as well be a thermostat or a pop-up toaster or (even less) a toy marble rolling to a stop in a board-game maze. You represent your feelings to your inner eye with pictures and images; you deploy the mental resources needful to preserve your sensations, perceptions, memories, fantasies, thoughts. The neuroscientific technician, however, knows—and can demonstrate that he knows—everything you think and feel before you think and feel it. Even if whatever in the organic substance of your brain that is identical to the experience of *consciousness* continues to operate, the scientist has access to its observable operations (although his experience cannot be yours). Implementing his mind-materializer, he is free to rob you of your conviction of private interiority, deprive you of what you had believed was your inviolate first-person subjectivity. Everything you think and feel, he knows in advance.

If the technology remains passive and the scientist observes but does not alter your brain activity, he is not yet *playing God.* Insofar as the passivity of the scientist makes his knowledge like the foreknowledge that an all-knowing but apparently passive God is believed to possess about human futures, we fear no violation of ourselves. That God knows the actions humans will perform in the unfolding of their lives does not necessarily dismay the believer, for many believers trust that God makes space for the exercise of will, agency being that which keeps the future open and makes history meaningful, agency being that which makes desires and intentions and performances worthy of attention and evaluation. *Playing God* means interfering with the subjective will of the human other in a dehumanizing way. It means exercising knowledge-derived power over the human pinned down as a cosmological object; it means exploiting the epistemological privilege to do practical, political work. Only when its operator uses it to relieve a human patient of its freedom, only when it becomes a body-controller does the mind-materializer become the device of a mad scientist.

Imagine waking up in a body still your own, but trapped inside it. The mad scientist now has taken over that part of your mind that makes your body move. You must taste the dog food the mad scientist makes you shovel into your mouth, chew and swallow for breakfast. You must feel the soaking as the vain scientist makes you walk to the office in a rainstorm, dressed in your silly threadbare yellow pajamas. You must feel despair when, in the lobby, you hear yourself shout at your boss and best friend, calling him or her a despicable oaf and bragging about having slept with his wife or husband (when you did not). When you are rerouted home, unemployed and overwhelmed, you are made to kick your dog and drink from the toilet and throw yourself against walls and place obscene phone calls to the police. His mind-materializer is working: you are the object of his experiment, impressively deprived of bodily freedom, living proof of the predictability of and controllability of materialized mind (which is to say, the reduction of mind to brain, and the reduction of body to machine obediently following braincase command).

Now I have tried to keep the example a little light, but the possibilities are bad: brainwashing, involuntary injections of mind-altering drugs, the irreversibly traumatizing effects of physical torture on personal memory come to mind as real-world analogues. We really fear being trapped as a consciousness inside an apparently comatose body; we fear being trapped in a persistent vegetative state. Such scenarios are hideously unfair. The scientistic scenario, however, is intended by the scientist—not the result of accident or illness. Consider what such an experiment would prove—consider what triumph it might trumpet as "empirical evidence" for the metaphysical presuppositions of an artificial intelligence or a cognitive neuroscience research programme. My point is not to accuse neuroscientists and AI workers of moral depravity. My point is to have all of us ask why *in the first place* we would ever desire to "prove" the materiality of our minds. Why, in this way or by any other scientific or experimental means, do we wish to prove the materiality of mind? What good could such proof provide? Why do we need such a scene of proof, when we have scenes of exchange—concrete exchanges of signs and objects, words and things—that only in the most clumsy and awkward ways can be reduced to such subject-object experimental models? It is not without reason that contemporary science is rigorously restrained in most institutional settings by Research Ethics boards. So should it be; otherwise we leave open the door to mad science.[2]

In the grotesque scenic inequality between the scientist using the mind-materializer and his control-deprived victim, we see a *reductio ad absurdum* of scientism—a vision that the esthetic revelation of the Frankenstein myth should embolden us to picture more clearly. The perception of that inequality opens up to a view of that which purveyors of religious wisdom, theologians, artists, and humanistic inquirers have always meant when they have insisted—against the metaphysical presuppositions of materialists—that the human being is *irreducibly social* and that scenic linguistic consciousness is a mystery it would be vain, mad to seek to dispel. The insistence nowadays emerges sometimes in opposition to the excesses of evolutionary psychologists.[3] At other times, it emerges in opposition to the excesses of genetic determinists.[4] At yet other times, it questions the kind of neuroscience that is propelled by eliminative materialism (see Tallis, *Aping; Why the Mind;* Searle, *Minds, Brains*). The human being is irreducible to a relation between bits of genetic material or an algorithm of evolutionary process. Recall our reflections in the section "Tortured Matter, Multiple Errors," from chapter One: if the image of the mad scientist torturing a supine body on a table into involuntary mock-human existence horrifies us, our horror follows from an intuition of the inequality between one human subject reducing another to *nothing but a material object.* For if we equalize the relations and make the knowledge and power doubly reversible, the moral horror evaporates. The madness *is* the scientist wishing to have (and quite happy when having) the epistemological privilege and the technological power all to and for himself.

Let us equalize the relations. Let us imagine now that there are two mind-materializers operating at the same time on the same two people, operated by the same two people. Two humans, equal to the other in expertise and power have each a machine that reads the mind of the other. Rather than one controlling and the other controlled, each has access to the other's mind-materialized. The relation no doubt would be strange and strained, but the moral horror would dissipate. It would dissipate because it would be a lot like—no reduction necessary—the mindful exchange of signs on the scene of representation. If Mary Shelley's Victor (or Moreau or Rossum or Tyrell) were on the table himself as the object of another Victor-figure, he would no longer be a Victor. To imagine a reciprocal face-off between equal possessors of mad technological power shows that the Victor-figure is fundamentally at war with moral and spiritual equality on the scene of representation between human beings. In the name of a purported cosmological truth, he seeks to justify a moral power. He exploits a law of "Nature" to rewrite moral laws in his own interest. Without the scenic and moral inequality, his experiment would be impractical and his technology worthless. But with the moral inequality, he is free and empowered to create with a clean, oblivious conscience nothing but impossible humans: monsters, beast people, robots, replicants, miserable clones, mechanical sex-dolls, suicidal cyborgs, judge-duping computer conversationalists.[5] The object of mad science is minimally exchangeable and maximally desacralizing. That which mad science desacralizes is reciprocal exchange between humans in the presence of a shared scenic center. I derive therefore a motto from the Frankenstein myth: *maybe you should ask why you want to prove the materiality of mind in the first place.*

A vain scientist plays God by trying to re-enact the origin of the human, and fails disastrously. If on the human scene, as generative anthropology proposes, linguistic signs exchanged now as in the beginning *first* acknowledge a sacred center of shared faith while *second* giving information about objective and indivisible "Nature" (meaning they give information only after the sacralisation and estheticization of the objects of pre-human appetitive attention), then human equality is not predicated upon a purely instrumental knowledge of any object, but particularly not upon knowledge of other humans as objects. When ordinary humanistic thinkers, including lay people outside the universities, imply that science tells us little or nothing about ultimate value, they acknowledge this priority of transcendent desire over practical cognition. They are presupposing the well-known human reality that appetite is not desire, that aggression is not violence, and that objects represented by signs on scenes of exchange are *different* from objects hypothetically detached from human purposiveness (wherever such objects might be hypothetically placed; they do not fit well on the scene of representation). The equality of humans is not predicated upon some objective truth that might be expressed in propositional form and matched to an indisputably real cosmological referent. We cannot perform an *experiment* to "prove" the spiritual equality of humans. That equality is predicated, on the contrary, on knowledge of the way that the exchange of signs that re-present things creates a sacred space where we can operate only as *subjects to and subjects for* one another. As Bernard Williams has written, "No one should make any claims about the importance of human beings to the universe: the point is about the importance of human beings to human beings" (*Ethics and the Limits* 118). To be bounded in a minimal but infinitely expandable nutshell: our scene of anthropology is not a scene of cosmology. Anthropology is not reducible to biology. It is mad science, or scientism, to reduce humans to nothing but objects of natural science. If such reductionism is mad, then why is it still desirable and respectable, if not required and prestigious, in the academy today?

Sharing Our Origin in Language

There is an anthropological truth latent in the Frankenstein myth, and I have tried in this book to make it explicit. To try to re-enact the origin of the human to prove that the human can be reduced to a material object is to court waste and ruin. Nothing shuts down exchange more quickly than the failure to recognize—because you believe you possess the "scientific truth" that will enable you to predict and control and describe in reductive terms another human person—the freedom that the other human has to defy the prediction you make about him or her. Despite the delusions of Victor and his heirs, other humans cannot be reduced to machines incapable of the free and equal exchange of signs; to try such a reduction is to destroy them as humans. Despite the quiet persistence of scientism in the modern era, humans cannot be reduced to language-less animals incapable of the exchange of symbols on a scene structured with a profane periphery and a sacred center; to try such a reduction is to start to destroy them as humans.

We may well, having absorbed the Frankenstein myth, revive the courage latent within us that gives us to believe what we know already, by now, after Auschwitz and Hiroshima: as members of one human community, each human person we name is as sacred as the one central Being our earliest ancestors began to name at the beginning. We may well do so even if the scientism kept more or less undisclosed in the scenes leading up to Auschwitz and Hiroshima has disenchanted the cosmos to such an extent that we can only think this Being as now *among* us, One in the center of our scenes of representation, rather than an agent outside the world who should have put an end to such horrors. Why expect from and demand of "God" a morality that we do not demand of ourselves? The question should anyway always be *what have I done to stop the violence?* If we can begin to affirm God as One *among* us mediating our every center of shared attention, invisible but present, rather than seeking to trap God as some inscrutably vague agent "out there" somewhere in the cosmos, whose existence needs to be proven by "reason and evidence," then we might learn what our earliest ancestors intuited at the beginning, that which still holds true. The one spiritual Other of humankind is *beyond representation,* but that does not make that One any the less real, as the often silent Being in/behind/of/at every desired and resented scenic center we share as competing and co-operating humans.

Our place in Nature is that of animals who have language. Our place in Nature is that of weird beings who use signs to represent things. The fundamental delusion of scientism is that we might somehow produce the best knowledge of our place in Nature if we could explain away the mystery of human consciousness, particularly this intractable linguistic consciousness. Few things are harder to do than to become truly aware of the mystery of language, for we are in language as worms are in earth, bats move among echoes, fish live in water. To explain the scenic exchange of signs away would reduce us humans, in turn, as sign-users, to nothing but parts of Nature. But it is scientism exclusively (scientific knowing severed from religion, from art, from trade) that makes parts of Nature into things nothing but material, into things predictable, disposable, manipulable but not sacred, beautiful, erotic or exchangeable. What human person desires to be objectified as just such a thing, just such an object of scientific cognition alone? Is our attachment to the resonant phrase *human being*—where *being* suggests something beyond quantification, beyond representation—neurotic or infantile? Or does it express an attachment to social exchange, an attachment calmly proud and quietly wise with the lived experience of our weird, language-using, scene-sharing animal kind over unthinkable millennia of history? We are not ourselves just objects because we alone on the planet exchange our representations of objects in scenic centers and buzzing, bustling markets. The exchangeability of objects on scenes of representation opens the ever-hopeful future of our human kind.

The cosmos existed once without us; it will "exist" without us once again. Nothing of significance follows. We live not in the ahistorical cosmos, but on historical scenes of significant representations, in communities with family, friends, co-workers, neighours and fellow citizens. Wherever you are now, steady reader, holding these pages in your hand, absorbing these words from a screen, you act and think and feel on such a scene with me and with every other person who reads these words. Do you wish for someone to convince you that you are not where you are? What value could we have for each other, if you and I were to reach out to one another as *nothing but* predetermined mechanical objects in the cosmos? Who got us to begin thinking science alone could let us or lead us to make sense of *all* our experience? Take away all your experiences of objects as sacred, esthetic, erotic, and economic. Take them away: in your world, the sacred, the esthetic, the erotic, and the economic have not *really* been. Try to forget every human and nonhuman object you have held and approached in such ways. Try. Reduce anthropology to materialist biology, biology to chemistry, chemistry to physics. Everything is really only physical. Leave yourself and your loved ones with your knowledge of how the mathematized and material cosmos works, devoid of anthropological historicity. Are you satisfied?[6]

Well may the ecological crisis frighten many of us, maybe most of us. But we may also have forgotten how massively indifferent Nature is to us as humans, prior to our making our home in it. Nature's sacred, beautiful indifference to us is an invitation to us (perhaps a Divine one, no less) to *count* for each other as humans, the most bizarre cosmological objects in all of Nature, so far as we can tell, for *we alone have language.* We are so filled with shame at our capacity for violence that we have forgotten that even "violence" is not all bad. We have forgotten how to enjoy being human, accepting our places on scenes of exchange as the gifts they are, our place in Nature as the gift it is—a gift given us not by "the universe" but by our human ancestors, parents, family, friends. We have forgotten how to rejoice in the mystery of our possession of the scene, together staring into the shared fire, saying nothing now, maybe saying something later, feeling good, a little grateful, maybe grateful to creation (or a Creator) for the burning wood that now warms us although once it lived as flesh of a growing tree. Should we not have cut it down and placed it in the center to become a source of light, heat, the flames that cook our food and call forth signs to convey our stories and plans? What else should the source of the warmth once have been? Who forces us to wish ourselves back into the prehistoric cold? Should we march offstage like the military robots at the end of the second-to-last act of *R.U.R.?* Should we hope to devolve into speechless creatures like the Beast People at the end of *Moreau,* as if language were not our thing, our very thing?

We are so filled with shame at our capacity for violence that we have forgotten how to love one another as humans. Maybe we can learn to do so again, like we did a little bit at the beginning, when we slipped out of prehuman animality onto the scene of representation, awestruck by the invention/discovery of the sign and the ever-self-giving Being in the scenic center that the sign gradually revealed. We hardly knew what was happening. But we knew something was happening when first we discovered/invented the sign. We still know we are choosing to make an event happen whenever we exchange signs. We do not need to merge into the Singularity[7] or torturously strive to become elite, pain-free, invulnerable post-human cyborgs. Information is not significance.[8] We are not nothing but computers.[9] We do not need to explain away our difference from the other mammals with whom we share a natural ancestry. The evolution of the human species is not the history of humankind. We are not nothing but animals.[10]

Mary Shelley, H. G. Wells, Karel Capek and Ridley Scott, and the other contributors to the Frankenstein myth—they have given us artworks which, when shared with care and consideration, offer access to experiences that will remind us of a truth we already know, but we are always in danger of forgetting. It is a blessed thing to be a real (not impossible) human. We do not need to re-enact the origin of the human, as Frankenstein and his heirs, fictional and real, think. We can learn again to be where we have always been, present to one another on the scene of representation.[11]

Notes

Notes to 1 / Minimal Anthropology

1. For a useful study of Shelley's *Frankenstein* and Wells' *The Island of Dr. Moreau* as science-fiction horror, see Hinckley. Significantly, in his study of robot films, Telotte excludes James Whale's *Frankenstein* (1931) and Erle C. Kenton's *The Island of Lost Souls* (1933)—a re-make of Well's *The Island of Dr. Moreau;* Telotte claims they could be legitimately "excepted from this study because of the way they *straddle the border between horror and science fiction*" (17) [emphasis added].

2. My choice of the phrase *violent origins* is a deliberate allusion to the title of the conference proceedings edited by Hamerton-Kelly and featuring contributions by Rene Girard, Walter Burkert, and Jonathan Z. Smith. Eric Gans' generative anthropology, as we will see, might be considered the completion of Girard's thought.

3. I owe this idea to Johnsen. See his Girardian analysis of Virginia Woolf's careful rejection of anger as incipient madness (113–19): "But for Woolf, being angry was like being mad" (116).

4. See Long's anthology of creation myths and accompanying commentary for evidence supporting such a characterization of the deities in many creation myths. As will become clearer as we proceed, however, the most important creator deity in the context of the Frankenstein myth is Yahweh or the God of the Hebrew and Christian scriptures.

5. For some of Gans' more important formulations of the idea of the victimary, see *Signs* 168–99. For the same phenomenon named differently and approached from perspectives other than those of generative anthropology, see Girard, *Things* 439–41; Siebers, 124–58; Good 24; Nisbet 303–307 (the entry "Victimology"); Robert Hughes.

6. Northrop Frye: "The tragic hero is very great as compared with us, but there is something else, something on the side of him opposite the audience, compared to which he is small. This something else may be called God, gods, fate, accident, fortune, necessity, circumstance, or any combination of these, but whatever it is the tragic hero is our mediator with it" (*Anatomy* 207). Again: ". . . tragic heroes are wrapped up in the mystery of their communion with that something beyond which we can see only through them, and which is the source of their strength and their fate alike" (208). In the Frankenstein myth, there is a double victimization—mad scientist and impossible human

may be destroyed as one unit, as if they were Siamese twins. However, the *something else* of which the impossible human alone is our mediator is the nonhuman condition into which the god-like mad scientist has thrown him. The condition of being an impossible human is *something on the side . . . opposite the audience.*

7. See in the MacDonald and Scherf edition of Shelley's *Frankenstein* (Primary Sources) excerpts from works by Erasmus Darwin and Humphry Davy collected in "Appendix B: The Education of Victor Frankenstein" as edited by Macdonald and Scherf (264–275). See also Butler; Eamon; Eichner.

8. Mary Shelley, in her introduction to the 1831 edition of the novel, recalling the famous nightmare in which the fiction of Victor Frankenstein's hubris germinated: "Frightful must it be; for supremely frightful would be the effect of any human endeavour *to mock the stupendous mechanism of the Creator of the World*" (Macdonald and Scherf 357) [emphasis added]. All further references to the text of *Frankenstein* come from this edition, identified by volume, chapter, and page number.

9. "And God saw that it was good." See Genesis 1:10; 1:12; 1:18; 1:21; 1:25; 1:31; 2:3.

10. For the theology to which I allude here, a theology inflected by the thought of Rene Girard, see Alison; Bailie; Anthony Bartlett; James Williams; Girard, *The Scapegoat* (1986). I take the capitalized naming of the Crucified from Anthony Bartlett.

11. My definition differs from the others I have come across, particularly because of its emphasis on reductionist materialism and because of my positing (within the discipline of generative anthropology) scenic symbolic exchange as the heart of *the anthropological.* For example, Sorell defines scientism as follows: "the belief that science, especially natural science, is much the most valuable part of human learning—much the most valuable part because it is the most authoritative, or beneficial or serious' (1); "What is crucial to scientism is not the identification of something as scientific or unscientific but the thought that the non-scientific is of negligible value" (9). While such a belief would certainly be *part* of scientism as I define it, Sorell's formulation does not catch the violence of scientism's dehumanization. It is mostly philosophers, theologians, and sociologists who have fought scientism. As many psychologists have succumbed to it as have opposed it (behaviorism is scientistic; cognitive neuroscience is frequently scientistic). Why so? The philosophers' respect for metaphysical questions; the theologians' presupposition of value in the question of God; and the sociologists' presupposition of the reality of irreducibly *social* facts (social reality is autonomous, its ontological level significantly distinct from the biological)—may offer a partial explanation.

12. More from Livingston: " . . . science is a selective and highly oriented attitude toward nature, and . . . it is neither always the best nor the only such attitude. It is moreover an attitude motivated by certain values, the foremost of which, is *truth*" (108). And later, with remarks on *events* that will become fully relevant when my argument about the need for a minimal model of human events emerges: "If science's orientation toward truth is very fundamental, the more specific direction in which its inquiry tends has to do with an additional attitude; the orientation toward truth does not explain, for example, why *science focuses not on single events, but on types of events;* not on particularities, but on classifiable, countable, and measurable phenomena. Nor does the goal of truth explain why science is most interested in lawlike regularities . . ." (109) [emphasis added]. Although I do not fully concur with the perhaps too-narrow goals Livingstone recommends for literary research at the end of the book (258–67), I strongly recommend *Literary Knowledge* as an introduction to the philosophy of science for scholars and students of literature and as a respectful refutation of the framework relativism idly perpetuated in many academic departments of literature.

13. On the Sokal hoax, see van Oort, "Science."

14. What gets denied is the following, from Rescher, *Limits:* "Man is a member not just of the *natural* but of the specifically *human* order of things. There is more to reality than science contemplates; in the harsh but stimulating school of life, we are set examinations involving problems for whose resolution our science courses by themselves do not equip us" (248). Yes: we need courses in matters such as the Frankenstein myth.

15. For references to this dictum, see Benton quoting Pinker 219; Steven Rose 254; Hurford x.

16. For the thinkers outside the discipline of generative anthropology whose work (in different ways, and from different positions) justifies the *I am not alone* claim, see in particular Belsey; Dupre; Eccles and Robinson; Evans; Fukuyama; Griesemer; Jones; Kass; Keller, "Master"; Krutch; Malik; Midgley; Olfason; Owen; Rose and Rose (the anthology); Stenmark; Tallis; Weizenbaum.

17. See Nagel, "Ethics without Biology." Compare these remarks by Dupre: " . . . evolution does indeed have momentous consequences for our view of ourselves and our place in the universe, but it does not have the kind of consequences most widely advertised today. In particular, *it is of limited use in illuminating human nature*" (*Darwin's* 2) [emphasis added].

18. For the criticism of "cultural studies" that underlies my taking a distance, see Stephen Adam Schwartz; van Oort, "The Critic as Ethnographer" in *End.*

19. By "environmentalism," Eccles and Robinson refer not to ideas organizing a reasonable concern for ecological crises but rather to this belief:

"What one does . . . or comes to want to believe or strive for is just what the environment has 'planted' in him" (6).

20. Rescher, *Limits:* "During most of the present century the number of American scientists has been increasing at 6 percent to yield an exponential growth rate with a doubling time of roughly twelve years. A startling consideration—one often but deservedly repeated—is that well over 80 percent of ever-existing scientists (in even the oldest specialities such as mathematics, physics, and medicine) are alive and active nowadays" (54).

21. Gans' major books on generative anthropology in order of publication would be listed thus: *The Origin of Language: A Formal Theory of Representation* (1981); *The End of Culture: Toward a Generative Anthropology* (1985); *Science and Faith: The Anthropology of Revelation* (1990); *Originary Thinking: Elements of Generative Anthropology* (1993); *Signs of Paradox: Irony, Resentment, and Other Mimetic Structures* (1997); *The Scenic Imagination: Originary Thinking from Hobbes to the Present Day* (2007); *A New Way of Thinking: Generative Anthropology in Religion, Philosophy, and Art* (2007). See also his *Chronicles of Love and Resentment,* now numbering over 400, available through the *Anthropoetics* website.

22. For work by some of those who are expanding and developing the discipline of generative anthropology, see the anthology edited by Katz. Places to start are the essays by Bertonneau, Eshelman, Fleming and O'Carroll, and Katz, "Question." For the clearest, most helpful book-length application of generative anthropology not authored by its founder, see van Oort, *End.* For a lucid exposition of the differences between the thought of Girard and that of Gans, see van Oort, "Mimetic." For longer introductions to generative anthropology, see Bartlett, "From First"; van Oort, "Cognitive." For concise introductions, see Bertonneau, "Gist"; Dennis 13–23; Goldman. The most accessible set of resources on generative anthropology is *Anthropoetics* (on the web).

23. For a succinct summary of Levi-Strauss' lack of interest in the diachronic dimension of myth, see Segal 118–199.

24. From a surprisingly large set of possible samples, this from one as erudite as Fernandez-Armesto is sufficiently representative: "The best available conclusion in our present state of knowledge is that there are many species with forms of communication specific to themselves, and it is unclear why language—even if it is in some sense a peculiarly human resource—should be treated as a basis for classifying the species that uses it apart from all others" (22). Here is another, from the philosopher Gray: "The calls of birds and the traces left by wolves to mark off their territories are no less forms of language than the songs of humans. What is distinctively human is not the capacity

for language. It is the crystallization of language in writing" (56). See also Grandin 280–81.

25. The following from philosopher of science Carnap is not irrelevant here: "It must not be forgotten that, both in the history of science and in the psychological history of a creative scientist, a theory has often first appeared as a kind of visualization, a vision that comes as an inspiration to a scientist long before he has discovered correspondence rules that may help in confirming his theory" (175). Likewise the philosopher of language Cooper, following Popper: "Hypotheses are not derived from observations; rather, observations serve to confirm or disconfirm hypotheses which alone tell us what to look for" (36).

26. I deliberately echo Peter Singer's disclosures about the frequency with which the mindless gesture toward the necessity of *further research* is repeated in the context of highly questionable, if not certainly unjustifiable experiments on nonhuman animals: *Animal Liberation* (25–94; esp. 30, 32, 35). Much experimentation on nonhuman animals seems to be exemplary of real-world mad science: scientific work the object-results of which are *maximally desacralizing and minimally exchangeable* (see below in this chapter, "Exchangeability and Desacralization"). Generative anthropology should lead us to be deeply respectful of nonhuman animals. One consequence of a widespread respect for the originary hypothesis might be the liberation of the many animals now used in "testing" to figure out what human language is or how it evolved. One wonders how many chimpanzees are sacrificed to the vain dream of trying to *prove* human language and animal communication are more, not less assimilable (when common sense, alas, would tell us they just are different in kind). It is ironic that research programmes designed to prove hypotheses to the effect that animals are more like than unlike us must inflict so much misery on those whose ontological status would presumably be elevated, were we to find evidence of linguistic prowess. If the animals had their "natural" way, they would not be living in laboratories. One might elevate animals better by letting them go and studying human language, religion, art, as they are (human) rather than what they are not (nonhuman).

27. I echo the title of the enjoyable book, *The World Without Us* (2007): having written an article for *Harper's* magazine about what happened to the ecological niche around Chernobyl after it was abandoned by humans, Weisman agreed to an editor's suggestion to expand his investigations to what would happen if on planet earth all humans were suddenly to disappear—not having done ourselves in with a nuclear holocaust, not having been wiped out by a virus, but just having disappeared. Expecting an anti-humanist tyrade against our capacity for ecological destructiveness, I was surprised by

the moderation and thoughtfulness of Weismann's projections. I was also amused by the irony that much of the book illustrates what would happen to our monumental technological achievements rather than how nonhuman life would flourish once liberated from our dominating presence. The world without us will still be, it seems, a world strongly marked by the material traces of our handiwork.

28. On the ability to navigate as a component in language evolution, see Wade 37. On prelinguistic hominids exchanging information with calls, see Hurford 329–330. On chimpanzee competitions for position, see Dunbar 9–34.

29. On fear as one of the fundamental animal emotions, see Grandin.

30. I hasten to add that to make such a claim is not to condone the oblivious cruelties against animals, cruelties which we seek to justify with a glib speciesism: to say that our reality transcends that of animals because of our possession of the sign is not to say that we cannot share the world responsibly with nonhuman animals and seek to welcome them with consideration onto the scene as we would welcome each other. In a strange way, our moral obligations to nonhuman animals go deeper than those we have to other humans—because animals cannot like normal humans *speak for themselves.* Therefore, we must speak for them—as we speak for human babies or people helplessly incapacitated by injury, age, addiction or the like. (The founding source for the concept of speciesism is Singer.)

31. Gans: "The nature of human action is distorted by any perspective that seeks to reduce [human action] to a set of predictable models—within which the model-maker's own activity both must and cannot escape accounting" ("Beyond").

32. Lobkowicz makes the point in different words: "Atheism presupposes a culture that believes or at least did believe in one transcendent God" (375).

33. Gans has defined love as *the deferral of resentment.* For a fuller explanation of these points, see his "Marcus Borg's Spiritual God" and "God is Love." See also my "Three Affirmations."

34. This is to say that, at the origin, the experience of the sacred for originary humans is very close to the experience of the desire-saturated subject as formulated thus by Girard: " . . . there is no desire except desire for absolute difference, and the subject always lacks this difference absolutely" (*Things Hidden* 389).

35. For a full exploration of this point, see my "Originary Personhood."

36. To posit the historical event of the recording in the studio as the referent of the famous song is deliberately naïve; however, it is the kind of referent a Beatles fan might desire as a concrete object (*how wonderful it would have*

been to witness the actual recording in person) and so it makes the point I am after. The "referent" of "Here Comes the Sun" is certainly more diffuse and irreducible than such a concrete thing.

37. See Reichmann 165–67 for a provoking illumination of the anti-humanism in Darwin's preference. Reichmann's careful, thorough book is an invaluable exploration for anyone interested in examining seriously the animal-human boundary without seeking to reduce to inconsequentiality the anomalous mystery of human language.

38. For some of the most important works that have developed this anthropology of the Cross, see (again) Alison; Bailie; Anthony Bartlett; James Williams. The most concise work by Girard offering access to this anti-sacrificial anthropology is *The Scapegoat* (trans. 1986).

39. For more on this, see my "Originary Object of Violence."

40. Midgley: "Dennett describes Darwin's 'dangerous idea'—that is, the idea of development by natural selection—as a 'universal acid . . . it eats through just about every traditional concept and leaves in its wake a revolutionised world-view, with most of the old land-marks still recognisable, but transformed in fundamental ways'" ("Why Memes?" 72). The quotation comes from Daniel Dennett's 1995 book *Darwin's Dangerous Idea: Evolution and the Meanings of Life.* Dennett is a champion of ultra-Darwinian reductionism.

41. Gans: "The human, from the perspective of the originary hypothesis, is *defined by the exchange of signs rather than the exchange of things;* the freedom and ease of the [exchange of signs] can only indirectly and over time be transferred to the [exchange of things]" ("Our Post-Postmodern") [emphasis added].

42. Since we can exchange as equals the name of God, we can be equal as humans before the divine Other who opposes us so as to create us as a single indivisible community. In the originary exchange of signs, each abortive gesture *intend*s both to share the referent as imaginary object and to defer the conflicted desire to consume it as real object. Putting into words now what could not have been expressed then, we might translate *sacred value* this way: the sacred center is equally forbidden, equally desired, equally inaccessible, equally resented for its inaccessibility. But the *name* of the central sacred being—the Creator, God, the One Other Being before whom all humans are equally valuable—is valuable in this practical world precisely because the Being named, thus kept in mind, is the One against whom the worth of each human as exchangeable object is valued (measured) as an absolute. We answer for our exchanges with other humans and for our treatment of human in the presence of this spiritual Other. We cannot *exchange* anything with God that

will make any of us equal to him; our universal equality, and therefore our minimal value as *nothing but* materially bound humans, is measured by one incalculably valuable spiritual difference: the difference between all of us us and our equally shared Divine Other. That is the difference—and it can only be thought as *immaterial*—that Frankenstein and his scientistic heirs want and try to *prove* is a nondifference, irrelevant, unreal, inconsequential. Scientism must fail or else concoct horrors because the difference between the transcendent and immanent is perfectly and utterly real to humans. We who know ourselves only as beings exchanging signs on the scene of representation.

43. Along the same lines, the theologian Errol Harris in his searching study of the tension between revelation and modern science, includes as one of the three types of "superstition" that he believes rampant in the modern world, the following, which he names "empiricism": " . . . contending that statements about God are devoid of factual significance because they cannot be supported by evidence derived from the senses" (38). See also Kaufman 86–87.

44. This differentiating includes the originary re-cognition that I myself (insofar as I become aware of my possession of my own body) am not *just* an object, because the *object* is always other than me and other than the locus my body occupies and leaves behind when its organic life ends. It would do no harm if, in our fitness-obsessed, image-saturated culture, we could re-learn something of the *unreality* of our own bodies as present to our consciousness. I am not my body: I am not just my body; I am other than my body alone; I observe my body, but I am not just it—just as the central locus of the scene of representation conceived as a Being is not to be confused with any of the represented objects that occupy the center and "represent" or make visible its Being. Individual human personhood is modeled on the transcendent personhood of the communally-known sacred Other.

45. For a cogent defense of the thesis that the most promising environmental politics will be based on recognition of this truth, see Scruton, *How to.*

46. Eric Cohen: "As experts, scientists know more than the rest of us; as citizens, they are just like the rest of us—sometimes wiser than most, and sometimes more foolish, especially when they falsely believe that their knowledge of scientific facts qualifies them to act as the lone moral arbiters of scientific research" (788).

47. Kolata: "In the 1960s, some intellectual leaders used cloning as a metaphor for the promise of science to allow humans to control their destiny and their evolution. By the end of the 1970s, in the context of the new movement to contain the awesome powers of biological scientists, cloning became a metaphor for the temptation of scientists to play God" (108).

48. Kass offers a list of the reasons many people find the idea of easily-accessible human cloning technology repugnant: "They recoil from [1] the prospect of mass production of human beings, with large numbers of look-alikes, compromised in their individuality; [2] the idea of father-son or mother-daughter twins; [3] the bizarre prospects of a woman giving birth to and rearing a genetic copy of herself, her spouse, or even her deceased father or mother; [4] the grotesqueness of conceiving a child as an exact replacement for another who has died; [5] the utilitarian creation of embryonic genetic duplicates of oneself, to be frozen away or created when necessary, in case of need for homologous tissues or organs for transplantation; [6] the narcissism of those who would clone themselves and the arrogance of others who think they know who deserves to be cloned or which genotype any child-to-be should be thrilled to receive; [7] the Frankensteinian hubris to create human life and increasingly control its destiny; [8] man playing God" (146–7) [enumeration added].

49. Kitcher: "If cloning human beings is undertaken in the hope of generating a particular kind of person, a person whose standards of what matters in life are imposed from without, then it is morally repugnant, not because it involves biological tinkering, but because it is continuous with other ways of interfering with human autonomy that we ought to resist" ("Human" 146). Meilander: "Even if we grant that a clone, reared in different circumstances than its immediate ancestor, might turn out to be a different person in some respects, the point of that person's existence would be grounded in our will and desire" (271–72). See also Caplan 66.

50. For arguments roughly similar to this one, see Agar 168–70; Habermas 60–66, 82–83. For dissenting views that see little or nothing to fear in the prospect of human cloning under controlled circumstances, see Dworkin; John Harris; McGee and Wilmut; Robertson. For a theologian who views the possibilities for therapeutic human genetic engineering favorably, see Cole-Turner.

51. The most relevant passage here is Gans' "The Two Varieties of Truth" in *Signs of Paradox* (1997) [chap. 4; 51–63].

52. I do not wish to imply that science is invested *exclusively* in the predictable; not so. As Kitcher demonstrates in his critique of Popper's falsifiability criterion, "In general, science is at least as concerned with reducing the number of unexplained phenomena as it is with generating correct predictions" ("Believing" 71).

53. For the Universal films, see Whale (listed deliberately among the secondary sources). For the Hammer films, much closer in spirit and content to the Frankenstein myth as this study defines it, see Fisher. Branagh's 1994

version foregrounds the egotism and recklessness of Victor. Connor's version, a Hallmark Hall of Fame production, is tame but more faithful to Mary Shelley's novel than any film version I know. In my opinion, any film of the story as told by Shelley must betray the essence of the novel and fall short of its power: to visualize the Monster is already to make him less frightening—less horrifyingly ugly—than Shelley wished him to be.

54. On creation myths, see Long, esp. ch. 4, "Creation from Nothing" (150–91). For the sensible but worth-pondering observation that creation from nothing is actually creation *from language,* see Campbell (86).

55. That the sign is not a material thing is one of the principles of generative anthropology. Here are a few representative remarks from Gans: "Signs are ideal entities that don't 'exist' at all. That such entities are not 'natural' does not make them 'supernatural,' which is only a mystified variety of 'natural.' Signs are not things but relations within a network of human communication. The idea of the linguistic sign is the most inexpressible of all ideas because it is the simplest—the minimal idea" ("Mind and Brain"). "Although signs are learned individually by individual minds, they don't subsist in those minds, but in the sphere of their communication. The point of the sign isn't to convey my or your experience, but to permit us to communicate about what is important to the linguistic community" ("Mind and Brain"). "There is nothing mystical about claiming that the sign has a different ontology from the elements of the real world. A sign is not a *thing;* it is a complex of things and the relations between them, mediated by the community that exchanges them" ("A New Mode of Being").

56. Rescher, *Limits:* "Thus, science ignores the individualized, affective, and person-linked dimension of human cognition: sympathy, empathy, feeling, insight, and personal reaction. The phenomena it takes as data for its theory projection and theory testing are publicly accessible" (244); "The limits of science inhere in the limits of its cognitive mission and mandate: the 'disinterested' depiction and rationalization of objective fact" (245); "It is no more a defect of science that it does not deal with *belle lettres* than it is a defect of dentistry that it does not deal with furniture repair" (247). Scientism, however, would argue that *belle lettres* are not really worth dealing with in the first place; scientism turns the limit of science into a defect by denying it is a limit.

57. Kuhn's remarks about the insulation of scientific communities resonate oddly with my contentions about the isolation of the Victor-figure: "The most esoteric of poets or the most abstract of theologians is far more concerned than the scientist with lay approbation of his creative work, though he may be even less concerned with approbation in general. That difference proves

consequential. Just because he is working only for an audience of colleagues, an audience that shares his values and beliefs, the scientist can take a single set of standards for granted. . . . Even more important, the insulation of the scientific community from society permits the individual scientist to concentrate his attention upon problems that he has good reason to believe he will be able to solve" (164).

58. In his judicious study *Modernism and Totalitarianism: Rethinking the Intellectual Sources of Naziism and Stalinism, 1945 to the Present* (2012), Shorten identifies three "totalitarian currents": utopianism, scientism, and revolutionary violence. Not a writer given to the rhetoric of frustration and grievance, Gans seeks whenever possible to minimize difference between interlocutors and dialectically to create spaces for dialogue rather than picking fights. And yet it bears remarking that Gans' work has consistently contained threads of wariness regarding those three currents—utopian thinking, revolutionary violence, and (less explicitly, thus this book), scientism. If I were forced to name three polemical opponents generative anthropology does have, I would name those three. Shorten's work is worth quoting at some length, especially given my wariness of sociobiology in all its forms: "But Darwinism did provide two concepts that became fixed in scientism's totalitarian current. First, it provided the idea of evolutionary progress. Second, it provided the idea of a violent struggle for existence. In each of these ways, Darwinism transcended the Enlightenment because it emphasized not free will and optimism, but fatalism and a kind of pessimism: the basis of human action was no longer conscious, rational choice but was, instead, a mixture of 'heredity' and 'environment'; and social progress would no longer follow automatically from the triumphant application of science, but progress now became conditional upon violent struggle. The second idea also legitimated projects in eugenics—projects controlling reproduction to improve the human species, albeit only once Darwin's original conception was inverted: artificial selection in the place of natural selection" (166–67).

59. On the maximally desacralizing and minimally exchangeable technology of nuclear weapons, see Lifton and Mitchell; Schell. See also my "Nuclear Warfare in the Movies."

60. Alexander's single-paragraph summary of Asilomar : "At least since the dawn of the Christian era there have been bio-Luddites. But the gene splicing era rekindled opposition to experimentation with the stuff of life. When word of recombinant DNA experiments in Paul Berg's Stanford lab leaked out, some scientists were concerned that Berg risked creating a kind of superbacteria that could unleash a plague. In one of the most important milestones in the history of biology, researchers gathered in 1973 at a

California conference center called Asilomar to decide on safety procedures for continued gene splicing experiments, and to assuage the public fears that science had gotten out of control. Today, the biologists involved look upon 'Asilomar,' as it is known, as a shining example of the responsible exercise of scientific caution" (126). For more on the Asilomar conference, see Appleyard 32–36; Evans 96.

Notes to Two / Shelley's *Frankenstein*

1. I use the 1818 edition of the novel in this book. The competition between the 1818 and 1831 editions for "textual authority" must be acknowledged, but I agree with the reasons given by the Broadview editors MacDonald and Scherf, reasons explained in greater detail by Marilyn Butler's "Introduction" to her version of the 1818 edition, for the preference of the less-doctored original text. Among other disadvantages, the 1831 edition presents a more "fatalistic view of human nature" and a less interesting, more "mechanistic view of non-human nature" than the 1818 edition (MacDonald and Scherf 39). An unquestioning obedience to the editorial principle that insists on the authority of any later revision as opposed to any earlier one on the basis of "final" authorial intentions has always puzzled me. What if the intentions of a later revision betray the insights and emphases of an earlier edition? Should one not judge each conflict between editions for textual authority on a case-by-case basis? Must one prefer the later edition even if it is, critics agree, *dull* by comparison with the earlier one? I believe not.

2. The scholarly basis for my biographical assertions in this paragraph, where not explicitly noted, may be found in Small (1972) and John Williams (1999).

3. Ketterer's study must be classed among those carefully wrought and intelligent monographs that leave one, initially, feeling that there is nothing more to say since the critic has done such a thorough job. With respect to the opposed directions I am charting, the following is also relevant: "the construction of the fictional world *Frankenstein,* the construction of the Monster, and the construction of the human world are offered as analogues of one another. *The extent to which the three areas of analogy are merely rough parallels and the extent to which they represent a fundamental identity is left open*" (96) [emphasis added]. What I will be calling the *doppelganger* approach would insist on the "fundamental identity" of the constructions, whereas the literalizing approach would take them as rough parallels and would scrupulously preserve the ontological differences between the three "areas of analogy."

4. Derrida's complete sentence reads: "The Jew is split, and split first of all, between the two dimensions of the letter: allegory and literality" (75).

5. For a rapprochement between generative anthropology and psychoanalytic thinking, see Gans, "Two Psychoanalytic Categories: Eros and the Unconscious"

(*Signs* 111–127). Girard, although he continues to be ignored by thinkers who find the anatomical invidiousness of psychoanalytic mythology infinitely encouraging, saw things accurately (in my opinion) long ago: "The main pitfall of psychoanalysis . . . is to assume that the individual being is rooted in scandal, according to an absurd and mythic thesis that presents parricidal and incestuous desire as the condition for the development of any form of consciousness" (*Things Hidden* 417). Gans says something similar: "To understand the originary form of desire is not suddenly to remember, as in a horror story, the figure one had not consciously noticed, but to become conscious of the alienation of desire in the present. . . . It is to be hoped that the producers of psychoanalytic discourse may . . . be cured of their infatuation with objectal cathexes by letting emerge within their own theoretical consciousness the basis of desire in the scene of human origin" (*Signs* 127). For other serious, searching critiques of psychoanalytic theory that have influenced this study, see Gellner; Grunbaum. For a valuable resituating of the mimetic theory of desire in the context of empirical and social psychology, see Livingston, *Models*.

6. I owe this analogy between the Monster's reality and the reality of Claudius' villainy to an illuminating analysis of *Frankenstein* by Richard van Oort, who has explored the ways the novel displays in particularly acute form the victimary rhetoric and self-centralizing paradoxes of the Romantic esthetic. Van Oort was the first to study Mary Shelley's novel through the lens of generative anthropology. Rather in the way that Charles Schug does in a related but not anthropological context, van Oort emphasizes the formal feature of the concentric narrative circles in the text, that organization foregrounding the intended comparison between the young sailor Walton, the wasted Victor, and the outcast Monster as they compete for victimary centrality in telling their tales of total woe. The merit of his work is that it both elucidates the implicit anthropology in Mary Shelley's text and refuses—quite unlike most *doppelganger* readings—to reduce the text to a symptom of her personal psychological experience. Inasmuch as he follows the path of the *doppelganger* myth by moving toward a certain *dissolving* of the *literal* difference between Victor and the Monster, his interpretation foregrounds the form of the novel at the expense of its content. My approach differs by trying to specify the novel's *mythic* content from the point of view of generative anthropology.

7. For arguments that find it appropriate and revealing to see Victor Frankenstein as a modern scientist, see Goodall; Manson and Stewart; Marcus; Stableford; Vasbinder; Ziolkowski. The opposing position is strongly taken by Ketterer in his 1986 review of Vasbinder; Ketterer cites, most importantly, the valuable study by Buchen. And yet although (unlike Ketterer) I have no quarrel with Vasbinder's work, I would also recommend Buchen as a source whose positions on Mary Shelley's use of alchemy seem reasonable. For a curious feminist analysis that ends up wishing Victor had used good, pre-modern, eco-friendly "spiritual alchemy" rather than bad, modern, masculinist "empirical alchemy," see Marie Roberts.

8. On this point, see Eric Gans, *Science and Faith* (1990), the chapter titled "The Christian Revelation"—particularly the reflections about the absence in the blinding light that stops Saul of the figure of the cross.

9. For elaborations of the idea that election is typical of the neoclassical and self-centralization typical of the romantic esthetic, elaborations in the framework of generative anthropology, see Gans, *Originary* 150–63; see also my "From First" 124–26.

10. Bowerbank tends to make this error in an essay arguing that Victor and the monster are "victims of society." She must categorize the universal responses as evidence of "prejudice" that could be unlearned: she writes, for example: "Victor shows the same prejudice against ugliness, albeit to a much lesser degree, in his reluctance to work with Professor Krempe" (426). But in the case of the Monster, it is not just a "prejudice," it is an ontological scandal, the result of a mad scientist playing God the creator. Elaine Graham invidiously attributes this ontological boundary to a "failure of imagination," which implies we could somehow be educated out of the visceral response: "The being's tragedy lies in the abuse of his gentle sensibilities by the *inflexibility and lack of imagination of human culture,* amounting to a powerful indictment of the *in*humanity of his detractors. Had the creature been willingly assimilated into human society, he could have developed a benign character" (67) [emphasis added]. Theodore Ziolkowski also underestimates the force of the ugliness: "If Victor Frankenstein had not been overcome by his *initial disgust,* if he had responded to his creature with love and understanding, it might have become an instrument of good rather than evil" (43) [emphasis added]. My contention is that it is a factual error to assert that the "disgust" Victor feels was or could have been only an "initial" response later altered. These critical responses are well-intentioned but ultimately victimary misreadings of the mythic content of the text.

11. With this remark, I have in mind, among other contexts, the following passage from Steve Fuller's discourse on the rivalry between philosophers of science Karl Popper and Thomas Kuhn: "Contemporary theories of knowledge rarely make reference to their religious roots. Nevertheless, these roots are indelibly marked in the philosophical tendency to think of *beliefs as compelled by evidence* rather than made by decision. Some philosophers even claim that it is psychologically impossible to decide to believe something. At best, such a decision is a pretence to belief (that is, to act 'as if' something were true); at worst, it is tantamount to wishful thinking. Clearly 'belief' is meant to be a rather profound state of mind, a partial *revelation of the truth,* no mere hypothesis adopted out of expedience for the sake of argument. The problem of knowledge then revolves around the search for some foolproof method, or criterion, for assessing the evidential quality of beliefs" (112) [emphasis added].

12. I part ways with Haynes, however, when she asserts: "Shelley thus suggests that claims of ignorance on the part of scientists for their failure to

foresee the consequences of their work are too glib to be credible and cannot be socially acceptable" (97). It makes more sense to hold Victor responsible for his willful self-isolation than to hold him responsible for his failure of foresight: the failure to foresee is the result of the isolation, rather than the expression of a conscious intention that he registers and then represses, later claiming ignorance of it.

13. In an otherwise fine study that defends Victor's scientific ambitions by placing them in the pre-scientific context of a repressive Calvinist ideology that produced the tortures of the reprobate's conscience, Goodall succumbs to the temptation in a particularly striking fashion: "the novel itself may be suggesting that the Creature is a natural being with elevated moral and social instincts and that, apart from some *acute cosmetic problems,* the experiment of which he is the result has been a profound success" (33) [emphasis added]. On the one hand, Goodall is quite correct to assert the "profound success" of Frankenstein's experiment. But on the other hand, it is a factual error to reduce the Monster's ugliness to "acute cosmetic problems."

14. I agree with Gigante: "The Creature's ugliness . . . constitutes a return of the repressed not linked to any particular childhood fixation. Instead the Creature appears as a return of what is universally repressed. . . . the horror at the core of all existence" (567). One might propose in originary terms that the "horror" is the truth of the exclusivity of the *human* mediation of the originary violence of the sparagmos, which can never quite be recuperated by the knowledge of the object that we gain by means of its (violent) consumption and distribution. The finality of knowledge of the economic-desacralized object always risks seeming horrible when set against the paradoxical dissatisfaction of the "knowledge" of the esthetic object. Recall the unstable structure that generates the esthetic effect, its oscillation between imaginary possession and recognized inviolability of the central object: merely imaginary possession risks no disillusionment in the way that real possession does; violation risks an experience of "horror" that the mediated recognition of inviolability does not. Science, like economic exchange, requires the "violation" of the erstwhile sacred untouchable object.

15. I have in mind the following remarks from Richardson's work on the "brain science" informing Shelley's fanciful thought experiment: "In gathering components for his gigantic science project, Victor has had recourse to the 'slaughterhouse' as well as the dissecting room. His experiment transgresses the 'wide chasm between man and the noblest animals' insisted upon by Coleridge but inexorably narrowed by materialist and 'corporealist' thinkers from La Mettrie and Herder to Darwin and Lawrence" (162); "Yet the Creature turns all too readily from sociability to savagery, from humankindness to ferocious rage. . . . . The reader is left to wonder whether Victor has 'endowed' his creation not only with human 'perceptions and passions' but with bestial ones as well" (162).

16. Clayton offers a useful summary of the pro-Monster tendency in postmodern criticism: "Today's positive interpretations [1990] draw on an undercurrent in *Frankenstein,* to which feminist critics of the 1970s and 1980s first drew attention: the novel's sympathy for the Creature. Contemporary advocates of AI emphasize many of the same things Shelley did: the emotional vulnerability of this new being, its abandonment in a hostile world, its sheer *creatureliness.* Turning its artificial origins from a liability to a virtue, writers and filmmakers focus on what humans owe to the things they create" (86). It is in the "turning its artificial origins from a liability to a virtue" that the victimary error initiates itself: the austerely frustrating, ironic structure of the vain scientist playing God and creating a *non*human gets set aside for more accessible satisfactions in which the Monster can be taken to represent this or that victimized human group whose revenge against their persecutors is legitimate in a "human" way.

17. Gans: "We note once more the relevance of Condillac's reference to the Genesis story. His originary sensation is not based on sensation but on 'needs.' The association of the children, not otherwise motivated, is cemented by their sharing of 'passions.' And however physical these 'needs' may be, Condillac describes the couple's mutual assistance in mimetic terms: 'He suffered by seeing the other suffer so miserably.' When he asserts that they act 'by instinct alone,' he refers of course to their prelinguistic, prereflective state, but this state is implicitly distinguished from similar states in animals by its proto-human mimeticism" (*Scenic Imagination* 42).

18. As Victor pursues him to the North Pole, the Monster learns how to manage dogs and a dog sled and takes up the technology of weapons: "A gigantic Monster, they said, had arrived the night before, armed with a gun and many pistols; putting to flight the inhabitants of a solitary cottage, through fear of his terrific appearance. He had carried off their store of winter food, and, placing it in a sledge, to draw which he had seized on a numerous drove of trained dogs, he had harnessed them, and the same night, to the joy of the horror-struck villagers, had pursued his journey across the sea in a directed that led to no land" (3.7.228).

19. This label, however crude, goes along with Victor's own self-descriptions and Walton's evaluations of his condition. Walton observes: "His limbs were nearly frozen, and his body dreadfully emaciated by fatigue and suffering. I never saw a man in so wretched a condition" (58). Walton calls him "a man on the brink of destruction" (58). Victor regrets his course: "'But I—I have lost everything, and cannot begin life anew'" (61). Toward the end of his long narrative, he offers this metaphor: "But I am a blasted tree; the bolt has entered my soul; and I felt then . . . what I shall soon cease to be—a miserable spectacle of wrecked humanity, pitiable to others, and abhorrent to myself'" (3.2.185).

20. The mystery Shelley invoked with the phrase "secret of life" remains an

enigma for biologists and philosophers of biology. Attempts to foreclose the discussion are premature. The philosopher of biology Dupre concedes: "It is true that the ultimate origins of life, the transition from primeval slime to the first living cell, remain little understood" (*Darwin's* 61). At the same time he cautions us that this is no reason to reject evolution (for creationism or intelligent design): "The situation points to an important aspect of inference to the best explanation. The best explanation available may well be, and perhaps often is, sketchy and partial" (62). See also Berlinski; Haught, esp. 55–76.

21. The Monster to Victor: "'God in pity made man beautiful and alluring, after his own image; but my form is a filthy type of your's [sic], more horrid from its very resemblance. Satan had his companions, fellow-devils, to admire and encourage him; but I am solitary and detested" (2.7.155). The Monster to Walton: "'Yet even the enemy of God and man had friends and associates in his desolation; I am quite alone'" (3.242).

22. "It is desolation": I am suggesting that the power of the Frankenstein myth lies partly in its creation of a new category of fictional victim that somehow exceeds or does not fit into previous categories. In this, the flourishing of the Frankenstein myth is like that of the Dracula myth. However, it is significantly unlike the vampire myth in that its victims, as victims of scientism, have real world analogues, while the victims of Dracula and his heirs do not. We do not fear real vampires in the real world of modern science; we do fear scientism in the real world of modern science. The exceeding of previous categories, the sense that something new emerges into view with such stories as Shelley's, follows from the literal non-human status of the victims, the *a priori* dehumanization of such strange human-like creatures as Victor's monster, the Beast People, Capek's robots, Tyrell's replicants, and the other impossible humans in texts contributing to the Frankenstein myth. "Desolation" is to refer to a category of human (or human-like) suffering that removes suffering from any sacrificial economy by virtue of which the suffering might be recuperated as "meaningful" for the victims or perpetrators or bystanders. The victims of Auschwitz and Hiroshima are the strongest examples of such desolation; they are the most powerful examples of such desolation. It would not work to think of them as victims of a romantic tragedy. Insofar as they were victims of dehumanizing scientism, we find in the victims of the Frankenstein myth their fictional counterparts.

23. "God or whatever means the Good/ Be praised that time can stop like this, / That what the heart has understood / Can verify in the body's peace / God or whatever means the Good." That is the second-to-last stanza of the love lyric "Meeting Point" by Louis MacNeice. It celebrates the kind of love the Monster hoped in vain to get from the aborted female companion that the vain scientist Victor at last had to tear to pieces.

Notes to Three / H. G. Wells's *Moreau*

1. On Thomas Henry Huxley as Charles Darwin's public promoter, see Caudill 3–45.

2. Suvin calls Wells "an apocalyptic writer'" ("Wells" 31), positioning him as "the central writer in the tradition of science fiction" (31) and as "the first significant writer who started to write science fiction from within the world of science, and not merely facing it" (32). Ruddick seconds the opinion: "science fiction in Britain arose through the Wellsian scientific romance" (*Ultimate* 61).

3. This complaint about the argumentative imbalance in chapter 14 is made, for example, by Fried: "[A]lthough Prendick is appalled by the pain involved in Moreau's procedures and obviously doesn't accept Moreau's characterization of him as a materialist, he also fails to give an account of his position that would distinguish it from Moreau's: he never gives his own views a name, and he is willing to accept vivisection if it is for a useful end" (109). Hinckley similarly complains: "Prendick is the representative of conventional morality, but he can find no arguments to resist Moreau's 'scientific' ideals, the brutal cruelty of 'pure reason' . . ." (90). Both comments assume that if Prendick presents no such arguments on the spot during the discourse of Moreau—after, I would note, he has been chased almost to death (as he believed) by an armed and dangerous hunting party—then he has no such arguments anywhere at his disposal. The assumption is ungrounded. Prendick spells out for us many of his objections to Moreau. He comes to see a sociopath in Moreau. For now, he correctly judges that it would be a waste of time to try to engage in debate with the vain vivisectionist anyway.

4. From an interview in *Young Man,* August 1897, during which Wells said: "I should say that *Moreau,* although it was written in a great hurry and is marred by many faults, is the best work I have done. It has been stupidly dealt with—as a mere shocker—by people who ought to have known better. The *Guardian* critic seemed to be *the only one who read it aright,* and who therefore succeeded in giving a really intelligent notice of it" (quoted in Parrinder, *Critical Heritage* 52) [emphasis added].

5. Hinckley, in a useful comparison between *Frankenstein* and *Moreau* as figurations of the mad scientist, has pointed to the inappropriateness of the analogy: "Wells' choice of the hyperrational Moreau as his deity figure is strangely at odds with a didactic critique of Christianity, whose ethos is 'Do unto others as you would have them do unto you,' a religion of sympathetic feeling rather than cold detachment. Moreau is in many respects the ultimate positivist . . . [. . . .] While the novel is, like *Frankenstein,* a radical critique of popular Christianity, the positivistic monster who plays God is also *an unappealing scientific alternative to religious rationality.* Reason, as Moreau defines it, seems more artificial than most religions" (90) [emphasis added]. In other words, the feebleness of the analogy of

Moreau- as-God makes the textual attack on "religion" likewise feeble.

6. Midgley makes somewhat the same point in her searching critique of the incoherent physicalist cosmologies of Jacques Monod and Steven Weinberg: "First, there is the tone of personal aggrievement and disillusion, which seems to depend, in both him and Monod, on failure to get rid of the animism or personification which they officially denounce. An inanimate universe cannot be hostile. To call it that is to reproach it for not being the divine parent of earlier belief" (*Evolution* 100).

7. Cheyette: "What is most disturbing about Wells' statements at this time [in the 1930s] is that in his *In Search of Hot Water,* for instance, he accurately foresaw the 'systematic attempt to exterminate' the Jew—'to exterminate him brutally and cruelly' (56)—while, at the same time, arguing that this was a logical consequence of the refusal of 'the Jews' to 'assimilate' and give themselves to 'the service of mankind.' This argument was once again made explicit in his *The Fate of Homo Sapiens* (1939), which was reprinted in *The Outlook for Homo Sapiens* (1942) during the War" (59). Neither does the following information, also reported by Cheyette, do Wells credit: "When faced with Jan Karski's eye-witness account of Belzec death camp in November 1942, Wells could only reply, in the words of Karski, that 'there is room for very serious research into the question of why antisemitism emerges in every country the Jews reside in'" (60). Coupland remarks that "an authoritarian elite without scruples about using violence was a long-established aspect of Wells' theory of revolutionary praxis" (545); he shows that Wells' main objection to Mussolini's fascism was only its nationalism, quoting from Wells' futuristic *The Shape of Things to Come* (1933): "'Except for the fundamentally important fact that these Fascisti were intensely nationalist, this control by self-appointed, self-disciplined elites was a distinct step towards our Modern State organisation'" (551). See also Wagar (19, 22). On the other side, certain facts that David C. Smith underlines should prevent us from leaping to condemnation: "This comment [that Wells helped fascism grow] seems risible today when one realizes that Wells called attention to the probable impact of Hitler and the Nazis as early as 1924, was banned from entering Italy by Mussolini, had his books burned at the 1933 Nuremberg rally, and was near the top of the list for execution if the Nazis has been successful in their proposed invasion of England" (231).

8. McConnell claims that Prendick is "arrogant about the naturalness of his own civilization" (105). Reed insinuates a similar complacency in his accusation that Prendick needs "an education in the nature of law, which he seems always to have taken on trust" (135). Haynes faults him for "self-righteous judgments" ("Unholy" 19) and goes so far as to damn his closing experiments in chemistry and astronomical observations as representing "a more subtle and hence, ultimately, a more insidious Moreau . . . none of his experiments is seen to have any relevance to society" ("Unholy" 23). Stover accuses Prendick of having a "preachy cast of mind" (*Island* 72, n.21), sneers at him as "the effete Prendick" (*Island* 67 n.16), and illogically

complains that he has not the courage to put into practice the very principles for which Moreau is, by Stover himself, condemned: "Moreau . . . represents vanguard over-man, whose lead Prendick by the end of the novel follows—but only up to the point of intellectually understanding the world-historical issues at stake. To act on it is beyond [Prendick]. That is for the propagandized reader of Wells' literature of power to do" (*Island* 159n144).

9. One strong reason *not* to take Moreau as a representative of Wells' own beliefs and opinions is that Moreau's self-isolated non-participation in the scientific community contradicts Well's belief in science as a shared pursuit, a fundamentally competitive, social enterprise. Reed points out that Wells was not naïve about the potential for science to be abused and misappropriated, that Wells remained "sensitive . . . to the uncontrolled nature of scientific discovery and the fallibility of scientific investigators" (*Natural* 97). Haynes emphasizes Wells' belief in science as organized *questioning* rather than rigid dogmatizing: "The same continued questioning of current hypotheses is one of the most striking characteristics of Wells' scientific romances in which current scientific assumptions are no more exempt from rigorous reconsideration than are political or religious ones, for [Wells] saw that those who accepted the 'authority' of science as absolute were in fact farthest from understanding the true scientific method" (*Discoverer* 37). Huntington has written that "the most complex alien, the isolated scientist, the exceptional human"—one who is "solitary against the crowd"—never simply met with Well's approval: "The figure will always attract *and bother* Wells" (70) [emphasis added].

10. Thus Huntington, in another foray into extra-textual moral superiority over the fictional Prendick: "Here the word 'humanity' has broken free from any question of biological descent: instead of a pun that poses a puzzle, the word here challenges the whole evolutionary criterion. *And yet Prendick's ethically motivated act is cold-hearted slaughter.* The ethical impulse is *contaminated by an almost spontaneous violence;* when Moreau complains that Prendick should not have killed the beast, Prendick evades both Moreau's point and the ethical one: "'I'm sorry,' said I, though I was not. 'It was the impulse of the moment'" (68) [emphasis added]. This comment strikes me as ungenerous. Prendick lies to Moreau to protect himself: the ironic point is that it was *not* for Prendick an *impulse of the moment.* He deceives Moreau, acting in solidarity with the beast folk against their "creator." It is strange for the reader to insinuate that Prendick there and then should have confessed his motive of euthanasia to Dr. Moreau, risking his own safety in what would have been a futile bid to influence a scientist for whom animal pain is a "little thing."

11. See Gans, "The Modernist Esthetic," in *Originary Thinking:* "The modernist solution to the discovery of the guilty violence at the origin of culture was to posit the guiltless violence of a precultural, prelinguistic human desire" (202). In the framework of generative anthropology, desire is mediated by language; there is no

such thing as a "prelinguistic desire." Properly speaking, animals have appetites and motives; but it creates considerable conceptual confusion to speak of animals having desires. The infinite yearnings that language makes possible are one with our self-awareness as participants in a drama governed by historical time. To the extent that animals do not share such historical self-awareness, animals neither require nor suffer *desire*. Appetite, motive, yes; desire, no. Desire, mediated by the scenic center, is uniquely human.

12. Here I agree with Stover's commentary: "Prendick is correct. While not a professional biologist, he has what Moreau lacks and even spurns: the common sense T. H. Huxley must have taught him. The simian larynx may be similar to man's, but *this anatomical detail is irrelevant to linguistic capacity* when the obviously greater differences in cranial capacity are compared. It is just this kind of gross error Huxley pilloried in *Man's Place in Nature*" (*Island* 136 n.101) [emphasis added].

13. To choose one such passage: commenting on Prendick's deception of the Beast People in his announcement of the mock-ascension of Moreau, Gomel writes: "He describes this move as a matter of self-preservation, but it is not difficult to see *in his contempt for Moreau's victims an acceptance of the vivisector's values.* His only regret is that he is, after all, too weak to emulate Moreau's self-transcendence . . ." (414). The evidence that counts against reducing Prendick's feelings toward the Beast People to *contempt* seems to me almost too massive to require elaboration. Gomel's fine study is marred by this self-righteous attack on the character within the text who—without question—explicitly, consistently challenges Moreau far more often than he "accepts" Moreau. Visceral feelings of horror and revulsion and disgust are not so easily assimilable to intellectual judgments of moral contemptibility. Prendick does not feel moral contempt for them; mostly, he pities them. Like him, they are victims of the deluded Moreau.

Notes to Four / Karel Capek's *R.U.R*

1. Harkins' remarks are typical of received critical opinion: "Capek has blended expressionism with realism. In fact, only the figure of the robot itself (and the final transformation of robots into men) is expresssionistic; otherwise, Capek's treatment is conventional and realistic" (*Karel* 86). Philmus reports the impact of the revisions Capek made for the 1921 opening performances: "many of the details he adds or removes in 1921 tend to give at least slightly greater prominence to *R.U.R.*'s Ibsenesque—as opposed to Expressionistic (and also its melodramatic) aspect" ("Matters" 22).

2. All further references to the edition we are using—translated by Novack (2004)—are parenthetical: "Pro." abbreviates "Comic Prologue"; 1, 2, 3 abbreviate Acts One, Two, Three respectively; page numbers follow. The reader

deserves to know there are considerable differences between English translations of R.U.R. currently available. My position is that the richest text is Novack's translation. I cannot imagine a contemporary theater company putting on the play that would not prefer Novack's text to Selver's, due simply to the greater richness of its content and the wider range of theatrical opportunities it presents. I also recommend Novack's earlier translation, only "slightly different" as the copyright page in the Penguin edition claims, in *Toward the Radical Center: A Karel Capek Reader* (ed. Peter Kussi), 34—109. Selver's earlier 1923 translation, done for the first London production of the play and based on the earlier 1920 text published *before* the first Czech performance of the play, is very different. Details are authoritatively explored in Philmus' "Matters."

3. As Alquist points out, it is only Domin who adds the corrupting utopian poison to the innocence of young Rossum's business plan: "[To abolish the necessity of human work] . . . was not the dream of the two Rossums. Old Rossum thought only of his godless hocus-pocus and young Rossum of his billions. And that wasn't the dream of your R.U.R. shareholders, either. They dreamed of the dividends" (2:54).

4. Naughton adds that the word is "derived from the Czech *robota* meaning 'curvee, forced labour, servitude' . . . related to the Russian *rabota* meaning simply 'work'" (73).

5. Perhaps it bears noting that Capek cannot be described as a Marxist: "The Czechoslovak Communist literary critics saw in Capek a representative of the ruling bourgeoisie, a defender of philosophical freedom, and a thinker who, in contrast to Marx, was firmly convinced of the unchangeability of social conditions. They accused him of near-earthiness, of escapism to the realm of metaphysics, and of futile humanitarianism" (Dresler 74).

6. Thus, Mr. Fabry the engineer and general technical director, to the uninitiated Helena: "Forgive me. It's great progress to give birth by machine. It's faster and more convenient. Any acceleration constitutes progress, Miss Glory. Nature had no grasp of the modern rate of work. From a technical point of view the whole of childhood is pure nonsense. Simply wasted time. An untenable waste of time" (Pro. 18). Thus Hallemeier, "head of the institute for Robot psychology and education" in a rapturous speech praising the mechanical value of precision: "When precision reigns, human law reigns, God's law reigns, the laws of the universe reign—everything reigns that should. The timetable is greater than the Gospels, greater than Homer, greater than all of Kant. The timetable is the most perfect manifestation of the human intellect. Mrs. Helena, I'll pour myself another" (1:44). That Hallemeier is getting drunk at the time may diminish a little the scandal of his blasphemy.

7. We might describe *economic* utopianism as that form of impossible dreaming that insists the moral reciprocity implicit in the exchange of *words* be replicated, be completely "incarnated," in the exchange of *things*. It is, of course, possible

to attempt such a replication; but not without implementing the considerable violence involved in sacrificing the relatively ethical to the absolutely moral—not without sacrificing the open, risky multiplication of human omnicentricity that we name freedom for the closed, "secure" compactness of human unity that we name communal solidarity. Certainly, the radical freedom implicit in modern market society is unsettling; but in the terms of generative anthropology, the freedom of choosing to use the sign and choosing to accept one's portion of the object as "equivalent" is also unsettling. Our very human-ness might well be qualified as such an unsettled openness to dangerous volunteering. Let us pause over the source of the unsettling effect of that openness: it is the *increase in desacralization of the object-as-sacred-whole* that, in the definition of originary science in chapter One, I described as operating in ratio with an *increase in exchangeability at the periphery.* Originary exchange is partly an exchange between the center and the periphery, for we experience the first piece of personal property we "own," our share of the central object, not as a gift from some person or other on the periphery but as a gift from the originary Being or God in/of/at the scenic center. Human exchange is *mediated by the scenic center:* that is why and how it differs from "material" exchanges between animals. The reciprocal exchange of signs is dependent upon the asymmetry of the human and the Divine. In keeping with the persistence of that dependency, the alluring, seemingly all-powerful concreteness of the "horizontal" exchange of things should not delude us into believing that the economic could ever be wholly detachable from the sacred and beautiful and erotic: "ritual and esthetic culture, far from being a mere 'superstructure' of the economic process, is the original locus of the valuation of goods and services" (Gans, "Culture vs. Exchange").

8. In the 1935 text, Capek takes pains to clarify this: "he [Old Rossum] created a new kind of matter by chemical synthesis; one which simply behaves a lot like the living; it [this new kind of matter] is an organic substance, different to that from which living cells are made; it is something like another alternative to life, a material substrate in which life could have evolved, if it had not, from the beginning, taken a different path" ("The Author of the Robots Defends Himself").

9. Another joke at the expense of Victor follows at this point in the script. Recall Victor's decision to make the creature eight feet tall. Domin tells Helena that young Rossum "assumed the role of God" (like his father) when he tried to manufacture "Superrobots. Working giants . . . twelve feet tall," but gave up when he found "those mammoths kept falling apart" (9). Twelve feet tall outdoes eight feet tall. More subtly, for Young Rossum the mere matter of engineered size, not the purportedly profound creative intention itself, is the basis of the blasphemy. His later humility is limited to his business decision to "make only Robots of normal human height and respectable human shape," now the standard practice in the factory.

10. For a very different view of Helena's role in the action, equally severe in its criticism of her, the men around her, and Capek (his failure to mount an overt feminist politics in the drama is thoroughly analyzed), see King.

11. Naughton reports the following: "He [Capek] stated that the idea of the robots had appeared to him *during a tram ride.* One day he had to go to Prague by suburban train, and it was uncomfortably full. It astonished him how modern conditions had made people disregard ordinary comfort in life. They were packed insiade and on the steps of the tramcar not like sheep, but like machines. He began to think about people not as individuals, but as machines, and on the way he searched for an expression which would denote *a person capable of working, but not of thinking.* This same horror of dehumanizing mass-anonymity and its potential for socio-political transformations and [for] human self-destruction is expressed in a letter to his future wife, Olga Scheifpflugov" (85) [emphasis added]. Bradbrook corroborates Naughton's account: "The happy end was also influenced by the author's promising start of his love relationship to Olga, to whom he wrote . . ." (*Karel* 45). The relevant comparable passage in Wells' *Moreau* reads: "Particularly nauseous were the blank expressionless faces of people in trains and omnibuses; they seemed no more my fellow-creatures than dead bodies would be, so that I did not dare to travel unless I was assured of being alone" (22:131).

12. Capek described this inspiration in a text where he argued that during the writing of the play he had been concerned primarily "not with Robots, but with people"; and that he intended it primarily as a meditation on the possibility of human extinction: "Imagine yourself standing over the grave of mankind; however jaundiced your view, you would surely realize the divine significance of the extinguished species and say—you too: It was a great thing to be a man" (from *Jeviste* [1921], quoted in Naughton 72).

13. I say "pseudo-freedom" for I agree with John Cohen's remarks in *Human Robots in Myth and Science* (1966): "Thirdly, suicide on the part of any future robot may have to be ruled out. A robot may be endowed with the capacity to bring about its own disorganization when conditions reach a given threshold of stress. But true suicide implies a foreknowledge of death and some idea of its significance, and this is a privilege of man" (139).

14. Thus I would have to distance myself from the position taken by Kinyon, which sentimentalizes the value of the "work" Radius himself does not really understand: "It is through labour, through the shaping of the object, that the slave gains superiority over the master. In Radius, the robot who challenges the master, one can see Hegel's master-slave dialectic at work. As Radius stresses, it is through active labor that he and other robots have become superior to humans" (382). I see no reason to accept Radius' perspective on his own superiority, unless one is firmly situated inside a discourse that *a priori* sides with the self-professed "challenger" and "slave'—that is, a victimary discourse of resentment. Such a perspective aims to humanize the essentially non-human robots. Kinyon outside

the text makes the same mistake that Helena makes inside the text.

15. In the Comic Prologue, Dr. Gall flirts with Helena by assuring her they would never send her away for anything: "Not for anything in the world, Miss Glory. Why would we send you away?" Helena's confessional reply is loaded with proleptic irony: "Because now you know—because—because I came to incite the Robots" (Pro. 17). Helena repeats this notion of having possessed "plans to instigate a r-revolt among your abominable Robots" (1:29) at a certain moment in her playful re-enactment with Domin of their courtship, a moment that significantly puts a halt to that lighthearted exercise—Domin leaps up in fear: "A revolt among the Robots!" (1: 29). Additional hints of such an affair appear in Nana's moralistic accusation: "He [God] knows very well why He didn't give you a child!" (1: 32). It is indicative of a particular intimacy between Gall and Helena that she rejects one of his attempts to comfort her with assurances of escape with uncharacteristic bluntness: "Oh, Gall, be quiet!" (2: 62). She does not shut down so abruptly any of the others who offer her equally false hopes (Fabry, Alquist); the resentment hints at an intimacy she does not share with those others.

16. On Capek's lifelong fear, suspicion, and contempt for the mob, see Matuska 23–24; 97; 100; 338; Naughton 85. For observations about "the streak of political neutralism in Capek" and the judgment that he was "an esthete first and foremost, and political man only second," see Harkins, "Real" 64.

17. It is no wonder that Wellek (among others) recognizes the radical centrism of Capek's politics: "Capek is a genuine democrat and has always advocated a humane, tolerant, and liberal government against extremists both on the right and on the left. Capek preaches—if one can call his lively papers preaching—civic duties, the right kind of nationalism, and ably defends his position against those who have accused him of a relativistic scepticism" (55).

18. I would argue that the comic power of the play has not received its due attention. I believe my approach is partly justified by the following words from Capek's article, "On Proletarian Art": "I . . . know that it [art] has other, loftier, more mysterious functions [than that of giving pleasure]. But this [mysterious function] is not its only goal. It must have been the mission of the primitive claw-type axe to contribute to the instrumental progress of man and to some day render testimony to man's beginning. But its more immediate and actual purpose was to kill a wolf or a bear. Likewise the actual purpose of art is to kill boredom, melancholy, and the greyness of life. If it can do more than this, so much the better, but if it fails to do this it is a poor ineffective clawhatchet because it will not protect us against the monsters that are devouring us" (quoted in Dresler 74).

19. I am tempted to see an inadvertent but intuitively correct mockery of the originary event in such a perfect-hindsight memory condemned to be incapable of generating real history: for the robots, the human community as center-of-resentment-and-desire has been wiped out in somewhat the same way that in the originary event, the sacrificial destruction and consumption of the central

food object is the necessary prerequisite of originary memory of the sacred center that mediates exchange. In Capek's prescient but perhaps unintended parody, humans become sacred to robots only after robots have annihilated them.

20. Following Gans, I understand the "erotic" as a cultural phenomenon, almost an esthetic effect, not a purely libidinal or "instinctual" force: "The model of the mimetic center confers on the other *an eroticism only possible in the human universe of the sign*" (Gans, *Signs* 113) [emphasis added]. The erotic "undisturbed" by the esthetic, as in the following formulation, would be a de-humanized erotic—not (properly speaking) an "erotic" at all: "In high art, the structure of originary experience is reproduced in an esthetic oscillation that is foregrounded at the expense of erotic fixation. The popular arts are less embarrassed to provide the material for imaginary wish fulfillment. But when the esthetic mechanism remains wholly occulted and the erotic reigns undisturbed, we fall altogether out of the domain of 'legitimate' culture into that of pornography" (*Signs* 113).

21. When I use the phrase "private sacred" or "erotic sacred," I am referring to that phenomenon designated by "private erotic scene" or "personal sacred" in the following passages, to be found in Gans' originary analysis of the minimal erotic (*Signs 112–120*): "Just as the esthetic becomes independent of the public scene of representation by internalizing the scene's mimetic structure in the subject's oscillation between sign and referent, the object of erotic desire is the incarnate sign of his/her own being, generating in the partner a *personal sacred* that lasts at least the time of a sexual encounter, and perhaps a lifetime" (114). "Because the erotic creates a microcosm of the human universe that requires no external transcendent figure, it is the privileged content of secular culture, which must arouse and purge desire without the benefit of the ritual reconstitution of collective presence. This privileged status only clearly emerges in the neoclassical era . . . Love creates a personal scene of representation homologous to the public one, with the beloved as its sacred center . . . Throughout the neoclassical era, the private erotic scene draws away the energies that classical forms concentrated on the public scene, until the romantics finally enshrine it as the authentic scene of origin" (114–115).

22. Gans: "The attachment of sacred meaningfulness to objects of economic value maintains peace in the community while permitting the consumption of these objects. Meaningfulness adds nothing directly to our material satisfactions; [meaningfulness] can only have arisen as a means to prevent conflict—the originary hypothesis" ("Culture vs. Exchange").

23. Gans: "What makes us want to imitate, and to continue to imitate, the aborted gesture-sign of the others is our judgment that the sign is indeed a sign, that it evokes its object without attempting to possess it. This is an esthetic judgment; the participant who makes it chooses what gives esthetic pleasure. The source of this pleasure is the temporary relief (or deferral) it provides from

originary resentment" (*Originary Thinking* 118). Consider also: "Freedom is the link between the vertical world and the horizontal one, the world of signs and the world of things, and this link cannot be expressed within the world of things alone"; " . . . a model of freedom is precisely what arises when a deterministic model is represented to its subject, who is then free to subvert it" ("Free Will"). One more remark: "The enunciation of our inalienable moral responsibility can . . . provide a convincing model of human freedom if it shows plausibly how [human freedom] emerged from its roots in prehuman necessity" ("Problem of the Subject"). Human freedom emerged in our first ancestors' choosing to exchange the ostensive sign with one another, choosing to re-present the object rather than simply consume it, to defer mimetic violence through representation. The originary human choice must have been the choice to be human as opposed to not being. If we are not prepared to concede the link between human linguistic consciousness and freedom, we might as well go to bed with the robots.

24. It must be conceded that in our time this claim (alas) has become debatable. For an earnest and respectable argument that humans and robots will in the near future come to have satisfying sex and fall in ontologically dizzying love, see Levy. Before moving humanoid robots, there were non-automated dolls: for a surprisingly even-keeled documentary film detailing the lucrative, flourishing sex doll industry in the West and Asia, see de Fren.

25. The encouragement and liberation of what I am calling the private sacred follows the "revelation" in consumer society of the "revolutionary anthropological truth" that Gans calls "the priority of desire over cognition." Gans, toward the end of his discussion of the psychoanalytic category of the unconscious: "The historical significance of consumer society has been difficult to evaluate in the face of the century of diatribes with which it has been greeted by the intelligentsia from Thorstein Veblen on down, incorrigible believers in the resentment-free utopia that will satisfy all their resentments. We are perhaps still insufficiently distanced from the Marxian-socialist illusion to carry out this evaluation. Nascent consumer society is clearly the efficient cause of modernism and its associated forms of political extremism, as well as of the decline of traditional metaphysics. The establishment of the priority of desire over cognition is not a temporary aberration engineered by the self-serving bourgeoisie, but the revelation of a revolutionary anthropological truth, the liberation of a genie that not even the most ruthless dictatorships have been able to put back in its bottle" (*Signs* 126). The worldwide rule of the robots is Capek's visionary foreseeing of what a "ruthless dictatorship" might look like. One tell-tale way of measuring such ruthlessness is to evaluate the extent to which a dictatorship abolishes the realm of the private sacred and the extent to which it prohibits the genuine erotic that nourishes the private sacred.

Notes to Five / Scott's *Blade Runner*

1. Almost every contributor to the collection of papers edited by Hilary Rose and Steven Rose, *Alas! Poor Darwin: Arguments against Evolutionary Psychology* (2000), attributes reductionism to evolutionary psychology, the offspring of sociobiology. See Benton 216–17, 223; Gould 92; Midgley, "Why?" 72; Nelkin 15, 19; Rose and Rose 7; Hilary Rose 116; Steven Rose 247–48. For equally relevant critiques from within Darwinian theory itself of what the author calls *ultra-Darwinism* (the genetic and sociobiological reductionism I have in mind), see Eldredge, *Reinventing* (1995), esp. 221–27.

2. For definition and interpretation of the postmodern era, I have relied much on Gans. I have also drawn on Bertens; Carroll; Chistopher Butler; Connor, "After"; Connor, *Postmodernist;* Lyon.

3. Scott, in an interview with Sammon printed in *Omni* magazine May 1982, quoted by Kolb in his "From Script": "I didn't want *Blade Runner* to be premonitory of *android* at all, because then people would think the film is about robots, when in fact it isn't. It was better that we come up with a new word altogether. *Replicant* was the choice. In a very real sense this was a stylistic decision" (147).

4. Senior: "the Frankensteinian theme of man crafting himself and experimenting with new forms provides both the conflict and the philosophical dialectic that run through the film and through cyberpunk fiction" (1). Bukatman in his helpful *BFI Film Classics* monograph makes a similar point: "Synthetic human narratives, from *Pygmalion* to *Pinocchio* to *Terminator 2,* have always challenged, or at least made explicit, definitions of 'natural' humanity and its role or function. Defining the human provides most of *Blade Runner*'s philosophical focus" (64). For one list of parallels between *Frankenstein* and *Blade Runner,* see Doll and Faller 96–97. For a detailed demonstration of *Blade Runner's* debt to the James Whale and Hammer Studio Frankenstein movies, see Heldreth (1991) 45–47. See also Rushing and Frentz (1989), 62–63.

5. Slade reports his impression that "Philip K. Dick aficionados . . . have generally responded to the cinematic adaptation with dismay" (13). For comparisons of the book and the 1982 movie, see Kerman, "Private"; Fitting; Fischer. It cannot be emphasised enough, however, that each of these studies would have to be substantively rewritten if they were comparing the 1992 Director's Cut to the book. The deletion of the voice-over and the drive-away happy ending significantly alter the content of the work. Kolb's "From Script" is packed with information relevant to the evolution from Dick's novel to Scott's film. Finally, Picart's quasi-feminist analysis is extraordinarily attentive to the evolution of the screenplay from draft to draft as the changes affect in particular

the construction of the movie's female figures Zhora, Pris, Rachael and Mary—the latter being an obscure character whose entire part was cut from all versions of the film released prior to the 2007 Final Cut Deluxe DVD box set.

6. Morrison's 1990 study of the film quotes Dick referring to the same revelation in a different context: "'There is among us something that is a bi-pedal humanoid, morphologically identical to the human being but which is not identical. It is not human to complain, as one SS man did in his diary, that starving children are keeping you awake'" (3). Five years before the 1981 interview with Sammon, Dick had made the same point in a public lecture: "A human being without the proper empathy or feeling is the same as an android built so as to lack it, either by design or mistake. We mean, basically, someone who does not care about the fate which his fellow living creatures fall victim to; he stands detached, a spectator, acting out by his indifference John Donne's theorem that 'No man is an island,' but giving that theorem a twist: that which is a mental and moral island *is not a man*" ("Man, Android and Machine" 202–203). However, a definition of the human cannot be constructed on the privileging of a particular emotion, even one as important as empathy is to biological survival and social welfare. The challenge is to link empathy to its roots in symbolic exchange. The originary hypothesis does that by situating our intuition of moral equality in the exchange of signs (prior to the exchange of things): we all have equal access to the meanings of language, and that is the originary source of our conviction that, concrete differences of wealth and power aside, we are beings worthy of equal moral consideration (empathy).

7. In a 1982 interview, Peoples reported the following positive impressions of Fancher's script when he first looked at it: "It was just wonderful! When I saw it, I was terrified! I thought that I was going to embarrass myself because it was just so good! The changes I made were really to make it more in line with Ridley's vision of that world" (L'Officier 32).

8. Scott (1982): "I found that one of my problems with the screenplay earlier on was that the man did no detection! He seemed to be told everything, and I wanted to see some detection work. So Peoples came in and started to take over Hampton's place in terms of helping me with those roots" (L'Officier 39).

9. From a September 1982 interview: "He, of course, was the reason it got made. We were in trouble because everyone said 'you've got to have a director, because this is too weird a film. You need a director to insure its value'" (L'Officier 28).

10. Scott: "I think *Blade Runner* is a good lesson for all serious film makers to 'stand by youg guns.' Don't listen to acclaim or criticism. Simply carry on" (Sammon 393).

11. Scott, interviewed in 1982: "It's a curious film. It runs on two levels. It's almost, in parts, philosophical. I hesitate to use the word with a commercial movie, but nevertheless, it is" (L'Officier 36).

12. Scott, as interviewed by Sammon: " . . . the American system is one containing a certain degree of optimism. I, on the other hand, tend to be a bit darker. To look to the dark side. Not because I'm a manic-depressive, but because I find darkness interesting" (390).

13. Scott himself was not immune to the temptation. In a 1982 interview, he declared: "One thinks one is dealing with a kind of Frankenstein's monster in character, but he [Roy Batty] really becomes more human than the average human as the film progresses, and therefore, in some respect, gains one's sympathy, possibly as much, or even *more* than the Deckard character!" (L'Officier 41). If we can sympathize with the creature and the creature is capable of nonviolence, so Scott's remark implies, then the creature is human. In "The Android as Doppelganger," Francavilla writes: "There is not only a confusion of identities and roles, but also a *reversal* of identities or roles. The characteristics of human life become attributed more and more to artificial life, and the replicants become, as the Tyrell slogan boasts, more human than human" (12). Slade: "The various killings provide more than kinetic pacing: they give substance to the replicants. The deaths of the replicants are particularly moving because they cling so to life: 'more human than human,' the Tyrell corporation advertises its products" (16). For Slade, the essence of the human seems to consist simply in the strength of a desire to cling to life. A remark from Gravett also belongs here: "Similarly, the replicants have lived up to the motto of the Tyrell Corporation; they are indeed 'more human than human,' often exhibiting greater compassion and empathy that the 'real' humans. If they could extend their lifespan, the replicants might eventually usurp humanity" (39). Forget about clinging to life; now Tyrell's slogan is about compassion and empathy and species supersession (during which the triumphant replicants presumably would show no compassion for the overridden human species). Vest contributes to the chorus: "The viewer, after witnessing Deckard's seduction of Rachael, can no longer be certain that the replicants are distinguishable from human beings. Their desire for life, love, and human experience makes the replicants appear more alive than their human oppressors. They become more human than human" (21).

14. Like Lev, Morrison argues that "this being [the replicant] not only mimics the truly human, but begins to exceed its creators in human passion and empathy" (3); "Roy is a nonhuman who is truly human in his striving, vitality, and development of emotions" (4). Morrison's criteria are emotional in content: passion, empathy, vitality, striving. But an ontology of the human founded simply on the fact of having feeling cannot withstand much scrutiny. A similar argument is presented by Heldreth: "The development of these emotions together with the possession of personal memories, however, eliminates a large number of the distinctions between human and replicant, for it is exactly the emotional responses and the individual memories that provide

personality" (49). If you have emotions and memories, you have personality; if you have personality, you are human. My West Highland terriers possess all three. I love them, but they are not humans.

15. In the context of my disagreements with Lev, it deserves to be noticed that Kolb (1991) presents evidence to show that People's revisions to the Hampton Fancher scripts are ultimately responsible for foregrounding "the question of what distinguishes us from beings that are the same in every respect *except their genetic origin*" [emphasis added]. Kolb proposes it is the Peoples' revisions that ultimately serve "to make them [the replicants] finally indistinguishable from humans" (133).

16. Ford in an interview in *Starburst* no. 53, as quoted in Kolb (1991): "Ford claims that he [Deckard the character] 'was anxious to make sure that this character represented an abhorrence of violence. And he does! He wanted to get out of the police force because he couldn't stand the killing. After every incident of having to kill someone, the character's revulsion is clearer. And, ironically, what he's killing are not human beings. That's what the thematic backbone of the film is. They're not really human beings! And yet, his empathy with something that *looks* like a human being—which is later to lead him into a romance with, basically, a machine—affects him'" (152).

17. There is little evidence to support the idea that their mutiny can be accurately described as a "revolution." Roy is the leader of a band of courageous souls, but nothing hints he is the flag-waving general of a mass movement or mobilized threat to the social order of Los Angeles 2019. Given the intensity of replicant suffering, the leftist critics' desires to nestle the replicant mission in the context of an imagined larger revolt story is understandable, but it has no explicit factual support in the script or on screen. *Blade Runner* is a horror film; it reads, in the words of Ridley Scott, like "a very dark novel." When Roy dies, the small revolt ends, and the little hope we have belongs only to the remnant, Rachael and Deckard.

18. For more nuanced interpretations that recognize the should-be-obvious fact that the differences between Christ and Roy outweigh the similarities, see Begley (187) and Bukatman.

19. In his 2007 book on Scott's whole career, Schwartz writes: " . . . he [Roy Batty] also dies a mass murderer, and his newly developed humanity does not excuse him from answering for his crimes. Ultimately, it is for this reason, and not because he is a replicant, that Batty's death remains a fitting ending to the plot. However justified from the replicants' point of view, the story cannot conclude with a murderer on the loose. So, even though we may sumpathize with Batty and begin to understand the perpetual terror experienced by those deemed subhuman, *in Scott's moral universe victimization does not legitimize murder.* And so, both the story's moral and narrative requirements hold Batty accountable" (47) [emphasis added].

20. In Sammon's *Future Noir,* Hauer reports that he convinced Scott to give

up an original plan to make the Batty-Deckard confrontation a martial-arts kind of conflict with Bruce Lee overtones: "So . . . I suggested , 'Why don't we make it more of a chase, based on the Game of Life? Try and make it a silly dance based on Batty's celebtation of his last drops of energy? Like a wicked game?' Ridley really liked that, and we sort of got that idea down on story boards before it was written up" (186).

Notes to Six / The Human Scene

1. See Gans' chapter "A Generative Taxonomy of Speech Acts" in *Originary Thinking* for a comparison of his theory of speech acts with that of Searle; see van Oort, "Three Models" for a consideration of the concept of fiction in Gans and Searle.

2. Eric Cohen: "The modern scientific method equips us with wonderful new powers and new knowledge, both to improve human life and to destroy it. But science itself does not tell us how to live or what to value in a world made new by scientific knowledge. The principles of biology cannot tell us whether to develop bioweapons or destroy human embryos for research. The principles of physics cannot tell us whether to build nuclear power plants or atomic weapons. Science needs to be governed by philosophical ethics and democratic politics" (786). And more: "For there is nothing in nature, at least not discoverable by modern scientific methods, that refutes the morality of despotism. Liberty and scientific liberty, it turns out, are not exactly the same. What is good for science, devoted first and foremost to the pursuit of knowledge, is not necessarily good for human beings, precisely because scientific knowledge of nature is not the highest good in human life" (787). Frankel: " . . . although it [irrationalism] points only clumsily at evil, that evil is there. The careful rational methods by which knowledge and technique have advanced have only rarely been used to examine the purposes to which this knowledge and intelligence are harnessed. It is natural that science, in such a setting, should seem to be a Frankenstein to those who are threatened by it" (93).

3. For critiques of evolutionary psychology, see especially Rose and Rose; Midgley; Tallis.

4. Dupre in *Darwin's Legacy:* "The central problem is that the role of genes in evolution has been grossly misrepresented. Genes are still widely described as carrying blueprints for the organism, recipes for putting together organisms, and such like. Consumers of science fiction are often treated to the idea that with sufficient skill it would be possible to read off the features of an organism simply from a knowledge of the sequence of base plans in a genome. At a more sophisticated level, many biologists still endorse the 'central dogma' that information about biological structure flows exclusively and unidirectionally from the genome. All of this can now be seen to be profoundly mistaken" (83).

See also Keller, *Century* 140.

5. For exemplary and well-worth reading variations on the Frankenstein myth deploying miserable clones, see Bay; Darnton; Ishiguro. For equally engaging variations with suicidal cyborgs, see Dantec; Piercy. For an account of the quest for judge-duping computer conversationalists, see Christian.

6. Next question: if you are not satisfied, then do you feel ready to cease wondering whether the possibility of the reduction of scenic anthropology to biological materialism might somehow be the most valuable truth about humans? Probably not. That we still feel we are probably not ready to cease reducing ourselves shows how dominant (I do not say oppressive) in our time is scientific naturalism, and how much hard work remains to be done to build the intellectual basis in our academic worlds for a self-respecting study of human reality based on a minimal but universal anthropology—an anthropology that founds itself on language, that takes as seriously as it should be taken the uniqueness and the mystery of human language. And I do not mean to despise language in the super-skeptical mode of resentful postmodern skepticism. I mean to love language in the affirmative mode of celebrating our human difference in all its horror and glory.

7. The singularity is that moment when, according to roboticist Hans Moravec, human machines will become immortal and human consciousness will achieve a new ontological level by being replaced by that of a super-network of computers. See Grossman for a recent account suggestive of its popularity and respectability.

8. Bavidge and Ground: "If we think of personal communication in terms of the contemporary concept of information transfer we need never refer to people or the subjective view at all" (163). Rescher, *Limits:* "While information is a matter of data, knowledge, by contrast, is something more select, more deeply issue-resolving—to wit, *significant and well-consolidated* information" (56); "The larger the body of information we have, the smaller will be the proportion of this information that represents real knowledge" (62).

9. For expressions of this belief from outside the framework of generative anthropology, consider the following. From Krutch: "When we think without reference to any preferences or 'values' we think like a machine. That means also thinking without reference to joy, or laughter, or love. Very often nowadays we are urged by certain sociologists, political propagandists, and even anthropologists to do just that although they prefer to call it 'thinking with detachment.' But the thing from which we are asked to detach ourselves is, nevertheless, the state of being human, and the result of such thinking would be a world fit for machines, not for men" (171). From Stanislaw Lem: "The most important thing about this is that computers never do anything consciously. We cannot attribute to them psychological reasons that would be understandable by way of intuitive feeling: they do not strive for power, they do not know egotism, for they have no ego and no personality. Whether as single units or as nets, they are, considered

anthropologically-psychologically, nobody at all" (324–25). From Fukuyama: "A moment's reflection will show that none of the key qualities that contribute to human dignity can exist in the absence of the others. Human reason, for example, is not that of a computer; it is pervaded by emotions, and its functioning is in fact facilitated by the latter. . . . We are social and political animals not merely because we are capable of game-theoretic reason, but because we are endowed with certain social emotions. Human sentience is not that of a pig or a horse, because it is coupled with human memory and reason" (172).

10. Gans, *Scenic:* "Every occurrence on the human scene, in distinction from the comings and goings of the animal world, is a unique event, a singularity that has its place in the series of singularities we call *history*" (2). Jaki: "He [Charles Darwin] did not suspect the extent to which the same familiarity [with scientific method] could also give rise to an unwarranted discrimination among various kinds of facts and to a shocking insensitivity to the countless facts of history which, unlike 'the facts' of science, *do not repeat themselves*" (7) [emphasis added]. Bernard Williams, in *Ethics and the Limits:* " . . . [social understanding] need not seek to join the natural sciences in providing an absolute conception of the world, but we need to have some reflective social knowledge, *including history,* that can command unprejudiced assent, if the better hopes for our self-understanding are to be realized" (199) [emphasis added].

11. Eric Gans: "Presence (on the scene of representation) as the deferral of (worldly) presence is the reality of language; on this point, as we might expect, the naïve experience of the ostensive is rendered self-aware by the declarative. Language is always already non-presence; we have no quarrel on this point with Derrida. But unlike the presence of metaphysics, which subsists within the context of the declarative, ostensive presence is not conceivable as the presence of the sign alone" (58). In other words, when we use language, we are not alone with language but present with and to each other as subjects on the scene of representation. Human linguistic exchange *cannot* be reduced to anything other than itself: it is *impossible.*

Works Cited

Primary Sources (Instantiations of the Frankenstein Myth)

Atwood, Margaret. 2003. *Oryx and Crake.* New York: Nan A. Talese / Doubleday, 2003.

Bay, Michael, dir. *The Island.* Perf. Ewen MacGregor, Scarlett Johansen, Sean Bean. Scr. Caspian Tredwell-Owen, Alex Kurtzman, Robert Orci. USA, 2005. DVD.

Branagh, Kenneth, dir. *Mary Shelley's Frankenstein.* Perf. Kenneth Branagh, Robert DeNiro, Helena Bonham Carter. UK, 1994. DVD.

Capek, Karel. *R.U.R. (Rossum's Universal Robots).* 1921. Trans. Claudia Novack. London: Penguin, 2004.

———. *R.U.R.* and *The Insect Play.* Trans. Paul Selver. Adapted by Nigel Playfair. Rpt. Oxford: U P, 1961.

———. *Toward the Radical Center: A Karel Capek Reader.* Ed. Peter Kussi. Foreword by Arthur Miller. New Haven: Catbird Press, 1990.

Connor, Kevin, dir. *Frankenstein.* USA, 2004. DVD.

Cook, Robin. *Mutation.* New York: G. P. Putnam's Sons, 1989.

Crichton, Michael. *Next.* New York: HarperCollins, 2006.

Dantec, Maurice G. *Babylon Babies.* Trans. Noura Wedell. New York: Semiotext(e), 2005.

Darnton, John. *The Experiment.* New York: Dutton, 1999.

Dear, Nick. *Frankenstein (Based on the Novel by Mary Shelley).* London: Faber and Faber, 2011.

Dick, Philip. "Impostor." Warrick and Greenberg 77–89.

———. "To Serve the Master." Warrick and Greenberg 167–77.

Fisher, Terence, dir. *The Curse of Frankenstein.* Perf. Peter Cushing, Christopher Lee. Scr. Jimmy Sangster. UK, 1957. DVD.

———. *The Evil of Frankenstein.* Perf. Peter Cushing, Francis Matthews. Scr. Jimmy Sangster. UK, 1964. DVD.

———. *Frankenstein Must Be Destroyed.* UK, 1969. Perf. Peter Cushing, Veronica Carlson, Freddie Jones. UK, 1969. DVD.

———. *The Revenge of Frankenstein.* Perf. Peter Cushing. UK, 1958. DVD.

Follett, Ken. *The Third Twin.* New York: Crown, 1996.

Hyman, Tom. *Jupiter's Daughter.* New York: Viking / Penguin, 1994.

Ishiguro, Kazuo. *Never Let Me Go.* Toronto: Random House, 2006.

James, Peter. *Perfect People.* London: Macmillan/ Pan, 2011.

Kenton, Erle C., dir. *Island of Lost Souls.* Perf. Charles Laughton, Richard Arlen, Bela Lugosi. USA, 1932. DVD.

Koontz, Dean and Kevin J. Anderson. *Dean Koontz's Frankenstein, Book One: Prodigal Son.* New York: Bantam/ Dell, 2005.

Koontz, Dean and Ed Gorman. *Dean Koontz's Frankenstein, Book Two: City of Night.* New York: Bantam/ Dell, 2005.

Mathews, Adrian. *Vienna Blood.* New York: HarperCollins, 1999.

Natali, Vincenzo, dir. *Splice.* Perf. Adrien Brody, Sarah Polley, Delphine Chaneac. Scr. Vincenzo Natali, Antoinette Terry Bryant, Doug Taylor. USA, 2009. DVD.

Nova, Craig. *Wetware.* New York: Random House, 2003.

Scott, Ridley, dir. *Blade Runner.* Perf. Joe Turkel, Rutger Hauer, Harrison Ford, Sean Young. Scr. Hampton Fancher and David Peoples. USA, 1982; Final Cut, 2007. DVD.

Shelley, Mary. *Frankenstein: or, the Modern Prometheus. The Original 1818 Text.* Ed. D. L. MacDonald and Kathleen Scherf. 2nd. ed. Peterborough, Ontario: Broadview Press, 1999.

Warrick, Patricia S., and Martin H. Greenberg, eds. *Robots, Androids, and Mechanical Oddities: The Science Fiction of Philip K. Dick.* Carbondale: Southern Illinois U P, 1984.

Wells, H. G. 1896. *The Island of Dr. Moreau.* Ed. Patrick Parrinder. New York: Penguin, 2005.

Wilson, F. Paul. *Sims.* New York: Tom Doherty Associates, 2003.

Secondary Sources

Abbot, Joe. "The 'Monster' Reconsidered: *Blade Runner*'s Replicant as Romantic Hero." *Extrapolation* 34 (1993): 340–50.

Agar, Nicholas. *Perfect Copy: Unravelling the Cloning Debate.* Cambridge: Ikon Books, 2002.

Aichison, Jean. "Recent Developments in Language Origin." *European Review* 12/2 (2004): 228–34. *Cambridge Journals Online.* Web.

Alexander, Brian. *Rapture: A Raucous Tour of Cloning, Transhumanism, and the New Era of Immortality.* New York: Basic Books, 2003.

Alison, James. *On Being Liked.* London: Longman, Darnton and Todd, 2004.

———. *Raising Abel: The Recovery of the Eschatological Imagination.* New York: Crossroad, 1996.

Anderson, Stephen R. *Doctor Doolittle's Delusion: Animals and the Uniqueness of Human Language.* New Haven and London: Yale U P, 2004.

Appleman, Philip, ed. *Darwin: Texts, Commentary.* 3rd ed. New York: Norton, 2001. Norton Critical Editions.

Appleyard, Bryan. *Brave New World: Staying Human in the Genetic Future.* New York: Viking, 1998.

Asimov, Isaac. *The Bicentennial Man and Other Stories.* Garden City, NY: Doubleday, 1976.

———. *The Robot Novels:The Cave of Steel [1953]; The Naked Sun [1957].* Garden City, NY: Doubleday, 1957.

———. *The Robots of Dawn.* Garden City, NY: Doubleday, 1983. Print

Augenstein, Leroy. *Come, Let Us Play God.* New York: Harper and Row, 1969.

Bailie, Gil. *Violence Unveiled: Humanity at the Crossroads.* New York: Crossroad, 1995.

Baldick, Chris. *In Frankenstein's Shadow: Myth, Monstrosity, and Nineteenth-Century Writing.* Oxford: Clarendon, 1987.

Barad, Judith. "*Blade Runner* and Sartre: The Boundaries of Humanity." *The Philosophy of Neo-Noir.* Ed. Mark T. Conrad. Lexington: U P of Kentucky, 2007. 21–34.

Barr, Marleen. "Metahuman 'Kipple' or, Do Male Movie Makers Dream of Electric Women? Speciesism and Sexism in *Blade Runner.*" Kerman, *Retrofitting* 25–31.

Barrett, William. *Death of the Soul: From Descartes to the Computer.* New York: Anchor Press / Doubleday, 1986.

Bartlett, Andrew. "Accusations of Playing God and the Anthropological Idea of God." Katz 299–350.

———. "From First Hesitation to Scenic Imagination: Originary Thinking with Eric Gans (Review Essay)." *Contagion: Journal of Violence, Mimesis and Culture* 15–16 (2008–2009): 89–172. *Project Muse.* 30 Aug. 2013. Web.

———. "Nuclear Warfare in the Movies." *Anthropoetics* 10/1 (Spring/ Summer 2004). Web.

———. "The Object of Originary Violence and the Second Person of the Trinity." *Anthropoetics* 14/2 (Winter 2009). Web.

———. "Originary Human Personhood." *Anthropoetics* 16/2 (Spring 2011). Web.

———. "Three Affirmations of the Being of God, Suggested by the Anthropological Idea of God." *Anthropoetics* 13/2 (Special GATE Issue, 2007). Web.

Bartlett, Anthony W. *Cross Purposes: The Violent Grammar of Christian Atonement.* Harrisburg, PA.: Trinity Press International / Continuum, 2001.

Bavidge, Michael, and Ian Ground. *Can We Understand Animal Minds?* New York: St. Martin's, 1994. Mind Matters.

Bayertz, Kurt. "Playing God." *Genethics: Technological Intervention in Human Reproduction as a Philosophical Problem.* Trans. Sarah L. Kirkby. Cambridge: U P, 1994. 174–97.

Beauchamp, Gorman. "*The Island of Dr. Moreau* as Theological Grotesque." *Papers on Language and Literature* 15 (1979): 408–417. *Academic Search Premier.* 30 Aug. 2013. Web.

Begley, Varum. "*Blade Runner* and the Postmodern: A Reconsideration." *Literature/Film Quarterly* 32/3 (2004): 186–92. *Academic Search Premier.* 30 Aug. 2013. Web.

Belsey, Catherine. "Biology and Imagination: The Role of Culture." Wells and McFadden 111–27.

Benton, Ted. "Social Causes and Natural Relations." Rose and Rose 206–24.

Bergonzi, Bernard. *The Early H. G. Wells: A Study of the Scientific Romances.* Manchester: U P, 1961.

Berlinski, David. "On the Origins of Life." *Commentary* 121 (February 2006): 222–33.

———. "On the Origins of the Mind." *Commentary* 118 (November 2004): 26–35.

Bertens, Hans. "Out of Left Field: The Politics of the Postmodern." Hoffman and Hornung 97–113.

Bertonneau, Thomas. "The Dispensations of Moira: Matter, Mind and Culture from Thales of Miletus to Walter Pater." Katz 213–48.

———. "The Gist of Eric L. Gans: From *The Origin of Language* to *The Scenic Imagination*." *Brussels Journal* 11 Nov. 2009. Web.

Bickerton, Derek. *Adam's Tongue: How Humans Made Language, How Language Made Humans*. New York: Hill and Wang, 2009.

Booker, M. Keith. *Alternate Americas: Science Fiction Film and American Culture*. Westport, Conn.: Praeger, 2006.

Boozer, Jack. "Crashing the Gates of Insight: *Blade Runner*." Kerman, *Retrofitting* 212–28.

Bowen, Roger. "Science, Myth and Fiction in H. G. Wells' *The Island of Dr. Moreau*. *Studies in the Novel* 8 (1976): 318–35.

Bowerbank, Sylvia. "The Social Order vs. the Wretch: Mary Shelley's Contradictory-Mindedness in Frankenstein." *English Literary History* 46 (1979): 418–31.

Bozzetto, Roger. "Moreau's Tragi-Farcical Island." Trans. Robert M Philmus and Russell Taylor. *Science-Fiction Studies* 20 (1993): 84–94. Web.

Bradbrook, Bohuslava R. *Karel Capek: In Pursuit of Truth, Tolerance and Trust*. Sussex: Academic Press, 1998.

Brome, Vincent. "H. G. Wells as a Controversialist." *University of Windsor Review* 2/2 (1967): 31–45.

Brooke, John Hedley. *Science and Religion: Some Historical Perspectives*. Cambridge: U P, 1991. Cambridge History of Science. Series ed. George Basalla.

Buchen, Irving. "Frankenstein and the Alchemy of Creation and Evolution." *Wordsworth Circle* 8 (1977): 103–112.

Bukatman, Scott. *Blade Runner*. London: British Film Institute, 1997. BFI Film Classics.

Burley, Justine and John Harris. "Human Cloning and Child Welfare." McGee 3rd ed. 234–50.

Butler, Christopher. "Postmodernism and Moral Philosophy." Hoffman and Hornung 69–86.

Butler, Marilyn. "Introduction." *Frankenstein: the 1818 Text* by Mary Shelley.

Oxford: U P, 1998. ix-li.

Campbell, Joseph. *The Masks of God: Primitive Mythology.* New York: Viking Penguin, 1959.

Capek, Karel. "The Author of the Robots Defends Himself." *Science Fiction Studies* 23/1 (Mar. 1996) [no. 23]. Web.

———. *The Absolute at Large.* Lincoln: U Nebraska P, 2005.

———. *War with the Newts.* Boston: Gregg Press, 1975.

Caplan, Arthur. "If Ethics Won't Work Here, Where?" In McGee 3rd. ed. 58–67.

Carnap, Rudolf. "The Nature of Theories." Klemke, Hollinger and Kline 162–77.

Carroll, Noel. "Moral Realism in the Age of Postmodernism." Hoffman and Hornung 87–96.

Caudill, Edward. *Darwinian Myths: The Legends and Misuses of a Theory.* Knoxville: U Tennessee P, 1997.

Cheyette, Bryan. "Beyond Rationality: H. G. Wells and the Jewish Question." *The Wellsian: Journal of the H. G. Wells Society* 14 (1991): 41–64.

Christian, Brian. "Mind vs. Machine." *Atlantic* Mar. 2011: 58–69.

Clayton, Jay. "*Frankenstein*'s Futurity: Replicants and Robots." Schor 84–99.

Clemit, Pamela. "Frankenstein, Matilda, and the Legacies of Mary Wollstonecraft." Schor 26–44.

Cohen, Eric. "The Permanent Limits of Modern Science: From Birth to Death." *Social Research* 73 (2006): 785–804. *Academic Search Premier.* 30 Aug. 2013. Web.

Cohen, John. *Human Robots in Myth and Science.* London: George Allen and Unwin, 1966.

Cole-Turner, Ronald. *The New Genesis: Theology and the Genetic Revolution.* Louisville, Kentucy: Westminster/ John Knox Press, 1993.

Columbus, Chris, dir. *The BiCentennial Man.* Perf. Robin Williams, Embeth Davidtz, Sam Neill. *USA, 1999.* DVD.

Connor, Stephen. "After Cultural Value: Ecology, Ethics, Aesthetics." Hoffman and *Hornung* 1–12.

———. *Postmodernist Culture: An Introduction to Theories of the Contemporary.* Oxford: Blackwell, 1997.

Cooper, David C. *Philosophy and the Nature of Language.* London: Longman, 1973.

Coupland, Philip. "H. G. Wells' 'Liberal Fascism.'" *Journal of Contemporary History* 35 (2000): 541–548.

Cranor, Carl F., ed. *Are Genes Us? The Social Consequences of the New Genetics.* New Brunswick, N.J.: Rutgers U P, 1994.

Cronenberg, David, *dir. The Fly.* Perf. Jeff Goldblum, Geena Davis. USA, 1986. DVD.

Darwin, Charles. *The Descent of Man.* 1871. [Excerpts.] Appleman 175–254.

———. *The Origin of Species.* 1859. [Excerpts.] Appleman 95–173.

Davis, John. *Exchange.* Buckingham: Open University Press, 1992. Concepts in Social Thought.

Deacon, Terrence W. *The Symbolic Species: The Co-evolution of Language and the Brain.* New York and Lond: Norton, 1997.

De Fren, Alison, dir. *The Mechanical Bride.* USA, 2012. DOXA Film Festival, Vancouver, British Columbia. 6 May 2013. Film.

Dempsey, Michael. Review of *Blade Runner. Film Quarterly* 36/2 (1982–83): 33–38. *JStor.* 30 Aug. 2013. Web.

Dennis, Ian. *Lord Byron and the History of Desire.* Newark: U Delaware P, 2009.

Derrida, Jacques. "Edmond Jabes and the Question of the Book." *Writing and Difference.* Trans. Alan Bass. Chicago: U P, 1978.

Dick, Philip K. *Blade Runner: Do Androids Dream of Electric Sheep?* 1968. Rpt. New York: Random House/ Ballantine, 1982.

———. "Man, Android, and Machine." *Science Fiction at Large.* Ed. Peter Nicholls. New York: Harper and Row, 1976. 199–224.

Doll, Susan and Greg Faller. "*Blade Runner* and Genre: Film Noir and Science Fiction." *Literature/Film Quarterly* 14/2 (1986): 89–100. *Academic Search Premier.* 30 Aug. 2013. Web.

Dresler, Jaroslav. "Capek and Communism." *The Czechoslovak Contribution to World Culture.* Ed. M. Recheigl Jr. The Hague: Mouton, 1964. 68–75.

Dunbar, Robin. *Grooming, Gossip, and the Evolution of Language.* London: Faber and Faber, 1996.

Dupre, John. *Darwin's Legacy: What Evolution Means Today.* Oxford: U P, 2003.

———. *The Disorder of Things: Metaphysical Foundations of the Disunity of Science.* Cambridge: Harvard U P, 1993.

Dutoit, Thomas. "Re-specting the Face as the Moral (of) Fiction in Mary Shelley's Frankenstein." *Modern Language Notes* 109/5 (Dec. 1994): 847–91.

Dworkin, Ronald. "Playing God: Genes, Clones, and Luck." *Sovereign Virtue: The Theory and Practice of Equality.* Cambridge, Mass.: Harvard U P, 2000. 426–52.

Eamon, William. "Secrets of Nature." *Encyclopedia of the Scientific Revolution from Copernicus to Newton.* Ed. Wilbur Applebaum. New York and London: Garland, 2000.

Eccles, John and Daniel N. Robinson. *The Wonder of Being Human: Our Brain and Our Mind.* Boston and London: Shambala/ New Science Library, 1985.

Eichner, Hans. "The Rise of Modern Science and the Genesis of Romanticism." *PMLA* 97 (1982): 8–30.

Elam, Keir. *The Semiotics of Theatre and Drama.* London and New York: Methuen, 1980. New Accents.

Eldredge, Niles. *Reinventing Darwin: The Great Debate at the High Table of Evolutionary Theory.* New York: Wiley, 1995.

Elshtain, Jean Bethke. *Democracy on Trial.* Toronto: House of Anansi, 1993. CBC Massey Lectures.

Eshelman, Raoul. "Originary Aesthetics and the End of Post-modernism." Katz 53–73.

Evans, John Hyde. *Playing God? Human Genetic Engineering and the Rationalization of the Public Bioethical Debate.* Chicago: U P, 2002.

Farrell, Kirby. "H. G. Wells and Neotony." *Cahiers Victoriens et Edouardiens* no. 46 (1997): 145–58.

Fernandez-Armesto, Felipe. *Humankind: A Brief History.* Oxford: U P, 2004.

Fitting, Peter. "Futurecop: The Neutralization of Revolt in *Blade Runner.*" *Science-Fiction Studies* 14/3 [no. 43] (Nov. 1987): 340–54. Web.

Fischer, William. "Of Living Machines and Living-Machines: *Blade Runner* and the Terminal Genre." *New Literary History* 20/1 (Autumn 1968): 187–98. *JStor.* 30 Aug. 2013. Web.

Flahaut, Francois. *Malice.* Trans. Liz Heron. London and New York: Verso, 2003. *Phronesis.* Series ed. Ernesto Laclau and Chantal Mouffe.

Fleming, Chris and John O'Carroll. "What is the Human? Eric Gans and the Structure of the Hypothesis." Katz 271–297.

Francavilla, Joseph. "The Android as Doppelganger." Kerman, *Retrofitting* 4–15.

Frankel, Charles. "The Nature and Sources of Irrationalism." *Science* 180 [no. 4089] (June 1, 1973): 727–31. *JStor*. 30 Aug. 2013. Web.

Fried, Michael. "Impressionist Monsters: H. G. Wells' *The Island of Dr. Moreau*." *Frankenstein, Creation, and Monstrosity*. Ed. Stephen Bann. London: Reaktion, 1994. 95–112.

Frye, Northrop. *Anatomy of Criticism: Four Essays*. Princeton: U P, 1957.

Fukuyama, Francis. *Our Posthuman Future: Consequences of the Biotechnology Revolution*. New York: Farrar, Strauss and Giroux, 2002.

Fuller, Steve. *Kuhn vs. Popper: The Struggle for the Soul of Science*. Cambridge: Ikon Books, 2003.

Gans, Eric. "Beyond 'Generative Anthropology': 2. Resistance to Mimetic Theory." *Chronicle* 145. 25 July 1998.

———. *Chronicles of Love and Resentment*. 1995—present. *Anthropoetics* homepage. Web.

———. "Culture vs. Exchange." *Chronicle* 193. 8 Jan. 2000.

———. "Culture against the Market." *Chronicle* 25. 13 Jan. 1996.

———. *The End of Culture: Toward a Generative Anthropology*. Berkeley: University of California Press, 1985.

———. "The Free Market." *Chronicle* 34. 16 Mar. 1996.

———. "Free Will and Cosmological Idiocy." *Chronicle* 88. 12 Apr. 1997.

———. "God is Love." *Chronicle* 119. 6 Dec. 1997.

———. "The Little Bang: The Early Origin of Language." *Anthropoetics* 5/1 (Spring/Summer 1999).

———. "Marcus Borg's Spiritual God." *Chronicle* 118. 22 Nov. 1997.

———. "Mind and Brain." *Chronicle* 133. 18 Apr. 1998.

———. "A New Mode of Being." *Chronicle* 355. 1 Mar. 2008.

———. *A New Way of Thinking: Generative Anthropology in Religion, Philosophy, Art*. Aurora, Co.: Davies Group Publishers, 2011.

———. *The Origin of Language: A Formal Theory of Representation*. Berkeley: University of California Press, 1981.

———. *Originary Thinking: Elements of Generative Anthropology*. Stanford: U P, 1993.

———. "Our Post-Postmodern Bobos." *Chronicle* 208. 20 May 2000.

———. "The Problem of the Subject." *Chronicle* 117. 15 Nov. 1997.

———. *Science and Faith: The Anthropology of Revelation.* Totowa, N.J.: Rowman and Littlefield, 1990.

———. *The Scenic Imagination: Originary Thinking from Hobbes to the Present Day.* Stanford: U P, 2007.

———. *Signs of Paradox: Irony, Resentment, and Other Mimetic Structures.* Stanford: U P, 1997.

Gellner, Ernest. *The Psychoanalytic Movement: The Cunning of Unreason.* 2nd. Ed. London: Fontana/ HarperCollins, 1985.

Gigante, Denise. "Facing the Ugly: The Case of Frankenstein." *English Literary History* 67/2 (2000): 565–87.

Girard, Rene. *"To Double Business Bound": Essays on Literature, Mimesis, and Anthropology.* Baltimore: Johns Hopkins U P, 1978.

———. *Things Hidden Since the Foundation of the World.* Trans. Stephen Bann and Michael Mettter. Stanford U P, 1987.

———. *The Scapegoat.* Trans. Patrick Gregory. Baltimore: Johns Hopkins U P, 1986.

Goldman, Peter. "Why Generative Anthropology?" *Chronicles of Love and Resentment* no. 445. 13 July 2103. Web.

Gomel, Elana. "From Dr. Moreau to Dr. Mengele" The Biological Sublime." *Poetics Today* 21/2 (Summer 2000): 393–421. *Academic Search Premier.* 30 Aug. 2013. Web.

Good, Graham. *Humanism Betrayed: Theory, Ideology, and Culture in the Contemporary University.* Montreal and Kingston: McGill-Queens U P, 2001.

Goodall, Jane. "Frankenstein and the Reprobate's Conscience." *Studies in the Novel* 31/1 (Spring 1999): 19–40.

Gould, Stephen Jay. "More Things in Heaven and Earth." Rose and Rose 85–105.

Graham, Elaine L. *Representations of the Post/Human: Monsters, Aliens and Others in Popular Culture.* New Brunswick, N.J.: Rutgers U P, 2002.

Graham, Gordon. "Playing God." *Genes: A Philosophical Inquiry.* London and New York: Routledge, 2002. 143–85.

Grandin, Temple, with Catherine Johnson. *Animals in Translation: Using the Mysteries of Autism to Decode Animal Behavior.* Orlando, Fl.: Harcourt/ Harvest Books, 2005.

Gravett, Sharon L. "The Sacred and the Profane: Examining the Religious Subtext of Ridley Scott's *Blade Runner.*" *Literature/Film Quarterly* 26/1 (1998): 38–45. *Academic Search Premier.* 30 Aug. 2013. Web.

Gray, John. *Straw Dogs: Thoughts on Human and Other Animals.* New York: Farrar, Strauss and Giroux, 2007.

Griesemer, James R. "Tools for Talking: Human Nature, Weismannism, and the Interpretation of Genetic Information." Cranor 69–88.

Gross, Paul R. and Norman Levitt. *Higher Superstition: The Academic Left and Its Quarrels with Science.* Baltimore: Johns Hopkins U P, 1994.

Grossman, Lev. "2045: The Year Man Becomes Immortal." *Time* 21 Feb. 2011: 42–49.

Grunbaum, Adolf. *The Foundations of Psychoanalysis: A Philosophical Critique.* Berkeley: U California P, 1984.

Gwaltney, Marilyn. "Androids as a Device for Reflection on Personhood." Kerman, *Retrofitting* 32–39.

Habermas, Jurgen. *The Future of Human Nature.* Cambridge: Polity Press, 2003.

Hamerton-Kelly, Robert, ed. *Violent Origins: Walter Burkert, Rene Girard, and Jonathan Z. Smith on Ritual Killing and Cultural Formation.* Intro. Burton Mack. Comm. by Renato Rosaldo. Stanford: U P, 1987.

Hammond, J. R. "*The Island of Dr. Moreau:* A Swiftian Parable." *The Wellsian: Journal of the H. G. Wells Society* no. 6 (1993): 30–41. Web.

Harkins, William E. *Karel Capek.* New York: Columbia U P, 1963.

———. "The Real Legacy of Karel Capek." *The Czechoslovak Contribution to World Culture.* Ed. M. Recheigl Jr. The Hague: Mouton, 1964. 60–67.

Harris, Errol E. *Revelation through Reason: Religion in the Light of Science.* New Haven: Yale U P, 1958.

Harris, John. *On Cloning.* London and New York: Routledge, 2004.

Harris, Marvin. *Cannibals and Kings: The Origins of Cultures.* New York: Random House, 1977.

Haught, John F. *Is Nature Enough? Meaning and Truth in the Age of Science.* Cambridge: U P, 2006.

Haynes, Roslynn. *From Faust to Strangelove: Representations of the Scientist in Western Literature.* Baltimore: Johns Hopkins U P, 1994.

———. *H. G. Wells, Discoverer of the Future: The Influence of Science on His Thought.* London and Basingstoke: MacMillan, 1980.

———. "The Unholy Alliance of Science in *The Island of Dr. Moreau.*" *The Wellsian: Journal of the H. G. Wells Society* no. 11 (1988): 13–24.

Hempel, Carl G. "Studies in the Logic of Explanation." Klemke, Hollinger and Kline 91–128.

Heldreth, Leonard G. "The Cutting Edges of *Blade Runner.*" Kerman, *Retrofitting* 40–52.

Heyd, David. "Introduction: Playing God." *Genethics: Moral Issues in the Creation of People.* Berkeley: U California P, 1992.

Hinckley, David. "Surgical Evolution, or, The Scalpel as Shortcut: The Doctor as Interface between Science Fiction and Horror." *No Cure for the Future: Disease and Medicine in Science Fiction and Fantasy.* Ed. Gary Westfahl and George Slusser. Westport, Conn.: Greenwood, 2002. 83–94. Contributions to the Study of Science Fiction and Fantasy 102.

Hitchcock, Susan Tyler. *Frankenstein: A Cultural History.* New York and London: Norton, 2007.

Hoeveler, Diane. "Frankenstein, Feminism, and Literary Theory." Schor 45–62.

Hoffman, Gerhard and Alfred Hornung, eds. *Ethics and Aesthetics: The Moral Turn of Postmodernism.* Heidelberg: Winter, 1996.

Holy Bible. Revised Standard Version. Camden, N.J.: Thomas Nelson and Sons, 1952.

Howard, Ted and Jeremy Rifkin. *Who Should Play God? The Artificial Creation of Life and What It Means for the Future of the Human Race.* New York: Delacorte Press, 1977.

Hughes, Robert. *The Culture of Complaint: The Fraying of America.* Oxford: U P, 1994.

Huntington, John. *The Logic of Fantasy: H. G. Wells and Science Fiction.* New York: Columbia U P, 1982.

Hurford, James R. *The Origin of Meaning: Language in the Light of Evolution I.* Oxford: U P, 2007. Studies in the Evolution of Language.

Instrell, Rick. "*Blade Runner:* The Economic Shaping of a Film." *Cinema and Fiction: New Modes of Adapting, 1950–1990.* Ed. John Orr and Colin Nicholson. Edinburgh: Edinburgh U P., 1992. 160–70.

Jackson, Kimberley. "Vivisected Language in H. G. Wells' *The Island of Dr. Moreau.*" *The Wellsian: Journal of the H. G. Wells Society* no. 29 (2006): 20–35.

Jacobus, Mary. "Is There a Woman in This Text?" *New Literary History* 14/1 (1982): 117–54. *JStor.* 30 Aug. 2013. Web.

Jaki, Stanley L. *The Savior of Science.* Grand Rapids: Eerdmans, 2000.

Johnsen, William A. *Violence and Modernism: Ibsen, Joyce, Woolf.* Gainesville: University Press of Florida, 2003.

Jones, Richard. *Reductionism: Analysis and the Fullness of Reality.* Lewisburg: Bucknell U P, 2000.

Juengel, Scott. "Face, Figure, Physiognomics: Mary Shelley's Frankenstein and the Moving Image." *Novel* 33 (2000): 353–76.

Kass, Leon. "The Wisdom of Repugnance." 1997. McGee and Caplan (2004) (4th ed.): 137–72.

Katz, Adam, ed. *The Originary Hypothesis: A Minimal Proposal for Humanistic Inquiry.* Aurora, Colorado: Davies Group Publishers, 2007. Critical Studies in the Humanities.

———. "The Question of Originary Method: The Generative Thought Experiment." Katz 101–37.

Kaufman, Gordon D. *God the Problem.* Harvard: U P, 1972.

Keller, Evelyn Fox. *The Century of the Gene.* Cambridge, Mass: Harvard U P, 2000.

———. "Master Molecules." Cranor 89–98.

Kellner, Douglas, Flo Leibowitz and Michael Ryan. "*Blade Runner:* A Diagnostic Critique." *Jump Cut: A Review of Contemporary Media* no. 29 (Feb. 1984): 6—8.

Kerman, Judith B. "Private Eye: A Semiotic Comparison of the Film *Blade Runner* and the Book *Do Androids Dream of Electric Sheep?*" *Patterns of the Fantastic, II.* Ed. Donald M. Hassler. Mercer Island, WA.: Starmont House, 1985. 69–75.

———, ed. *Retrofitting* Blade Runner: *Issues in Ridley Scott's* Blade Runner *and Philip K Dick's* Do Androids Dream of Electric Sheep? Bowling Green, Ohio: Popular Press, 1991.

Ketterer, David. *Frankenstein's Creation: The Book, the Monster, and Human Reality.* Victoria, B.C.: Department of English at U Victoria, 1979. English Literary Studies no. 16.

———. "Alchemy and Science in *Frankenstein.*" Review of *Scientific Attitudes in Mary Shelley's Frankenstein* by Samuel Holmes Vasbinder. *Science Fiction Studies* 40 (Nov. 1986). Web.

———. *Karel Capek: Life and Work.* Trans. Norma Comrada. New Haven: Catbird Press, 2002.

King, Sharon D. "A Better Eve: Women and Robots in Capek's *R.U.R.* and Pavlovsky's *El Robot*." *Women in Theatre*. Cambridge: U P, 1989. 99–107.

Kinyon, Kamila. "The Phenomenology of Robots: Confrontations with Death in Karel Capek's *R.U.R.*" *Science-Fiction Studies* 26/2 [no. 79] (July 1999): 379–400. Web.

Kitcher, Philip. "Believing Where We Cannot Prove." Klemke, Hollinger and Kline 55–77.

———. "Human Cloning." McGee 3rd ed. 136–53.

Klemke, E.D., Robert Hollinger and A. David Kline, eds. *Introductory Readings in the Philosophy of Science*. Buffalo: Prometheus Books, 1988.

Klima, Ivan. Introduction. *R.U.R. (Rossum's Universal Robots)*. Trans. Claudia Novack. London: Penguin, 2004. Vii-xxv.

Kolata, Gina. *Clone: The Road to Dolly and the Path Ahead*. New York: William Morrow and Co., 1998.

Kolb, William M. "Script to Screen: *Blade Runner* in Perspective." Kerman, *Retrofitting* 132–53.

Krutch, Joseph Wood. *The Measure of Man: On Freedom, Human Values, Survival and the Modern Temper*. New York: Grosset and Dunlap, 1953.

Kuhn, Thomas. *The Structure of Scientific Revolutions*. 2nd. ed, enlarged. Chicago: U P, 1970.

Kuper, Adam. *The Chosen Primate: Human Nature and Cultural Diversity*. Cambridge, Mass.: Harvard U P, 1994.

Lederer, Susan E. *Frankenstein: Penetrating the Secrets of Nature (An Exhibition by the National Library of Medicine)*. New Brunswick, N. J.: Rutgers U P, 2002.

Lem, Stanislaw. "Robots in Science Fiction." Trans. Frans Rottensteiner. *SF: The Other Side of Realism*. Ed. Thomas D. Clareson. Bowling Green, Kentucky: Bowling Green U Popular P, 1971. 307–25.

Lev, Peter. "Whose Future? *Star Wars, Alien,* and *Blade Runner*." *Literature/Film Quarterly* 26/1 (1998): 30–37. *Academic Search Premier*. 30 Aug. 2013.

Levine, George. "The Ambiguous Heritage of Frankenstein." Levine and Knoepflmacher 3–30.

———, and U. C. Knoepflmacher, eds. *The Endurance of Frankenstein: Essays on Mary Shelley's Novel*. Berkeley: U California P, 1979.

Levy, David. *Love and Sex with Robots: The Evolution of Human-Robot Relationships.* New York: Harper Perennial, 2007.

Lifton, Robert Jay and Greg Mitchell. *Hiroshima in America: A Half-Century of Denial.* New York: Avon, 1995.

Lipking, Laurence. "Frankenstein, the True Story; or, Rousseau Judges Jean-Jacques." *Mary Shelley: Frankenstein.* Ed. J. Paul Hunter. New York: Norton, 1996. 313–31. Norton Critical Editions.

Livingston, Paisley. *Literary Knowledge: Humanistic Inquiry and the Philosophy of Science.* Ithaca: Cornell U P, 1988.

———. *Models of Desire: Rene Girard and the Psychology of Mimesis.* Baltimore: Johns Hopkins U P, 1994.

Lobkowicz, Nicholas. "Christianity and Culture." *Review of Politics* 53/2 (1991): 378–89. *JStor.* 30 Aug. 2013. Web.

L'Officier, Randy and Jean Marc L'Officier. "Blade Runner." *Science-Fiction FilmMaking in the 1980s: Interviews with Actors, Directors, Producers and Writers.* Ed. Lee Goldberg, Jean-Marc L'Officier, and William Rabkin. Jefferson, N.C.: McFarland, 1995.

Long, Charles H. *Alpha: The Myths of Creation.* New York: Collier, 1963.

Lowe, E. J. *An Introduction to the Philosophy of Mind.* Cambridge: U P, 2000.

Lyon, David. *Postmodernity.* Minnesota: U Minnesota P, 1994. Concepts in Social Thought.

MacNeice, Louis. "Meeting Point." *The New Penguin Book of Love Poetry.* Ed. Jon Stallworthy. London: Penguin, 2001.

Malik, Kenan. "What Science Can and Cannot Tell Us about Human Nature." Wells and McFadden 164–83.

Manson, Michael and Robert Scott Stewart. "Heroes and Hideousness: *Frankenstein* and Failed Unity." *SubStance* 22 (1993): 228–42. *JStor.* 30 Aug. 2013. Web.

Marcus, Steven. "Frankenstein: Myths of Scientific and Medical Knowledge and Stories of Human Relations." *Southern Review* 38 (2002): 188–201. *Academic Search Premier.* 30 Aug. 2013. Web.

Marder, Elissa. "*Blade Runner*'s Moving Still." *Camera Obscura* 27 (Sep. 1991): 88–107.

Matuska, Alexander. *Man Against Destruction [or] Karel Capek, an Essay.* Trans. Cathryn Alan. London: George Allen and Unwin, 1964.

Maugham, Somerset. *The Magician.* 1908. Rpt. Harmondsworth: Penguin, 1983.

Mayr, Ernst. *One Long Argument: Charles Darwin and the Genesis of Modern Evolutionary Thought.* Cambridge, Mass.: Harvard U P, 1991.

McConnell, Frank. *The Science Fiction of H. G. Wells.* Oxford: U P, 1981.

McGee, Glen, ed. *The Human Cloning Debate.* 3rd. ed. Berkeley: Berkeley Hills Books, 2002.

McGee, Glen, and Arthur Caplan, eds. *The Human Cloning Debate.* 4th. ed. Berkeley: Berkeley Hills Books, 2004.

McGee, Glen, and Ian Wilmut. "A Model for Regulating Cloning." McGee, 3rd ed. 221–34.

McNamara, Kevin R. "*Blade Runner*'s Post-Invidividual Workspace." *Contemporary Literature* 38/3 (Autumn 1997): 422–46. *Jstor.* 30 Aug. 2013. Web.

Meilander, Gilbert. "Human Cloning Would Violate the Dignity of Children." McGee 3rd., ed. 268–75.

Mellor, Anne K. "Making a 'Monster': An Introduction to Frankenstein." Schor 9–25.

Midgley, Mary. *Beast and Man: The Roots of Human Nature.* 1979. Rev. ed. London and New York: Routledge, 1995.

———. *Evolution as a Religion: Strange Hopes and Stranger Fears.* 1985. Rev. ed. London and New York: Routledge Classics, 2002.

———. "Why Memes?" Rose and Rose 67–84.

Moers, Ellen. "Female Gothic." Levine and Knoepflmacher 77–87.

Morrison, Rachela. "Casablanca Meets Star Wars: The Blakean Dialectics of *Blade Runner.*" *Literature/Film Quarterly* 18/1 (1990): 2–10. *Academic Search Premier.* 30 Aug. 2013. Web.

Morton, Timothy. "Mary Shelley as Cultural Critic." Schor 259–73.

Nagel, Thomas. "Ethics without Biology." *Mortal Questions.* Cambridge: U P, 1979.

Napoli, Donna Jo. "Do Animals Have Language?" *Language Matters: A Guide to Everyday Thinking about Language.* New York: Oxford U P, 2003. 72–86.

Naughton, James D. "Futurology and Robots: Karel Capek's *R.U.R.*" *Renaissance and Modern Studies* 28 (1984): 72–86.

Nelkin, Dorothy. "Less Selfish than Sacred? Genes and the Religious Impulse in Evolutionary Psychology." Rose and Rose 14–27.

Nielsen, Kai. *Ethics Without God.* Buffalo: Prometheus Books, 1973.

Nisbet, Robert K. *Prejudices: A Philosophical Dictionary.* Cambridge, Mass.: Harvard U P, 1982.

Nossal, G.J.V. and Ross L. Coppell. "Scientists Playing God." *Reshaping Life: Key Issues in Genetic Engineering.* 3rd. ed. Cambridge: U P, 2002.

Olafson, Frederick A. *Naturalism and the Human Condition.* London and New York: Routledge, 2001.

Orwell, George. *Nineteen Eighty-four.* 1949. Harmondsworth: Penguin, 1969. Penguin Modern Classics.

Owen. D. R. G. *Scientism, Man, and Religion.* Philadelphia: Westminster Press, 1952.

Parrinder, Patrick, ed. *H. G. Wells: The Critical Heritage.* London: Routledge and Kegan Paul, 1972.

———. *Science Fiction: Its Criticism and Teaching.* London and New York: Methuen, 1980. New Accents.

Paul, Diane B. "Eugenic Anxieties, Social Realities, and Political Choices." Cranor 142–54.

Peoples, David, scr. *Soldier.* Dir. Poul Anderson. Perf. Kurt Russell. USA, 1998. DVD.

Peters, Ted. *Playing God? Genetic Determinism and Human Freedom.* New York and London: Routledge, 1997.

Philmus, Robert. "Introducing Moreau." *The Island of Dr. Moreau: A Variorum Text.* Ed. Philmus. Athens and London: U Georgia P, 1993. xi-xlviii.

———. "Matters of Translation: Karel Capek and Paul Selver." *Science-Fiction Studies* 28/3 [no. 83] (Mar. 2001): 7–32. Web.

———. *Visions and Re-visions: (Re)Constructing Science Fiction.* Liverpool: U P, 2005.

Philmus, Robert, and David Y. Hughes, eds. *H. G. Wells: Early Writings in Science and Science Fiction.* Berkeley: U Cal P, 1975.

Picart, Caroline Joan S. *Remaking the Frankenstein Myth on Film: Between Laughter and Horror.* Albany: SUNY Press, 2003.

Piercy, Marge. *He, She, and It.* New York: Fawcett Crest, 1993.

Pinker, Steven. "The Biology of Fiction." Wells and McFadden 27–39.

Popper, Karl. "Science: Conjectures and Refutations." Klemke, Hollinger and Kline 19–27.

Pyle, Forest. "Making Cyborgs, Making Humans: Of Terminators and Blade Runners." *Film Theory Goes to the Movies*. Ed. Jim Collins, Hilary Radner, and Ava Preacher Collins. New York: Routledge, 1993.

Reed, John R. *The Natural History of H. G. Wells*. Athens: Ohio U P, 1982.

———. "The Vanity of Law in *The Island of Dr. Moreau*." *H. G. Wells Under Revision*. Ed. Patrick Parrinder and Christopher Rolfe. London: Associated University Presses, 1990. 134–44.

Reichmann, James B., S. J. *Evolution, Animal "Rights," and the Environment*. Washington, D. C.: Catholic University of America Press, 2000.

Rescher, Nicholas. *Human Interests: Reflections on Philosophical Anthropology*. Stanford: U P, 1990. Stanford Studies in Philosophy.

———. *The Limits of Science*. Rev. ed. Pittsburgh: U P, 1999.

Restivo, Sal. "Modern Science as a Social Problem." *Social Problems* 35/3 (1988): 206–225. *JStor*. 30 Aug. 2013.

Richardson, Alan. "Embodied Universalism, Romantic Discourse, and the Anthropological Imagination." *British Romanticism and the Science of the Mind*. Cambridge: U P, 2001. 151–180.

Riechardt, Jasia. "Artificial Life and the Myth of Frankenstein." *Frankenstein, Creation and Monstrosity*. Ed. Stephen Bann. London: Reaktion, 1992. 136–57.

Roberts, Marie Mulvey. "The Male Scientist, the Man-Midwife, and Female Monster: Appropriation and Transmutation in *Frankenstein*." *A Question of Identity: Women, Science and Literature*. Ed. Marina Benjamin. New Brunswick, N. J.: Rutgers U P, 1993. 59–73.

Robertson, John. "Cloning as a Reproductive Right." McGee and Caplan 177–92.

Rose, Hilary. "Colonising the Social Sciences?" Rose and Rose 106–128.

Rose, Hilary and Steven Rose, eds. *Alas, Poor Darwin: Arguments against Evolutionary Psychology*. London: Jonathan Cape, 2000.

Rose, Steven. "Escaping Evolutionary Psychology." Rose and Rose 247–65.

Ruddick, Nicholas. "Annotations, Appendices, Adaptations: Recent Work on H. G. Wells' Scientific Romances." *Science Fiction Studies* 32 (2005): 316–22.

———. *Ultimate Island: On the Nature of British Science Fiction*. Westport,

Conn." Greenwood, 1993. Contributions to the Study of Science Fiction and Fantasy 55.

Rushing, Janice Hocker and Thomas S. Frentz. "The Frankenstein Myth in Contemporary Cinema." *Critical Studies in Mass Communications* 6 (1989): 61–68. *Academic Search Premier.* 30 August 2013.

Rust, Joshua. *John Searle.* London: Continuum, 2009.

Sammon, Paul. *Future Noir: The Making of Blade Runner.* New York: HarperCollins, 1996.

Schanze, Jens, dir. & scr. *Plug and Pray.* Germany, 2010. 29th Vancouver International Film Festival. 1 Oct. 2010. Film.

Schell, Jonathan. *The Fate of the Earth.* New York: Knopf, 1982.

Schor, Esther, ed. *The Cambridge Companion to Mary Shelley.* Cambridge: U P, 2003.

Schug, Charles. "The Romantic Form of Mary Shelley's Frankenstein." *Studies in English Literature* 17 (1997): 607–19. *JStor.* 30 Aug. 2013. Web.

Schwartz, Richard A. *The Films of Ridley Scott.* Westport, Conn. And London: Praegar, 2001.

Schwartz, Stephen Adam. "Everyman an Ubermensch: The Culture of Cultural Studies." *Substance* 29/1 (2000): 104–138. *JStor.* 30 Aug. 2013. Web.

Scruton, Roger. *How to Think Seriously about the Planet: The Case for anEnvironmental Conservatism.* Oxford: U P, 2012.

Searle, John. *Freedom and Neurobiology:Reflections on Free Will, Language and Political Power.* New York: Columbia U P, 2007. Columbia Themes in Philosophy.

———. *Minds, Brains, and Science.* Cambridge, Mass.: Harvard U P, 1984.

———. *Mind, Language, and Society: Philosophy in the Real World.* London: Weidenfeld and Nicolson, 1999.

———. *The Rediscovery of the Mind.* Cambridge, Mass.: MIT Press, 1992.

Segal, Robert A. *Myth: A Very Short Introduction.* Oxford: U P, 2004.

Senior, W. A. "*Blade Runner* and Cyberpunk Humanity." *Film Criticism* 21/1 (Fall 1996): 1–12. *Academic Search Premier.* 30 Aug. 2013.

Shapiro, Michael J. "'Manning ' the Frontiers: The Politics of (Human) Nature in *Blade Runner.*" *In the Nature of Things: Language, Politics, and the Environment.* Ed. Jane Bennett and William Chaloupka. Minneapolis: U Minnesota P, 1993. 65–84.

Sherborne, Michael. "Two Cheers for Stoverism." *The Wellsian: Journal of the H. G. Wells Society* 19 (1996): 50–53.

Sherman, Kenneth. "Are We Not Men?" *Queen's Quarterly* 102 (1995): 869–75.

Shorten, Richard. *Modernism and Totalitarianism: Rethinking the Intellectual Sources of Nazism and Stalinism, 1945 to the Present.* Houndsmills, Basingstoke: Palgrave Macmillan, 2012.

Showalter, Elaine. "The Apocalyptic Fables of H. G. Wells. *Fin de Siecle / Fin du Globe: Fears and Fantasies of the Late Nineteenth Century.* Ed. John Stokes. New York: St. Martin's Press, 1992. 69–84.

Siebers, Tobin. *The Ethics of Criticism.* Ithaca: Cornell U P, 1988.

Silverman, Kaja. "Back to the Future." *Camera Obscura* 27 (Sep. 1991): 109–32.

Singer, Peter. *Animal Liberation.* New York: Random House, 1990.

Skal, David J. *Screams of Reason: Mad Science and Modern Culture.* New York and London: Norton, 1998.

Slade, Joseph W. "Romanticizing Cybernetics in Ridley Scott's *Blade Runner.*" *Literature/Film Quarterly* 18/1 (1990): 11–18. *Academic Search Premier.* 30 Aug. 2013. Web.

Small, Christopher. *Ariel Like A Harpy: Shelley, Mary, and Frankenstein.* London: Victor Gollancz, 1972.

Smith, David C. "H. G. Wells and His Contemporary Critics: A Developing Reputation." *Cahiers Victoriens et Edouardiens* no. 47 (1997): 213–44.

Snow, C. P. *The Two Cultures and the Scientific Revolution.* New York: Cambridge U P, 1961.

Sobchack, Vivian. *Screening Space: The American Science Fiction Film.* New York: Ungar, 1987.

Sorell, Tom. *Scientism: Philosophy and the Infatuation with Science.* Routledge: London and New York, 1991.

Spielberg, Steven, dir. *AI: Artificial Intelligence.* USA, 2001. Perf.. Haley Joel Osment, Jude Law, William Hurt. DVD.

———. *Jurassic Park.* Perf. Sam Neill, Laura Dern, Jeff Goldblum. USA, 1993. DVD.

Splicing Life: A Report on the Social and Ethical Issues of Genetic Engineering with Human Beings. President's Commission for the Study of Ethical Problems in Medicine and Biomedical and Behavioral Research. Washington, D.C.: The Commission, 1983.

Stableford, Brian. "*Frankenstein* and the Origins of Science Fiction." *Anticipations: Essays on Early Science Fiction and Its Precursors.* Ed. David Seed. Liverpool: U P, 1995. 46–57.

Steffen-Fluhr, Nancy. "The Definitive Moreau." *Science-Fiction Studies* no. 61 (1993): 433–39. Web.

Stenmark, Mikael. *Scientism: Science, Ethics and Religion.* Aldershot: Ashgate, 2001.

Sterrenburg, Lee. "Mary Shelley's Monster: Politics and Psyche in Frankenstein." Levine and Knoepflmacher 143–71.

Stover, Leon. "Introduction." *The Island of Dr. Moreau: A Critical Text of the 1896 London First Edition, with an Introduction and Appendices.* Jefferson, N. C.: McFarland, 1996.

Suvin, Darko. *Metamorphoses of Science Fiction: On the Poetics and History of a Literary Genre.* New Haven: Yale U P, 1979.

———. "Wells as the Turning Point of the Science Fiction Tradition." 1979. Rpt. in *Critical Essays on H. G. Wells.* Ed. John Huntington. Boston: G. K. Hall, 1991. 23–33.

Swift, Jonathan. *Gulliver's Travels.* 2nd ed. Ed. Robert Greenberg. New York: Norton, 1970. Norton Critical Editions.

Tallis, Raymond. *Aping Mankind: Neuromania, Darwinitis, and the Misrepresentation of Humanity.* Durham: Acumen, 2011.

———. *Why the Mind is Not a Computer: A Pocket Lexicon of Neuromythology.* 1994. Exeter: Imprint Academic, 2004.

Telotte, J. P. *Replications: A Robotic History of the Science Fiction Film.* Urbana: U Illinois P, 1995.

Tiitsman, Jenna. "If Only You Could See What I've Seen with Your Eyes: Destabilized Spectatorship and Creation's Chaos in *Blade Runner.*" *Cross Currents* 54.1 (Spring 2004): 32–47. *Academic Search Premier.* 30 Aug. 2013.

Tomasello, Michael. *The Cultural Origins of Human Cognition.* Cambridge, Mass.: Harvard U P, 2001.

Turney, Jon. *Frankenstein's Footsteps: Science, Genetics and Popular Culture.* New Haven and London: Yale U P, 1998.

Vallorani, Nicoletta. "Hybridizing Science: The 'Patchwork Biology' of Dr. Moreau." *Cahiers Victoriens et Edouardiens* no. 46 (1997): 245–61.

van Oort, Richard. *The End of Literature: Essays in Anthropological Aesthetics.* Aurora, Colorado: Davies Group Publishers, 2009.

———. "Mimetic Theory and its Rivals: A Reply to Pablo Bandera." *Contagion: Journal of Violence, Mimesis and Culture* 17 (2010): 189–204.

———. "Science and Culture, or How Real is Real? Reflections on the Sokal Debate." Guest columnist. *Chronicles of Love and Resentment* no. 45. 1 June 1996. Web.

———. "Three Models of Fiction: The Logical, the Phenomenological, and the Anthropological (Searle, Ingarden, Gans)." *New Literary History* 29 (1998): 439–65. *JStor.* 30 Aug. 2013. Web.

Vasbinder, Samuel Holmes. *Scientific Attitudes in Mary Shelley's Frankenstein.* Ann Arbor, Michigan: UMI Research Press, 1984. Studies in Speculative Fiction 8.

Verhey, Allen D. "Playing God." *Genetic Ethics: Do the Ends Justify the Means?* Ed. John F. Kilner et al. Grand Rapids, Michigan: Eerdmans, 1997. 60–74.

Vernier, J. P. "Evolution as a Literary Theme in H. G. Wells' Science Fiction." *H. G. Wells and Modern Science Fiction.* Ed. Darko Suvin and Robert Philmus. Lewisburg: Bucknell U P, 1977. 70–89.

Vest, Jason P. *Future Imperfect: Philip K. Dick at the Movies.* Westport, Conn.: Praeger, 2007.

Wade, Nicholas. *Before the Dawn: Recovering the Lost History of our Ancestors.* New York: Penguin Press, 2006.

Wagar, W. Warren. *H. G. Wells: Traversing Time.* Middletown, Conn.: Wesleyan U P, 2004.

Weizenbaum, Joseph. *Computer Power and Human Reason.* 1976. Rpt. London: Penguin, 1984.

Wellek, Rene. "Karel Capek." 1936. Rpt. *Essays on Czech Literature.* The Hague: Mouton, 1963.

Wells, H. G. *The Complete Science Fiction Treasury of H. G. Wells.* Orig. pub. *Seven Famous Novels.* 1934. Rpt. New York: Avenell, 1978.

———. *The First Men in the Moon.* 1898. *Complete* 391–528.

———. "Human Evolution, An Artificial Process." 1896. Rpt. *H. G. Wells: Early Writings in Science and Science Fiction.* Ed. Robert Philmus and David Y. Hughes. Berkeley: U California P, 1975. 211–19.

———. *The Invisible Man.* 1897. *Complete* 161–262.

———. *The Time Machine.* 1895. *Complete* 3–66.

———. *The War of the Worlds.* 1898. *Complete* 265–388.

Wells, Robin Headham and Johnjoe McFadden, eds. *Human Nature: Fact and Fiction.* London and New York: Continuum, 2006.

———. "Introduction." Wells and McFadden 1–24.

Whale, James. dir. *Frankenstein.* Perf. Colin Clive, Boris Karloff, Mae Clarke. Scr. John A. Balderston. USA, 1931. DVD.

———. *Bride of Frankenstein.* Perf. Colin Clive, Boris Karloff, Elsa Lancaster. Scr. William Hurlbut. USA, 1935. DVD.

White, Eric. "The End of Metanarratives in Evolutionary Biology." *Modern Language Quarterly* 51/1 (March 1990): 63–81. *Academic Search Premier.* 30 Aug. 2013. Web.

Wiesman, Alan. *The World without Us.* New York: HarperCollins, 2007.

Williams, John. *Mary Shelley: A Literary Life.* London: Macmillan, 2000.

Williams, James G. *The Bible, Violence and the Sacred: Liberation from the Myth of Sanctioned Violence.* New York: HarperCollins, 2001.

Wilson, Eric G. "Movie-going and Golem-Making: The Case of *Blade Runner.*" *Journal of Film and Video* 57/3 (Fall 2005): 31–43. *JStor.* 30 Aug. 2013. Web.

Winner, Langdon. "Resistance is Futile: The Posthuman Condition and Its Advocates." *Is Human Nature Obsolete? Genetics, Bioengineering, and the Future of the Human Condition.* Ed. Harold W. Baillie and Timonty K. Casey. Cambridge and London: MIT Press, 2005. 385–411.

Wood, Robin. "Papering the Cracks: Fantasy and Ideology in the Reagan Era." *Hollywood from Vietnam to Reagan . . . and Beyond.* Rev. ed. New York: Columbia U P, 2003. 144–67.

Ziman, John. "What Is Science?" Klemke, Hollinger and Kline 28–33.

Ziolkowski, Theodore. "Science, Frankenstein, and Myth." *Sewanee Review* 89 (1981): 34–56. *JStor.* 30 Aug. 2013. Web.

INDEX

Printed in Dunstable, United Kingdom

91221467R00201